Power Systems Analysis

Power Systems Analysis

Second Edition

P.S.R. Murty

Butterworth-Heinemann
An imprint of Elsevier

Butterworth-Heinemann is an imprint of Elsevier
The Boulevard, Langford Lane, Kidlington, Oxford OX5 1GB, United Kingdom
50 Hampshire Street, 5th Floor, Cambridge, MA 02139, United States

British Library Cataloguing-in-Publication Data
A catalogue record for this book is available from the British Library

Library of Congress Cataloging-in-Publication Data
A catalog record for this book is available from the Library of Congress

ISBN: 978-0-08-101111-9

For Information on all Butterworth-Heinemann publications
visit our website at https://www.elsevier.com/books-and-journals

Working together
to grow libraries in
developing countries

www.elsevier.com • www.bookaid.org

Publishing Director: Joy Hayton
Senior Editorial Project Manager: Kattie Washington
Production Project Manager: Mohana Natarajan
Designer: Mark Rogers

Typeset by MPS Limited, Chennai, India

To
My Students

Contents

Preface

In the current edition, the content in the introduction is brought up-to-date. The subject matter is enlarged to widen the scope of the book and its utilization.

In chapter 3 "Incidence Matrices" is reorganized and expanded to include "augmented cut-set incidence matrix" and augmented loop incidence matrix. Chapter 4, Network Matrices, is created exclusively to deal with network matrices. Symmetrical components are presented separately in Chapter 6, Symmetrical Components. Analysis of three-phase networks is presented in Chapter 7, Three-Phase Networks. Theory and modeling of synchronous machines is included as Chapter 8, Synchronous Machine. Line, transformer, and load modeling is added as Chapter 9, Lines and Loads. It is hoped that inclusion and addition of more background material will enable both the teacher and the student to cope with problems arise in power system analysis more effectively in teaching and in understanding.

P.S.R. Murty

INTRODUCTION

Power is an essential prerequisite for the progress of any country. The modern power system has features unique to itself. It is the largest man-made system in existence and is the most complex system. The power demand is more than doubling every decade.

Planning, operation, and control of interconnected power system poses a variety of challenging problems, the solution of which requires extensive application of mathematical methods from various branches.

Thomas Alva Edison was the first to conceive an electric power station and operate it in New York in 1882. Since then, power generation originally confined to steam engines expanded using (steam turbines) hydroelectric turbines, nuclear reactors, and others.

The inter connection of the various generating stations to load centers through extra high voltage (EHV) and ultra high voltage (UHV) transmission lines necessitated analytical methods for analyzing various situations that arise in operation and control of the system.

Power system analysis is the subject in the branch of electrical power engineering which deals with the determination of voltages at various buses and the currents that flow in the transmission lines operating at different voltage levels.

1.1 THE ELECTRICAL POWER SYSTEM

The electrical power system is a complex network consisting of generators, loads, transmission lines, transformers, buses, circuit breakers, etc. For the analysis of a power system in operation, a suitable model is needed. This model basically depends upon the type of problem on hand. Accordingly, it may be algebraic equations, differential equations, transfer functions, etc. The power system is never in steady state as the loads keep changing continuously.

However, it is possible to conceive a quasistatic state during which period the loads could be considered constant. This period could be 15–30 min. In this state, power flow equations are nonlinear due to the presence of product terms of variables and trigonometric terms. The solution techniques involve numerical (iterative) methods for solving nonlinear algebraic equations. Newton–Raphson method is the most commonly used mathematical technique. The analysis of the system for small load variations, wherein speed or frequency and voltage control may be required to maintain the standard values, transfer function, and state variable models is better suited to implement proportional, derivative, and integral controllers or optimal controllers using Kalman's feedback coefficients. For transient stability studies involving sudden changes in load or circuit condition due to faults, differential equations describing energy balance over a few half-cycles of time period are required. For studying the steady state performance a number of matrix models are needed.

Power Systems Analysis. DOI: http://dx.doi.org/10.1016/B978-0-08-101111-9.00001-X

Consider the power system shown in Fig. 1.1. The equivalent circuit for the power system can be represented as in Fig. 1.2. For study of fault currents the equivalent circuit in Fig. 1.2 can be reduced to Fig. 1.3 up to the load terminals neglecting the shunt capacitances of the transmission line and magnetizing reactances of the transformers.

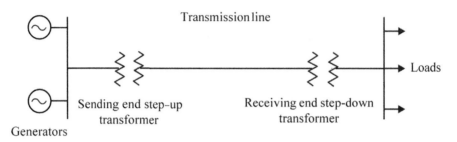

FIGURE 1.1

Sample power system.

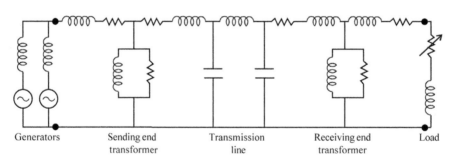

FIGURE 1.2

Equivalent circuit for sample power system.

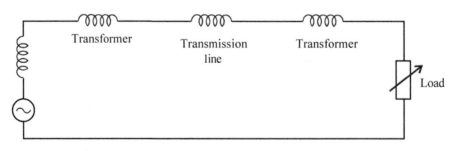

FIGURE 1.3

Reduced equivalent circuit.

While the reactances of transformers and lines which are static do not change under varying conditions of operation, the machine reactances may change and assume different values for different situations. Also, composite loads containing 3-phase motors, 1-phase motors, d-c motors, rectifiers, lighting loads, heaters, welding transformers, etc. may have very different models depending upon the composition of its constituents.

The control of a turbo generator set to suit to the varying load requirement requires a model. For small variations a linearized model is convenient to study. Such a model can be obtained using transfer function concept and control can be achieved through classical or modern control theory. This requires modeling of speed governor, turbo generator, and power system itself as all these constitute the components of a feedback loop for control. The ultimate objective of power system control is to maintain continuous supply of power with acceptable quality. Quality is defined in terms of voltage and frequency.

1.2 **NETWORK MODELS**

Electrical power network consists of large number of transmission lines interconnected in a fashion that is dictated by the development of load centers. This interconnected network configuration expands continuously. A systematic procedure is needed to build a model that can be constantly up-graded with increasing interconnections.

Network solutions can be carried out using Ohm's law and Kirchoff's laws.
Either

$$e = Zi$$

or

$$i = Ye$$

model can be used for steady state network solution. Thus it is required to develop both Z-bus and Y-bus models for the network. To build such a model, graph theory and incidence matrices will be quite convenient.

1.3 **FAULTS AND ANALYSIS**

Study of the network performance under fault conditions requires analysis of a generally balanced network as an unbalanced network. Under balanced operation, all the three-phase voltages are equal in magnitude and displaced from each other mutually by 120° (elec.). It may be noted that unbalanced transmission line configuration is balanced in operation by transposition, balancing the electrical characteristics.

Under fault conditions the three-phase voltages may not be equal in magnitude and the phase angles too may differ widely from 120° (elec.) even if the transmission and distribution networks are balanced. The situation changes into a case of unbalanced excitation.

Network solution under these conditions can be obtained by using transformed variables through different component systems involving the concept of power invariance.

1.4 THE PRIMITIVE NETWORK

Network components are represented either by their impedance parameters or admittance parameters. Fig. 1.4 represents the impedance form, the variables are currents and voltages. Every power system element can be described by a primitive network. A primitive network is a set of unconnected elements.

a and b are the terminals of a network element a—b. V_a and V_b are voltages at a and b.

V_{ab} is the voltage across the network element a—b.
e_{ab} is the source voltage in series with the network element a—b
z_{ab} is the self-impedance of network element a—b.
j_{ab} is the current through the network element a—b.

From Fig. 1.4, we have the relation

$$V_{ab} + e_{ab} = z_{ab} i_{ab} \tag{1.1}$$

In the admittance form the network element may be represented as in Fig. 1.5.

y_{ab} is the self-admittance of the network element a—b
j_{ab} is the source current in parallel with the network element a—b

From Fig. 1.5, we have the relation

$$i_{ab} + j_{ab} = y_{ab} v_{ab} \tag{1.2}$$

The series voltage in the impedance form and the parallel source current in the admittance form are related by the equation.

$$-j_{ab} = y_{ab} e_{ab} \tag{1.3}$$

A set of unconnected elements that are depicted in Fig. 1.4 or 1.5 constitute a primitive network. The performance equations for the primitive networks may be either in the form

$$\underline{e} + \underline{v} = [z]\underline{i} \tag{1.4}$$

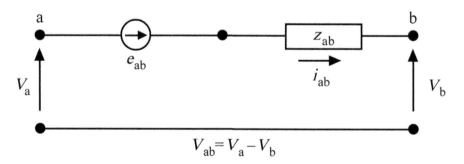

FIGURE 1.4

Primitive network in impedance form.

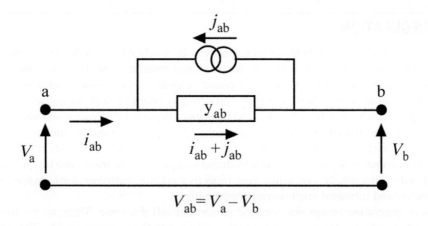

FIGURE 1.5

Primitive network in admittance form.

or in the form

$$\underline{i} + \underline{j}[y]\underline{v} \tag{1.5}$$

In Eqs. (1.4) and (1.5) the matrices $[z]$ or $[y]$ contain the self-impedances or self-admittances denoted by $z_{ab,ab}$ or $y_{ab,ab}$. The off-diagonal elements may in a similar way contain the mutual impedances or mutual admittances denoted by $z_{ab,cd}$ or $y_{ab,cd}$ where ab and cd are two different elements having mutual coupling. If there is no mutual coupling, then the matrices $[z]$ and $[y]$ are diagonal matrices. While in general $[y]$ matrix can be obtained by inverting the $[z]$ matrix, when there is no mutual coupling, elements of $[y]$ matrix are obtained by taking reciprocals of the elements of $[z]$ matrix.

Analysis of any specific power system requires comprehensive knowledge about the components that comprise the power system. Analysis of the system to monitor its operation at a first level includes steady state operation as well as system under faulted condition.

1.5 POWER SYSTEM STABILITY

Power system stability is a word used in connection with alternating current power systems denoting a condition wherein, the various alternators in the system remain in synchronous with each other. Study of this aspect is very important, as otherwise, due to a variety of changes, such as, sudden load loss or increment, faults on lines, short circuits at different locations, circuit opening, reswitching, etc., occuring in the system continuously some where or other may create blackouts.

Study of simple power systems with single machine or a group of machines represented by a single machine, connected to infinite bus gives an insight into the stability problem.

At a first level, study of these topics is very important for electrical power engineering students.

1.6 **DEREGULATION**

The growth of technology in all fields of human activity is closely related to the consumption of electric power. The basic components of electric power system remained essentially the same in spite of great changes taking place in every aspect of power generation, transmission distribution, and consumption of electrical energy. New strategies introduced in operation and control resulted in better economy, higher security, greater reliability, and increased customer participation and satisfaction.

Deregulation was introduced all over the world decomposing erstwhile monolithic vertically integrated power system into gencos, transcos, and discos (discoms) to deal with generation, transmission, and distribution separately. This introduced competitive electricity markets, cheaper electric energy, efficient capacity expansion, cost reflective pricing, customer participation, improved service benefits, and increased employment opportunities.

Gencos or generating companies generate power and sell the same. Transcos are transmitting companies which own the transmission networks and operate the respective grids. They operate to transmit power from gencos to customers. For the service they render, they collect transmission charges. Discos or distribution companies own and operate local distribution networks. They buy and sell electric energy to their customers in different ways.

In a deregulated system an entity, independent system operator (ISO) is created and is given the responsibility of maintaining system security and reliability. ISO is not allowed in buy−sell transactions. He is permitted to own some generation to provide reserve power when needed in some cases. Another entity, market operator is given the responsibility of receiving bids for supply of electric energy for sale and to settle the market operations. In general, transcos operate as ISOs. Practices differ from country to country. Two distinct types of electricity markets came into existence, namely (1) pool type markets and (2) bilateral markets. The ISO plays an important role in pool type markets while his role in bilateral markets is limited.

1.7 **RENEWABLE ENERGY RESOURCES**

The rise of oil prices and green house gas emissions influenced the quest for renewable energy resources and their integration with the electric power grid. The oil prices are increasing by 1.5% per annum over the last two decades. The green house gas emission increased in a short span of about 12 years from 1990 to 2012 by 5.2%, and this is seen as an alarming situation. Relentless efforts are on all over the globe to increase the proportion of green energy as much as possible, so that the limited fossil fuels are conserved, pollution level is controlled and environment is protected.

These developments introduced several new problems in power system analysis. In the context of multiple players in deregulated markets with open access to transmission network, the computation of profits and losses, the determination of tariffs for different players careful, and complicated calculation are required. Marginal cost characteristics are needed to be obtained by the use of Lagrange multipliers. Voltage-Var support from a vast variety of flexible alternating current transmission system (FACT) devices enter into several decision taking processes. Demand side management has attained prominence. Transmission system congestion management also gained importance. The emergence of smart grid concept for realization is revolutionizing the electric power system scenario.

GRAPH THEORY

2.1 INTRODUCTION

Graph theory has many applications in several fields such as engineering, physical, social and biological sciences, linguistics, etc. Any physical situation that involves discrete objects with interrelationships can be represented by a graph. In *Electrical Engineering*, graph theory is used to predict the behavior of the network in analysis. However, for smaller networks, node or mesh analysis is more convenient than the use of graph theory. It may be mentioned that Kirchoff was the first to develop theory of trees for applications to electrical network. The advent of high-speed digital computers has made it possible to use graph theory advantageously for larger network analysis. In this chapter a brief account of graph theory is given that is relevant to power transmission networks and their analysis.

2.2 DEFINITIONS

Element of a graph: Each network element is replaced by a line segment or an arc while constructing a graph for a network. Each line segment or arc is called an *element*. Each potential source is replaced by a short circuit. Each current source is replaced by an open circuit.

Node or vertex: The terminal of an element is called a *node* or a *vertex*.

Edge: An element of a graph is called an *edge.*

Degree: The number of edges connected to a vertex or node is called its *degree.*

Graph: An element is said to be incident on a node, if the node is a terminal of the element. Nodes can be incident to one or more elements. The network can thus be represented by an interconnection of elements. The actual interconnections of the elements give a graph.

Rank: The *rank of a graph* is $n-1$ where n is the number of nodes in the graph.

Subgraph: Any subset of elements of the graph is called a *subgraph.* A subgraph is said to be proper if it consists of strictly less than all the elements and nodes of the graph.

Path: A path is defined as a subgraph of connected elements such that not more than two elements are connected to any one node. If there is a path between every pair of nodes then the graph is said to be connected. Alternatively, a graph is said to be connected if there exists at least one path between every pair of nodes.

Planar graph: A graph is said to be planar, if it can be drawn without-out cross over of edges. Otherwise, it is called nonplanar (Fig. 2.1).

Closed path or loop: The set of elements traversed starting from one node and returning to the same node form a closed path or loop.

Power Systems Analysis. DOI: http://dx.doi.org/10.1016/B978-0-08-101111-9.00002-1

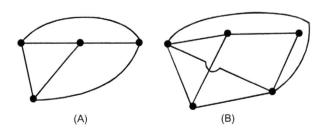

FIGURE 2.1

(A) Planar graph. (B) Nonplanar graph.

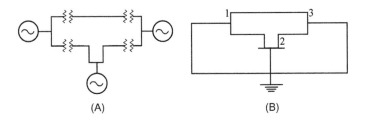

FIGURE 2.2

(A) Power system single-line diagram. (B) Positive sequence network diagram.

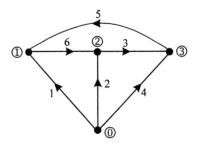

FIGURE 2.3

Oriented connected graph.

Oriented graph: An oriented graph is a graph with direction marked for each element Fig. 2.2A shows the single-line diagram of a simple power network consisting of generating stations, transmission lines, and loads. Fig. 2.2B shows the positive sequence network of the system in Fig. 2.2A. The oriented connected graph is shown in Fig. 2.3 for the same system.

2.3 TREE AND COTREE

Tree: A tree is an oriented connected subgraph of an oriented connected graph containing all the nodes of the graph, but, containing no loops. A tree has $(n-1)$ branches where n is the number of

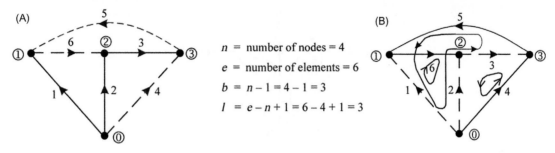

FIGURE 2.4

(A) Tree for the system in Fig. 2.3. (B) Cotree for the system in Fig. 2.3.

nodes of graph G. The branches of a tree are called twigs. The remaining branches of the graph are called links or chords.

Cotree: The links form a subgraph, not necessarily connected called cotree. Cotree is the complement of tree. There is a cotree for every tree.

For a connected graph and subgraph:

1. there exists only one path between any pair of nodes on a tree;
2. every connected graph has at least one tree;
3. every tree has two terminal nodes and;
4. the rank of a tree is $n-1$ and is equal to the rank of the graph.

The number of nodes and the number of branches in a tree are related by

$$b = n - 1 \tag{2.1}$$

If e is the total number of elements then the number of links l of a connected graph with branches b is given by

$$1 = e - b \tag{2.2}$$

Hence, from Eq. (2.1), it can be written that

$$1 = e - n + 1 \tag{2.3}$$

A tree and the corresponding cotree of the graph for the system shown in Fig. 2.3 are indicated in Fig. 2.4A and B.

2.4 BASIC LOOPS

A loop is obtained whenever a link is added to a tree, which is a closed path. As an example to the tree in Fig. 2.4A if the link 6 is added, a loop containing the elements 1−2−6 is obtained. Loops which contain only one link are called *independent loops* or *basic loops*.

It can be observed that the number of basic loops is equal to the number of links given by Eq. (2.2) or (2.3). Fig. 2.5 shows the basic loops for the tree in Fig. 2.4A.

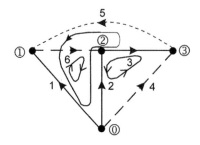

FIGURE 2.5

Basic loops for the tree in Fig. 2.4A.

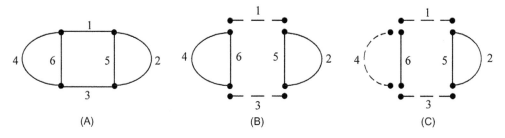

FIGURE 2.6

Graph (A), subgraph (B), and cutest (C).

2.5 CUT-SET

A cut set is a minimal set of branches K of a connected graph G, such that the removal of all K branches divides the graph into two parts. It is also true that the removal of K branches reduces the rank of G by one, provided no proper subset of this set reduces the rank of G by one when it is removed from G.

Consider the graph in Fig. 2.6A.

The rank of the graph = (no. of nodes $n-1$) = $4 - 1 = 3$. If branches 1 and 3 are removed, two subgraphs are obtained as in Fig. 2.6B. Thus 1 and 3 may be a cut set. Also, if branches 1, 4, and 3 are removed, the graph is divided into two subgraphs, as shown in Fig. 2.6C. Branches 1, 4, and 3 may also be a cut set. In both the above cases the rank both of the subgraphs is $1 + 1 = 2$. It can be noted that (1, 3) set is a subset of (1, 4, 3) set. The cut set is a minimal set of branches of the graph, removal of which cuts the graph into two parts. It separates nodes of the graphs into two graphs. Each group is in one of the two subgraphs.

2.6 BASIC CUT-SETS

If each cut-set contains only one branch, then these independent cut-sets are called basic cut-sets. In order to understand basic cut-sets, select a tree. Consider a twig b_k of the tree. If the twig is

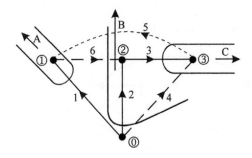

FIGURE 2.7

Cut-set for the tree in Fig. 2.4A.

removed, the tree is separated into two parts. All the links which go from one part of this disconnected tree to the other, together with the twig b_k constitutes a cut-set called basic cut-set. The orientation of the basic cut-set is chosen as to coincide with that of the branch of the tree defining the cut-set. Each basic cut-set contains at least one branch with respect to which the tree is defined which is not contained in the other basic cut-set. For this reason the n-1 basic cut-sets of a tree are linearly independent.

Now consider the tree in Fig. 2.4A.

Consider node (1) and branch or twig 1. Cut-set A contains the branch 1 and links 5 and 6 and is oriented in the same way as branch 1. In a similar way, C cut-set cuts the branch 3 and links 4 and 5 and is oriented in the same direction as branch 3. Finally, cut-set B cuts branch 2 and also links 4, 6, and 5 and is oriented as branch 2, and the cut-sets are shown in Fig. 2.7.

WORKED EXAMPLES

2.1. *For the network shown in figure below, draw the graph and mark a tree. How many trees will this graph have? Mark the basic cut-sets and basic loops.*

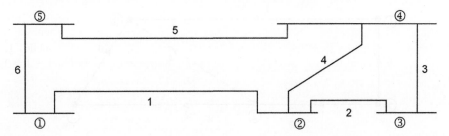

FIGURE E.2.1

Network for 2.1.

Solution:
   Assume that bus (1) is the reference bus

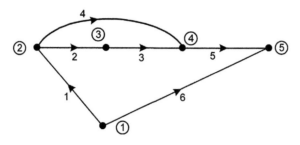

Number of nodes $n = 5$
Number of elements $e = 6$
The graph can be redrawn as

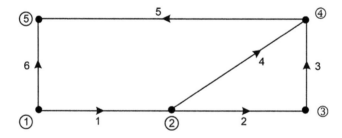

   Tree: A connected subgraph containing all nodes of a graph, but no closed path is called a *tree*.

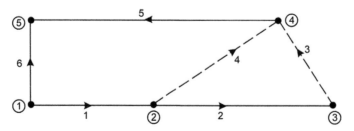

Number of branches $n - 1 = 5 - 1 = 4$
Number of links $= e - b = 6 - 4 = 2$
(*Note:* number of links = number of cotrees).

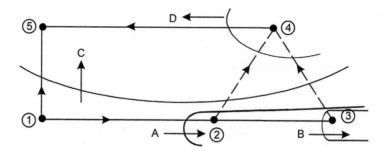

FIGURE E.2.5

The number of basic cut-sets = no. of branches = 4; the cut-sets A, B, C, and D, are shown in figure.

2.2. Show the basic loops and basic cut-sets for the graph shown below and verify any relations that exist between them.

(Take 1−2−3−4 as tree 1.)

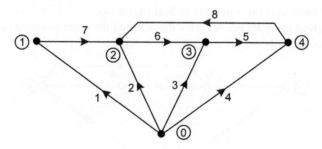

FIGURE E.2.6

Graph for 2.2.

Solution:

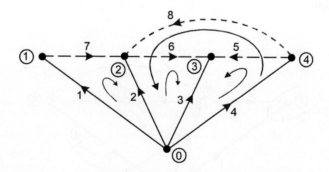

FIGURE E.2.7

Tree and cotree for the graph.

If a link is added to the tree, a loop is formed, loops that contain only one link are called *basic loops*.

Branches,

$$b = n - 1 = 5 - 1 = 4$$
$$1 = e - b = 8 - 4 = 4$$

The four loops are shown in the following figure

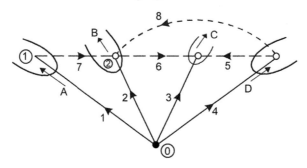

FIGURE E.2.8

Basic cut-sets A, B, C, and D.

The number of basic cuts (4) = number of branches b (4).

2.3. For the graph given in figure below, draw the tree and the corresponding cotree. Choose a tree of your choice and hence write the cut-set schedule.

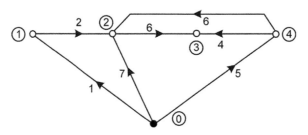

FIGURE E.2.9

Oriented connected graph.

Solution:

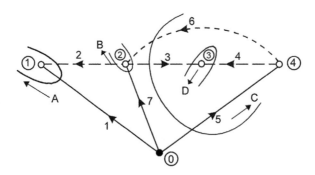

FIGURE E.2.10

Basic cut-sets A, B, C, and D.

The f-cut set schedule (fundamental or basic)

A: 1, 2
B: 2, 7, 3, 6
C: 6, 3, 5
D: 3, 4

2.4. *For the power systems shown in the figure, draw the graph, a tree, and its cotree.*

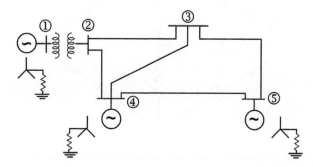

FIGURE E.2.11

Network for 2.4.

Solution:

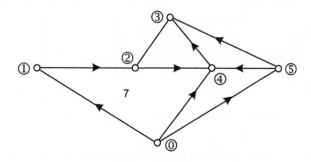

FIGURE E.2.12

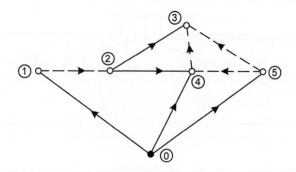

FIGURE E.2.13

Tree and cotree 2.4.

PROBLEMS

P.2.1. Draw the graph for the network shown. Draw a tree and cotree for the graph.

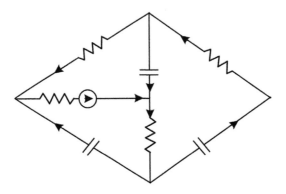

FIGURE P.2.1

Network for P.2.1.

P.2.2. Draw the graph for the circuit shown.

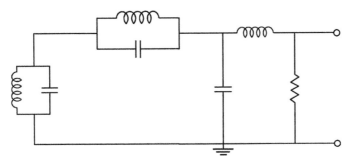

FIGURE P.2.2.

Circuit for P.2.2.

P.2.3. Draw the graph for the network shown.

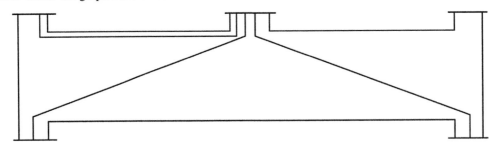

NETWORK FOR P.2.3.

Circuit for P.2.2.

Mark basic cut-sets, basic loops, and open loops.

QUESTIONS

2.1. Explain the following terms:
1. Basic loops
2. Cut set
3. Basic cut-sets

2.2. Explain the relationship between the basic loops and links; basic cut-sets and the number of branches.

2.3. Define the following terms with suitable example:
1. Tree
2. Branches
3. Links
4. Cotree
5. Basic loop

2.4. Write down the relations between the number of nodes, number of branches, number of links, and number of elements.

2.5. Define the following terms.
1. Graph
2. Node
3. Rank of a graph
4. Path

INCIDENCE MATRICES

There are several incidence matrices that are important in developing the various networks matrices, like bus impedance matrix, branch admittance matrix, etc., using singular or nonsingular transformation.

These various incidence matrices are basically derived from the connectivity or incidence of an element to a node, path, cut-set, or loop.

Incidence matrices

The following incidence matrices are of interest in power network analysis.

1. Element-node incidence matrix
2. Bus incidence matrix
3. Branch-path incidence matrix
4. Basic cut-set incidence matrix
5. Augmented cut-set incidence matrix
6. Basic loop incidence matrix
7. Augmented loop incidence matrix

Each of these incidence matrices will now be explained with reference to the power network described in Fig. 2.2A.

3.1 ELEMENT-NODE INCIDENCE MATRIX

Element-node incidence matrix $\overline{A}$ shows the incidence of elements to nodes in the connected graph. The incidence or connectivity is indicated by the operator as follows:

$\alpha_{pq} = 1$ if the pth element is incident to and directed away from the qth node.

$\alpha_{pq} = -1$ if the pth element is incident to and directed towards the qth node.

$\alpha_{pq} = 0$ if the pth element is not incident to the qth node.

The element-node incidence matrix will have the dimension $e \times n$ where "e" is the number of elements and "n" is the number of nodes in the graph. It is denoted by $\overline{A}$.

The element-node incidence matrix for the graph of Fig. 2.3 is shown in Fig. 3.1.

It is seen from the elements of the matrix that

$$\sum_{q=0}^{3} \alpha_{pq} = 0; \quad p = 1, 2, \ldots, 6 \tag{3.1}$$

Power Systems Analysis. DOI: http://dx.doi.org/10.1016/B978-0-08-101111-9.00003-3

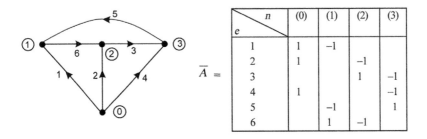

The element-node incidence matrix $\overline{A}$:

$\displaystyle{}^{n}_{e}$	(0)	(1)	(2)	(3)
1	1	−1		
2	1		−1	
3			1	−1
4	1			−1
5		−1		1
6		1	−1	

FIGURE 3.1

Element-node incidence-matrix for the graph of Fig. 2.3.

$A =$

$\displaystyle{}^{e}_{\text{Bus}}$	(1)	(2)	(3)
1	−1		
2		−1	
3		+1	−1
4			−1
5	−1		+1
6	1	−1	

FIGURE 3.2

Bus incidence matrix for graph in Fig. 2.3.

It can be inferred that the columns of $\overline{A}$ are linearly independent. The rank of $\overline{A}$ is less than n the number of nodes in the graph.

3.2 BUS INCIDENCE MATRIX

The network in Fig. 2.2B contains a reference reflected in Fig. 2.3 as a reference node. In fact any node of the connected graph can be selected as the reference node. The matrix obtained by deleting the column corresponding to the reference node in the element-node incidence matrix $\overline{A}$ is called *bus incidence matrix A*. Thus the dimension of this matrix is $e(n-1)$ and the rank will therefore be, $n-1 = b$, where b is the number of branches in the graph. Deleting the column corresponding to node (0) from Fig. 3.1 the bus incidence matrix for the system in Fig. 2.2A is obtained. This is shown in Fig. 3.2.

If the rows are arranged in the order of a specific tree, the matrix A can be partitioned into two submatrices A_b of the dimension $b(n-1)$ and A_1 of dimension $l(n-1)$. The rows of A_b correspond to branches and the rows of A_1 correspond to links. This is shown in Fig. 3.3 for the matrix in Fig. 3.2.

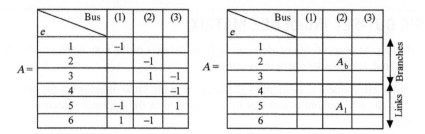

FIGURE 3.3

Partitioning of matrix A.

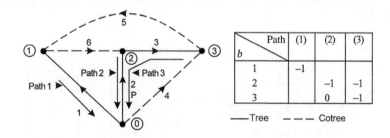

FIGURE 3.4

Branch-path incidence matrix for network.

3.3 BRANCH-PATH INCIDENCE MATRIX K

Branch-path incidence matrix, as the name itself suggests, shows the incidence of branches to paths in a tree. The elements of this matrix are indicated by the operators as follows:

$K_{pq} = 1$ If the pth branch is in the path from qth bus to reference and oriented in the same direction.

$K_{pq} = -1$ If the pth branch is in the path from qth bus to reference and oriented in the opposite direction.

$K_{pq} = 0$ If the pth branch is not in the path from the qth bus to reference.

For the system in Fig. 2.4A the branch-path incidence matrix K is shown in Fig. 3.4. Node (0) is assumed as reference.

While the branch-path incidence matrix relates branches to paths, the submatrix A_b of Fig. 3.3 gives the connectivity between branches and buses. Thus the paths and buses can be related by $A_b K^t = U$ where U is a unit matrix.

Hence

$$K^t = A_b^{-1} \tag{3.2}$$

3.4 BASIC CUT-SET INCIDENCE MATRIX

This matrix depicts the connectivity of elements to basic cut-sets of the connected graph. The elements of the matrix are indicated by the operator as follows:

$\beta_{pq} = 1$ if the pth element is incident to and oriented in the same direction as the qth basic cut-set.
$\beta_{pq} = -1$ if the pth element is incident to and oriented in the opposite direction as the qth basic cut-set.
$\beta_{pq} = 0$ if the pth element is not incident to the qth basic cut-set.

The basic cut-set incidence matrix has the dimension $e \times b$. For the graph in Fig. 2.3A, the basic cut-set incidence matrix B is obtained as in Fig. 2.7 (Fig. 3.5).

It is possible to partition the basic cut-set incidence matrix B into two submatrices U_B and U_l corresponding to branches and links, respectively. For the example on hand the partitioned matrix is shown in Fig. 3.6.

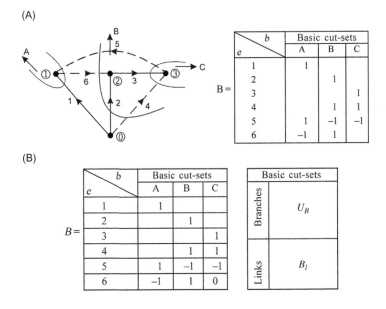

(A)

$B =$

e \ b	Basic cut-sets A	B	C
1	1		
2		1	
3			1
4		1	1
5	1	−1	−1
6	−1	1	

(B)

$B =$

e \ b	Basic cut-sets A	B	C
1	1		
2		1	
3			1
4		1	1
5	1	−1	−1
6	−1	1	0

Basic cut-sets		
Branches	U_B	
Links	B_l	

FIGURE 3.5

Basic cut-set incidence matrix for the graph in (A) drawn and shown.

$$
\begin{bmatrix} 0 & 0 & -1 \\ -1 & 0 & 1 \\ 1 & -1 & 0 \end{bmatrix} \cdot \begin{bmatrix} -1 & 0 & 0 \\ 0 & -1 & 0 \\ 0 & -1 & -1 \end{bmatrix} = \begin{bmatrix} 0 & 1 & 1 \\ 1 & -1 & -1 \\ -1 & 1 & 0 \end{bmatrix}
$$

FIGURE 3.6

Illustration of equation $A_1 K^t = B_1$

The identity matrix U_b shows the one-to-one correspondence between branches and basic cut-sets.

It may be recalled that the incidence of links to buses is shown by submatrix A_1 and the incidence of branches to buses by A_b. There is a one-to-one correspondence between branches and basic cut-sets. Since the incidence of links to buses is given by

$$B_1 A_b = A_1 \tag{3.3}$$

Therefore

$$B_1 = A_1 A_b^{-1} \tag{3.4}$$

However from Eq. (3.2) $K^t = A_b^{-1}$
Substituting this result in Eq. (3.1)

$$B_1 = A_1 K^t \tag{3.5}$$

This is illustrated in Fig. 3.6.

3.5 AUGMENTED CUT-SET INCIDENCE MATRIX $\tilde{B}$

In the basic cut-set incidence matrix the number of cut-sets is equal to the number of branches. The number of cut-sets can be made equal to the number of elements by introducing fictitious or imaginary cut-sets called tie-cut-sets. Each tie cut-set contains only one link of the connected graph. The tie cut-sets for the system in Fig. 2.7 are shown in Fig. 3.7.

An augmented cut-set incidence matrix is formed by adjoining to the basic cut-set incidence matrix columns corresponding to the tie-cut-sets additionally. Just as in the case of basic cut-sets here too, the tie cut-set is oriented in the same direction as the associated link.

The augmented cut-set incidence matrix for the system in Fig. 3.7 is shown in Fig. 3.8.

Matrix $\overline{B}$ may be partitioned as shown in Fig. 3.3 (Fig. 3.9).

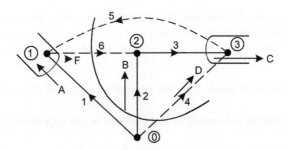

FIGURE 3.7

Tie cut-sets D, E, and F are for each link added.

$\tilde{B} =$

e \ Bus	Basic cut-sets				Tie-sets		
1	1						
2		1					
3			1				
4		1	1	1			
5	1	−1	−1		1		
6	−1	1				1	

FIGURE 3.8

Augmented cut-set incidence matrix for Fig. 3.7.

$\tilde{B} =$

e \ e	Basic cut-sets			Tie-sets		
1	1					
2		1				
3			1			
4		1	1	1		
5	1	−1	−1		1	
6	−1	1				1

e \ e	Basic cut-sets	Tie-sets
1		
2	U_b	O
3		
4		
5	B_l	U_l
6		

FIGURE 3.9

Partitioning of augmented cut-set incidence matrix of Fig. 3.1.

3.6 BASIC LOOP INCIDENCE MATRIX

In Section 2.3 basic loops are defined and in Fig. 3.10 basic loops for the sample system under discussion are shown. Basic loop incidence matrix C shows the incidence of the elements of the connected graph to the basic loops. The incidence of the elements is indicated by the operator as follows:

$\gamma_{pq} = 1$ if the pth element is incident to and oriented in the same direction as the qth basic loop.

$\gamma_{pq} = -1$ if the pth element is incident to and oriented in the opposite direction as the qth basic loop.

$\gamma_{pq} = 0$ if the pth element is not incident to the qth loop.

The basic loop incidence matrix has the dimension $e \times l$ and the matrix is shown in Fig. 3.11. It is possible to partition the basic loop incidence matrix as in Fig. 3.12. The unit matrix U_l shown the one-to-one correspondence of links to basic loops.

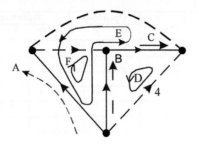

FIGURE 3.10

Basic loops (D, E, F) and open loops (A, B, C).

e \\ 1	D	E	F
1		−1	1
2	−1	1	−1
3	−1	1	
4	1		
5		1	
6		0	1

$$C =$$

FIGURE 3.11

Basic loop incidence matrix for Fig. 3.10.

e \\ b	Basic loops		
1			1
2	−1		−1
3	−1	1	
4	1		
5		1	
6		1	

$$C =$$

e \\ l	Basic loops
Branches	C_b
Links	U_l

FIGURE 3.12

Partitioning of basic loop incidence matrix.

3.7 AUGMENTED LOOP INCIDENCE MATRIX

It has been indicated that the number of basic loops correspond to the number of links. It is possible to augment the matrix by introducing open loops, such that the total number of loops correspond to the number of elements. An open loop is defined as a path between adjacent nodes connected by a

$$\tilde{C} =$$

e / e	Open loops			Basic loops		
	A	B	C	D	E	F
1	1					1
2		1		−1		−1
3			1	−1	1	
4				1		
5					1	
6						1

FIGURE 3.13

Augmented loop incidence matrix.

$$\tilde{C} =$$

	A	B	C	D	E	F
1	1					1
2		1		−1		−1
3			1	−1	1	
4				1		
5					1	
6						1

$$=$$

	Open loops	Basic loops
Branches	U_b	C_b
Links	0	U_l

FIGURE 3.14

Augmented loop incidence matrix.

branch. Thus the number of open loops will be $(e-1)$ equal to b, the number of branches. The open loops are shown in Fig. 3.10 as A, B, and C. The augmented loop incidence matrix is shown in Fig. 3.13.

The augmented loop incidence matrix is a square matrix of dimension $e \times e$ and is nonsingular. The matrix $\tilde{C}$ can be partitioned as shown in Fig. 3.14.

The utility of these augmented incidence matrices will be seen when nonsingular transformation method of obtaining network matrices is discussed.

3.8 NETWORK PERFORMANCE EQUATIONS

The power system network consists of components such as generators, transformers, transmission lines, circuit breakers, capacitor banks, etc., which are all connected together to perform specific function. Some are in series and some are in shunt connection.

Whatever may be their actual configuration, network analysis is performed either by nodal or by loop method. In case of power system, generally, each node is also a bus. Thus in the bus frame of reference the performance of the power network is described by $(n-1)$ independent nodal equations, where n is the total number of nodes. In the impedance form the performance equation, following Ohm's law will be

$$\overline{V} = [Z_{BUS}]\overline{I}_{BUS} \tag{3.6}$$

where $\overline{V}_{BUS}$ = vector of bus voltages measured with respect to a reference bus; $\overline{I}_{BUS}$ = vector of impressed bus currents; $[Z_{BUS}]$ = bus impedance matrix.

The elements of bus impedance matrix are open circuit driving point and transfer impedances. Consider a 3-bus or 3-node system. Then

$$[Z_{BUS}] = \begin{matrix} & \begin{matrix} (1) & (2) & (3) \end{matrix} \\ \begin{matrix} (1) \\ (2) \\ (3) \end{matrix} & \begin{bmatrix} z_{11} & z_{12} & z_{13} \\ z_{21} & z_{22} & z_{23} \\ z_{31} & z_{32} & z_{33} \end{bmatrix} \end{matrix}$$

The impedance elements on the principal diagonal are called driving point impedances of the buses and the off-diagonal elements are called transfer impedances of the buses. In the admittance frame of reference

$$\overline{I}_{BUS} = [Y_{BUS}] \cdot \overline{V}_{BUS} \tag{3.7}$$

where $[Y_{BUS}]$ = bus admittance matrix whose elements are short circuit driving point and transfer admittances.

By definition

$$[Y_{BUS}] = [Z_{BUS}]^{-1} \tag{3.8}$$

In a similar way, we can obtain the performance equations in the branch frame of reference. If b is the number of branches, then b independent branch equation of the form

$$\overline{V}_{BR} = [Z_{BR}] \cdot \overline{I}_{BR} \tag{3.9}$$

describe network performance. In the admittance form

$$\overline{I}_{BR} = [Y_{BR}]\overline{V}_{BR} \tag{3.10}$$

where $\overline{I}_{BR}$ = vector of currents through branches; $\overline{V}_{BR}$ = vector of voltages across the branches; $[Y_{BR}]$ = branch admittance matrix whose elements are short circuit driving point and transfer admittances of the branches of the network; $[Z_{BR}]$ = branch impedance matrix whose elements are open circuit driving point and transfer impedances of the branches of the network.

Like wise, in the loop frame of reference, the performance equation can be described by l independent loop equations where l is the number of links or basic loops. In the impedance from

$$\overline{V}_{LOOP} = [Z_{LOOP}] \cdot \overline{I}_{LOOP} \tag{3.11}$$

and in the admittance form

$$\overline{I}_{LOOP} = [Y_{LOOP}] \cdot \overline{V}_{LOOP} \tag{3.12}$$

where $\overline{V}_{LOOP}$ = vector of basic loop voltages; $\overline{I}_{LOOP}$ = vector of basic loop currents; $[Z_{LOOP}]$ = loop impedance matrix; $[Y_{LOOP}]$ = loop admittance matrix.

WORKED EXAMPLES

3.1. *Obtain the oriented connected graphs for the given power network shown below. Hence obtain C and C̃.*

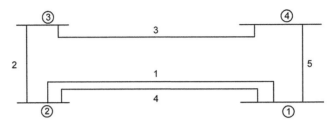

FIGURE E.3.1

Power network.

Solution:

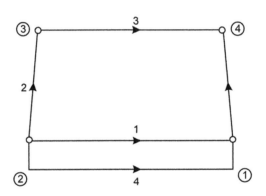

FIGURE E.3.2

Oriented connected graph.

The oriented connected graphs is shown in figure above. The basic and open loops identified are shown below for the tree and cotree selected.

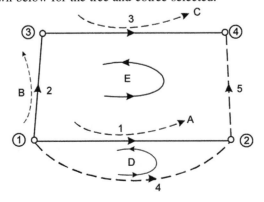

FIGURE E.3.3

Basic and open loops.

The basic loop incidence matrix C is given by

$$
C =
\begin{array}{c|c|c|c}
\diagdown & 1 & D & E \\
e & & & \\
\hline
1 & & -1 & 1 \\
2 & & & -1 \\
3 & & & -1 \\
4 & & 1 & \\
5 & & & 1 \\
\end{array}
$$

FIGURE E.3.4

Basic loop incidence matrix.

The augmented loop incidence matrix $\tilde{C}$ is given by

$$
\tilde{C} =
\begin{array}{c|c|c|c|c|c|c}
\diagdown & 1 & A & B & C & D & E \\
e & & & & & & \\
\hline
1 & & 1 & & & -1 & 1 \\
2 & & & 1 & & & -1 \\
3 & & & & 1 & & -1 \\
4 & & & & & 1 & \\
5 & & & & & & 1 \\
\end{array}
$$

FIGURE E.3.5

Augmented loop incidence matrix.

3.2. *For the network shown in figure form the bus incidence matrix, A, branch-path incidence matrix K, and loop incidence matrix C.*

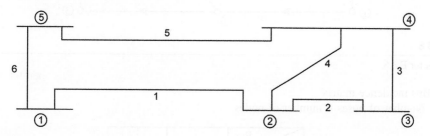

FIGURE E.3.6

Network for 3.2.

Solution:

For the tree and cotree chosen for the graph shown below the basic cut-sets are marked. Bus (1) is taken as reference.

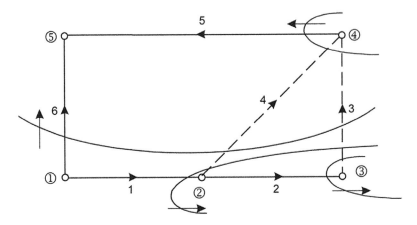

FIGURE E.3.7

Basic cut-sets for E.3.6.

The basic loops are shown in the following figure.

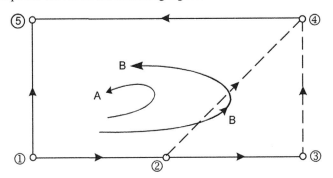

FIGURE E.3.8

Basic loops for E.3.6.

1. Bus incidence matrix
   Number of buses = number of nodes

$A =$

e \ Bus	(2)	(3)	(4)	(5)
1	−1	0	0	0
2	1	−1	0	0
3	0	1	−1	0
4	1	0	−1	0
5	0	0	1	−1
6	0	0	0	−1

FIGURE E.3.9

Bus incidence matrix A for E.3.6.

Branches/Link	Bus e	Bus (2)	(3)	(4)	(5)
1		−1	0	0	0
2		1	−1	0	0
5		0	1	1	−1
6		0	0	0	−1
3		0	1	−1	1
4		1	0	−1	0

$A =$

$=$

	Bus e	Buses
B branches		A_b
L links		A_1

FIGURE E.3.10

Partitioning of matrix A.

2. Branch-path incidence matrix (K):

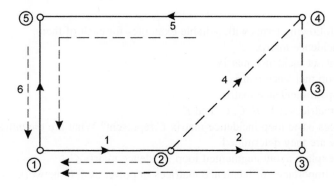

FIGURE E.3.11

Branches and the paths.

Path e	(2)	(3)	(4)	(5)
1	−1	−1		
2		−1	0	
5			1	
6			−1	−1

$K =$

FIGURE E.3.12

Branch path incidence matrix K.

3. Basic loop incidence matrix C:

$$C = \begin{array}{c|c|c|c} \diagdown I \;\; \diagup & & & \\ e & & A & B \\ \hline 1 & 1 & 1 \\ 2 & 0 & 1 \\ 3 & 0 & 1 \\ 4 & 1 & 0 \\ 5 & 1 & 1 \\ 6 & -1 & -1 \end{array}$$

$\diagup I$ e	A	B
1	1	1
2	0	1
3	0	1
4	1	0
5	1	1
6	-1	-1

$C =$

$\diagup I$ e	A	B
1	1	1
2	0	1
5	1	1
6	-1	-1
3	0	1
4	1	0

$= \begin{bmatrix} C_b \\ \\ C_l \end{bmatrix}$

FIGURE E.3.13

Basic loop incidence matrix.

QUESTIONS

3.1. Define the following terms with suitable examples for each of them:
 1. Node incidence matrix
 2. Basic cut-set incidence matrix
 3. Basic loop incidence matrix
 4. Branch-path incidence matrix
3.2. Define the matrices $A, V, \tilde{B}, C, \tilde{C}$, and K
3.3. 1. What does basic loop incidence matrix C represent? What are the entries of this matrix and how are they determined.
 2. Briefly explain about augmented loop incidence matrix $\tilde{C}$
3.4. What is the importance of incidence matrices in power system network solution? Explain.

PROBLEMS

P.3.1. The transpose of the bus incidence matrix of a power system network is given by

$$[A]^t = \begin{array}{c} \\ 1 \\ 2 \\ 3 \end{array} \begin{bmatrix} 1 & 2 & 4 & 5 \\ -1 & 1 & 1 & 0 \\ 0 & 0 & -1 & -1 \\ 0 & -1 & 0 & 1 \end{bmatrix}$$

Draw its oriented graph
P.3.2. Obtain the $B, \tilde{B}, C, \tilde{C}$, and K matrices with usual notation for the graph of P.3.1.
P.3.3. Prove the following relations for the graphs of P.3.1.
 1. $A_b k^t = U$
 2. $\tilde{C}\tilde{B}^t = U$
 3. $C_b = -B_1^t$
 4. $b_1 = A_1 k^t$

P.3.4. For the network shown in figure taking bus 1 as reference prove the following.

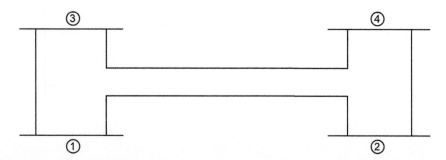

FIGURE P.3.4

Network for P.3.4.

1. $C_b = -B_1^t$
2. $\tilde{C}\tilde{B} = U$
3. $AC_bA_bK^t = U$
4. $B_1 = A_1k^t$

NETWORK MATRICES

4.1 INTRODUCTION

It is indicated in Chapter 1, Introduction, that network solution can be carried out using Ohm's Law and Kirchoff's Law. The impedance model given by

$$e = Z \cdot i$$

or the admittance model

$$i = Y \cdot e$$

can be used depending upon the situation or the type of problem encountered. In network analysis, students of electrical engineering are familiar with nodal analysis and mesh analysis using Kirchoff's laws. In most of the power network solutions, the bus impedance or bus admittance is used. Thus it is necessary to derive equations that relate these various models.

4.2 NETWORK MATRICES

Network matrices can be formed by two methods:
   Viz.

1. Singular transformation
2. Nonsingular transformation.

4.2.1 NETWORK MATRICES BY SINGULAR TRANSFORMATIONS

The network matrices that are used commonly in power system analysis that can be obtained by singular transformation are:

1. Bus admittance matrix
2. Bus impedance matrix
3. Branch admittance matrix
4. Branch impedance matrix
5. Loop impedance matrix
6. Loop admittance matrix.

Power Systems Analysis. DOI: http://dx.doi.org/10.1016/B978-0-08-101111-9.00004-5

4.2.1.1 Bus Admittance Matrix and Bus Impedance Matrix

The bus admittance matrix Y_{BUS} can be obtained by determining the relationship between the variables and parameters of the primitive network described in Section 2.1 to bus quantities of the network using bus incidence matrix. Consider Eq. (1.5)

$$\bar{i} + \bar{j} = [y]\,\bar{v}$$

Premultiplying by $[A^t]$, the transpose of the bus incidence matrix

$$[A^t]\bar{i} + [A^t]\bar{j} = A^t[y]\bar{v} \tag{4.1}$$

Matrix A shows the connections of elements to buses. $[A^t]i$ thus is a vector, wherein, each element is the algebraic sum of the currents that terminate at any of the buses. Following Kirchoff's current law, the algebraic sum of currents at any node or bus must be zero. Hence

$$[A^t] = 0 \tag{4.2}$$

Again $[A^t]$ term indicates the algebraic sum of source currents at each of the buses and must equal the vector of impressed bus currents. Hence,

$$\bar{I}_{BUS} = [A^t]\bar{j} \tag{4.3}$$

Substituting Eqs. (4.2) and (4.3) into Eq. (4.1)

$$\bar{I}_{BUS} = [A^t][y]\bar{v} \tag{4.4}$$

In the bus frame, power in the network is given by

$$[\bar{I}^*_{BUS}]^t V_{BUS} = P_{BUS} \tag{4.5}$$

Power in the primitive network is given by

$$(\bar{j}^*)^t \bar{v} = P \tag{4.6}$$

Power must be invariant, for transformation of variables to be invariant. That is to say, that the bus frame of referee corresponds to the given primitive network in performance. Power consumed in both the circuits is the same.

Therefore

$$[\bar{I}^*_{BUS}]\bar{V}_{BUS} = [\bar{j}^*]\bar{v} \tag{4.7}$$

Conjugate transpose of Eq. (4.3) gives

$$[\bar{I}^*_{BUS}]^t = [\bar{j}^*]^t A^* \tag{4.8}$$

However, as A is real matrix $A = A^*$

$$[\bar{I}^*_{BUS}]^t = (\bar{j}^*)^t[A] \tag{4.9}$$

Substituting Eq. (4.9) into Eq. (4.7)

$$(\bar{j}^*)^t[A] \cdot \bar{V}_{BUS} = (\bar{j}^*)^t\bar{v} \tag{4.10}$$

i.e.,

$$[A]\,\bar{V}_{BUS} = \bar{v} \tag{4.11}$$

Substituting Eq. (4.11) into Eq. (4.4)

$$\bar{I}_{\text{BUS}} = [A^{\text{t}}][y][A]\bar{V}_{\text{BUS}} \tag{4.12}$$

From Eq. (3.7)

$$\bar{I}_{\text{BUS}} = [\bar{Y}_{\text{BUS}}]\,\bar{V}_{\text{BUS}} \tag{4.13}$$

Hence

$$[Y_{\text{BUS}}] = [A^{\text{t}}][y][A] \tag{4.14}$$

Once $[Y_{\text{BUS}}]$ is evaluated from the above transformation, (Z_{BUS}) can be determined from the relation:

$$Z_{\text{BUS}} = Y_{\text{BUS}}^{-1} = \{[A^{\text{t}}][y][A]\}^{-1} \tag{4.15}$$

4.2.1.2 Branch Admittance and Branch Impedance Matrices

In order to obtain the branch admittance matrix Y_{BR}, the basic cut-set incidence matrix $[B]$ is used. The variables and parameters of primitive network are related to the variables and parameters of the branch admittance network.

For the primitive network

$$\bar{i} + \bar{j} = [y]\bar{v} \tag{4.16}$$

Premultiplying by B^{t}

$$[B]^{\text{t}}i + [B]^{\text{t}}j = [B]^{t}[y]\,\bar{v} \tag{4.17}$$

It is clear that the matrix $[B]$ shows the incidence of elements to basic cut-sets.

Each element of the vector $[B^{\text{t}}]\bar{i}$ is the algebraic sum of the currents through the elements that are connected to a basic cut-set. Every cut-set divides the network into two connected subnetworks. Thus each element of the vector $[B^{\text{t}}]\bar{i}$ represents the algebraic sum of the currents entering a subnetwork which must be zero by Kirchoff's law.

Hence

$$[B^{\text{t}}]\bar{i} = 0 \tag{4.18}$$

$[B^{\text{t}}]\bar{j}$ is a vector in which each element is the algebraic sum of the source currents of the elements incident to the basic cut-set and represents the total source current in parallel with a branch.

$$[B^{\text{t}}]\bar{j} = \bar{I}_{\text{BR}} \tag{4.19}$$

therefore,

$$\bar{I}_{\text{BR}} = [B^{\text{t}}][y]\bar{v} \tag{4.20}$$

For power invariance:

$$\bar{I}_{\text{BR}}^{*t} \cdot \bar{V}_{\text{BR}} = \bar{j}^{*t}\bar{v} \tag{4.21}$$

conjugate transpose of Eq. (4.19) gives $j^{*t}[B]^{*} = \bar{I}_{\text{BR}}^{*t}$. Substituting this in Eq. (4.21)

$$(j)^{*t}[B]^{*}\bar{V}_{\text{BR}} = (j^{*})^{t}\bar{v}$$

As $[B]$ is a real matrix

$$[B]^* = [B] \tag{4.22}$$

Hence

$$(J^*)^t [B] \overline{V}_{BR} = (j^*)^t \overline{v} \tag{4.23}$$

i.e.,

$$\overline{v} = [B] \overline{V}_{BR} \tag{4.24}$$

Substituting Eq. (4.24) into Eq. (4.20)

$$\overline{I}_{BR} = [B]^t [y][B] \overline{V}_{BR} \tag{4.25}$$

However, the branch voltages and currents are related by

$$\overline{I}_{BR} = [Y_{BR}] \cdot \overline{V}_{BR} \tag{4.26}$$

comparing Eqs. (4.25) and (4.26)

$$[Y_{BR}] = [B]^t [y][B] \tag{4.27}$$

Since, the basic cut-set matrix $[B]$ is a singular matrix the transformation $[Y_{BR}]$ is a singular transformation of $[y]$. The branch impedance matrix, then, is given by

$$[Z_{BR}] = [Y_{BR}]^{-1}$$
$$[Z]_{BR} = [Y]_{BR}^{-1} = \{[B^t][y][B]\}^{-1} \tag{4.28}$$

4.2.1.3 Loop Impedance and Loop Admittance Matrices

The loop impedance matrix is designated by $[Z_{LOOP}]$. The basic loop incidence matrix $[C]$ is used to obtain $[Z_{LOOP}]$ in terms of the elements of the primitive network.

The performance equation of the primitive network is

$$\overline{v} + \overline{e} = [Z] \overline{i} \tag{4.29}$$

Premultiplying by $[C^t]$

$$[C]^t [\overline{v}] + [C]^t \overline{e} = [C]^t [z] \overline{i} \tag{4.30}$$

As the matrix $[C]$ shows the incidence of elements to basic loops, $[C^t] \overline{v}$ yields the algebraic sum of the voltages around each basic loop.

By Kirchoff's voltage law, the algebraic sum of the voltages around a loop is zero. Hence, $[C^t] \overline{v} = 0$. Also $[C^t] \overline{e}$ gives the algebraic sum of source voltages around each basic loop, so that,

$$\overline{V}_{LOOP} = [C^t] \overline{e} \tag{4.31}$$

From power invariance condition for both the loop and primitive networks

$$(\overline{I}_{LOOP}^*)^t \cdot \overline{V}_{LOOP} = (\overline{i}^*)^t \overline{e} \tag{4.32}$$

for all values of $\overline{e}$.

Substituting $\overline{V}_{LOOP}$ from Eq. (4.31)

$$(\overline{I}_{LOOP}^*)^t[C^t]\overline{e} = [i^*]^t\overline{e} \tag{4.33}$$

Therefore,

$$\overline{i} = [C^*]^t\overline{I}_{LOOP} \tag{4.34}$$

However, as $[C]$ is a real matrix $[C] = [C^*]$
Hence

$$\overline{i} = [C]\overline{I}_{LOOP} \tag{4.35}$$

From Eqs. (4.31), (4.32), and (4.30)

$$\overline{V}_{LOOP} = [C^t][z][C]\overline{I}_{LOOP} \tag{4.36}$$

However, for the loop frame of reference the performance equation from Eq. (3.11) is

$$\overline{V}_{LOOP} = [Z_{LOOP}]\overline{I}_{LOOP} \tag{4.37}$$

comparing Eqs. (4.57) and (4.58)

$$[Z_{LOOP}] = [C^t][z][C] \tag{4.38}$$

$[C]$ being a singular matrix the transformation Eq. (4.38) is a singular transformation of $[z]$. The loop admittance matrix is obtained from

$$[Y_{LOOP}] = [Z_{LOOP}^{-1}] = \{[C]^t[z][C]\}^{-1} \tag{4.39}$$

Summary of Singular Transformations
$[z]^{-1} = [y]$
$[A^t][y][A] = [Y_{BUS}];$
$[Y_{BUS}]^{-1} = [Z_{BUS}]$
$[B^t][y][B] = [Y_{BR}];$
$[Y_{BR}]^{-1} = [Z_{BR}]$
$[C^t][Z][C] = [Z_{LOOP}];$
$[Z_{LOOP}]^{-1} = [Y_{LOOP}]$

4.2.2 NETWORK MATRICES BY NONSINGULAR TRANSFORMATION

The augmented incidence matrices are nonsingular and hence, the network matrices obtained by using these incidence matrices are through nonsingular transformations.

4.2.2.1 Branch Admittance Matrix

Consider the augmented cut-set incidence matrix obtained in Section 3.5. The augmented network is obtained by connecting a fictitious branch in series with each link of the original network.

As the fictitious branch cannot be allowed to alter the interconnected network, its admittance is set equal to zero. The current source of the fictitious branch carries the same current as the current source of the link so that the voltage across the fictitious branch is zero. A tie cut-set, thus, can be treated as a fictitious branch in series with a cut-set containing a link (Fig. 4.1).

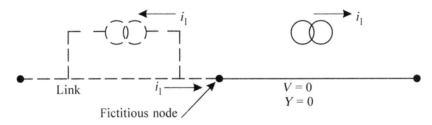

FIGURE 4.1

Tie cut-set.

For the branch frame of reference the performance Eq. (3.10) gives

$$\bar{I}_{BR} = [\bar{Y}_{BR}] \cdot \bar{V}_{BR}$$

However, the elements of Y_{BR} can be obtained directly from the admittance matrix $\bar{Y}_{BR}$ of the augmented network. A relationship will now be established between the matrices $[\bar{Y}_{BR}]$ and $[y]$ of the primitive network.

In the following treatment, B, C, Y, y, U, Z, z are all matrices, i, j, v, V are all vectors unless otherwise stated.

The performance equation for the primitive network is given by Eq. (1.5) as

$$\underline{i} + \underline{j} = [y]\,(\underline{v})$$

Premultiplying it by $[\hat{B}^{-t}]$, we get

$$[\hat{B}^t]\underline{i} + [\hat{B}^t]\underline{j} = [\hat{B}^t][y]\,\underline{v} \tag{4.40}$$

Eq. (4.40) can be put in the partitioned form (refer Fig. 3.9)
Since

$$\begin{array}{|c|c|} \hline U_b & 0 \\ \hline B_1 & U_1 \\ \hline \end{array} = \hat{B}$$

$$\begin{bmatrix} \{U_b\} & \{B_l^t\} \\ [0] & [U_l] \end{bmatrix} \begin{bmatrix} i_b \\ i_l \end{bmatrix} + \begin{bmatrix} \{U_b\} & \{B_l^t\} \\ [0] & [U_l] \end{bmatrix} \begin{bmatrix} j_b \\ j_l \end{bmatrix} = \begin{bmatrix} \{U_b\} & \{B_l^t\} \\ [0] & [U_l] \end{bmatrix} [y] \cdot [v] \tag{4.41}$$

$$\begin{bmatrix} i_b + \{B_l^t\}i_l \\ i_l \end{bmatrix} + \begin{bmatrix} i_b + \{B_l^t\}j_l \\ j_l \end{bmatrix} = \text{R.H.S. of Eq.(4.41)} \tag{4.42}$$

$$i_b + \{B_l^t\}i_l = \{B^t\}\underline{i} \tag{4.43}$$

and also

$$i_{j_b} + \{B_1^t\}j_l = \{B^t\}\underline{j} \tag{4.44}$$

It is seen that

$$[B^t]i = 0 \tag{4.45}$$

and

$$\{B\}^t \underline{j} = \underline{I}_{BR} \tag{4.46}$$

The L.H.S. of Eq. (4.41) becomes

$$\left[\frac{[o]}{i_l} \right] + \left[\frac{I_{BR}}{j_l} \right] = \left[\frac{I_{BR}}{i_l + \underline{j}_l} \right] \tag{4.47}$$

Thus Eq. (4.47) becomes

$$\hat{I}_{BR} = \left[\frac{I_{BR}}{i_l + \underline{j}_l} \right] \tag{4.48}$$

$(i_l + j_l)$ is equal to the algebraic sum of the source currents of a fictitious branch and its associated link. From Eqs. (4.40) and (4.48):

$$\hat{I}_{BR} = \{\hat{B}^t\}[y]\underline{v} \tag{4.49}$$

But, the voltage across the fictitious branches is zero. So the voltage vector of the augmented network becomes

$$\hat{V}_{BR} = \left[\frac{V_{BR}}{\overline{0}} \right] \tag{4.50}$$

the voltages across the original network are given by Eq. (4.25) as

$$\underline{v} = [B]\overline{V}_{BR}$$

However, since

$$[B]\underline{V}_{BR} = [\hat{B}]\hat{V}_{BR} \tag{4.51}$$

$$\underline{v} = [\hat{B}]\hat{\underline{V}}_{BR} \tag{4.52}$$

Substituting Eq. (4.53) into Eq. (4.48)

$$\hat{\underline{I}}_{BR} = [\hat{B}^t][y][\hat{B}]\hat{\underline{V}}_{BR} \tag{4.53}$$

the performance equation of the augmented network is

$$\hat{\underline{I}}_{BR} = \hat{Y}_{BR} \cdot \hat{V}_{BR} \tag{4.54}$$

written in partitioned form, Eq. (4.75) becomes

$$\begin{bmatrix} [Y_A] & [Y_B] \\ [Y_C] & [Y_D] \end{bmatrix} = \begin{bmatrix} [U_b] & [B_l]^1 \\ [0] & [U_l] \end{bmatrix} \begin{bmatrix} y_{bb} & y_{bl} \\ y_{lb} & y_{ll} \end{bmatrix} \begin{bmatrix} [U_b] & [0] \\ B_l & [U_l] \end{bmatrix} \tag{4.55}$$

where

$$[y] = \begin{bmatrix} [y_{bb}] & [y_{bl}] \\ [y_{lb}] & [y_{ll}] \end{bmatrix} \tag{4.56}$$

and

y_{bb} = Primitive admittance matrix of branches
$[y_{bl}] = [y_{lb}]^t$ = Primitive admittance matrix of the network whose elements are the mutual admittances between branches and links
$[y_{ll}]$ = Primitive admittance matrix of links.

From Eq. (4.56), after performing the operation

$$\begin{bmatrix} [Y_A] & [Y_B] \\ [Y_C] & [Y_D] \end{bmatrix} = \begin{bmatrix} [y_{bb} + B_l' y_{lb}] & [y_{bl} + B_l' y_{ll}] \\ [y_{lb}] & [y_{ll}] \end{bmatrix} \begin{bmatrix} [U_b] & [0] \\ [B_l] & [U_l] \end{bmatrix}$$

$$= \left[\begin{array}{c|c} [y_{bb} + B_l^t + y_{lb}B_l + B_l' y_{ll}B_l] & [y_{bl} + B_l' y_{ll}] \\ \hline [y_{lb} + y_{ll}B_l] & [y_{ll}] \end{array} \right] \quad (4.57)$$

From which equation we obtain

$$[Y_A] = [Y_{bb}] + [B_l^t][y_{lb}] + [y_{bl}][B_l^t][y_{ll}][B_l] \quad (4.58)$$

Again

$$[Y_{BR}] = [B^t][y][B] \quad (4.59)$$

$$= [\{U_b\}\{B_l^t\}] \begin{bmatrix} \{y_{bb}\} & \{y_{bl}\} \\ \{y_{lb}\} & \{y_{ll}\} \end{bmatrix} \cdot \begin{bmatrix} \{U_b\} \\ \{B_l\} \end{bmatrix} \quad (4.60)$$

$$= [\{U_b\}\{y_{bb}\} + \{B_l^t\}\{y_{lb}\}\{U_{bl}\}\{y_{bl}\} + \{B_l^t\}\{y_{ll}\}] \begin{bmatrix} \{U_b\} \\ \{B_l\} \end{bmatrix} \quad (4.61)$$

$$= [y_{bb}] + [B_l^t][y_{lb}] + [y_{bl}][B_l] + [B_l^t][y_{ll}][B_l] \quad (4.62)$$

Thus, we find that from Eqs. (4.58) and (4.62)

$$[Y_A] = [Y_{BR}] \quad (4.63)$$

The branch impedance can be found from

$$[Z_{BR}] = [Y_A]^{-1} \quad (4.64)$$

4.2.2.2 Loop Impedance and Loop Admittance Matrices

Augmented loop incidence matrix $\hat{C}$ is used to relate the variables and parameters of the primitive network to those of an augmented network. In order to obtain the loop impedance matrix $[Z_{LOOP}]$, a fictitious link is connected in parallel with each branch of the original network to obtain the augmented network. The impedance of each fictitious link is made zero. A voltage source is put in the parallel fictitious link with voltage set equal and opposite to the voltage across the associated branch. This preserves the performance of the network unchanged (Fig. 4.2).

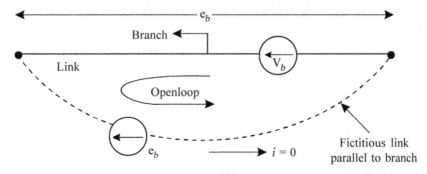

FIGURE 4.2

Fictitious link to augmented network.

It follows that the current through the link is zero. Thus an open loop is formed. An open loop contains a branch and a fictitious link carrying no current. The performance equation in the loop frame of reference from Eq. (3.11) is

$$\underline{V}_{\text{LOOP}} = [Z_{\text{LOOP}}] \cdot \underline{I}_{\text{LOOP}}$$

For the augmented network, it will be

$$\hat{V}_{\text{LOOP}} = [\hat{Z}_{\text{LOOP}}] \cdot \hat{I}_{\text{LOOP}} \tag{4.65}$$

The primitive network performance Eq. (1.4) is

$$\underline{v} + \underline{e} = [z]\underline{i}$$

Premultiplying by $[\hat{C}]^t$

$$[\hat{C}]^t \underline{v} + [\hat{C}]^t \underline{e} = [\hat{C}]^t [z]\underline{i} \tag{4.66}$$

From Fig. (3.14) equation (4.66) can be partitioned as

$$\begin{bmatrix} U_b & 0 \\ \hline C_b^t & U_l \end{bmatrix} \begin{bmatrix} \underline{v}_b \\ \underline{v}_l \end{bmatrix} + \begin{bmatrix} U_b & 0 \\ \hline C_b^t & U_l \end{bmatrix} \begin{bmatrix} e_b \\ e_l \end{bmatrix} = \begin{bmatrix} U_b & 0 \\ \hline C_b^t & U_l \end{bmatrix} [z] [i] \tag{4.67}$$

The partitioning is done to separate out the voltages associated with branches and links. Performing the multiplication

$$\begin{bmatrix} \underline{v}_b \\ C_b^t \underline{v}_b + \underline{v}_l \end{bmatrix} + \begin{bmatrix} e_b \\ C_b^t e_b + e_l \end{bmatrix} = \begin{bmatrix} U_b & 0 \\ C_b^t & U_l \end{bmatrix} \begin{bmatrix} z_{bb} & z_{bl} \\ z_{lb} & z_{ll} \end{bmatrix} [i] \tag{4.68}$$

where

$[z_{bb}]$ = Primitive impedance matrix of branches
$[z_{bl}] = [z_{lb}]^t$ = Primitive impedance matrix whose elements are mutual impedance between branches and links
$[Z_{ll}]$ = Primitive impedance matrix of links

$$[C]_b^t \underline{v}_b + \underline{v}_l = [C^t]\underline{v}$$
$$[C_b^t]\underline{e}_b + \underline{e}_l = [C^t]\underline{e}$$

Further, since the algebraic sum of voltages around a closed loop is zero,

$$[C^t]\,\underline{v} = 0$$

Because the sum of the source voltages in a loop is the loop voltage vector

$$[C^t] = \underline{e} = \underline{V}_{\text{LOOP}}$$

Eq. (4.68) becomes

$$\begin{bmatrix} \underline{v}_b \\ 0 \end{bmatrix} + \begin{bmatrix} \underline{e}_b \\ \underline{V}_{\text{LOOP}} \end{bmatrix} = \begin{bmatrix} \underline{v}_b + \underline{e}_b \\ \underline{V}_{\text{LOOP}} \end{bmatrix} = \begin{bmatrix} U_b & 0 \\ C_b^t & U_l \end{bmatrix} \begin{bmatrix} z_{bb} & z_{bl} \\ z_{lb} & z_{ll} \end{bmatrix} [i] \tag{4.69}$$

$\underline{v}_b + \underline{e}_b$ is a vector in which each element is equal to the algebraic sum of the source voltages in an open loop. The augmented loop voltage vector

$$[\hat{V}_{\text{LOOP}}] = \begin{bmatrix} \underline{e}_b + \underline{v}_b \\ \underline{V}_{\text{LOOP}} \end{bmatrix} \tag{4.70}$$

From Eqs. (4.66) and (4.70)

$$[\hat{C}^t][z]\underline{i} = \hat{V}_{\text{LOOP}} \tag{4.71}$$

Noting that the currents in the open loop are zero, the current vectors of the augmented network

$$\bar{\hat{I}}_{\text{LOOP}} = \begin{bmatrix} \bar{0} \\ \hat{I}_{\text{LOOP}} \end{bmatrix} \tag{4.72}$$

The currents through elements of the original network are given by Eq. (4.36)

$$\underline{i} = [C]\bar{I}_{\text{LOOP}} \tag{4.73}$$

It is seen that

$$[C]\underline{I}_{\text{LOOP}} = [\hat{C}]\hat{I}_{\text{LOOP}} \tag{4.74}$$

so that

$$\underline{i} = [\hat{C}]\bar{\hat{I}}_{\text{LOOP}} \tag{4.75}$$

substitution from Eq. (4.75) into Eq. (4.70) and using Eq. (4.71)

$$\hat{V}_{\text{LOOP}} = [\hat{C}]^t[z][\hat{C}]\hat{I}_{\text{LOOP}} \tag{4.76}$$

However, the performance equation for the augmented loop frame is

$$\hat{V}_{\text{LOOP}} = [\hat{Z}]_{\text{LOOP}} \cdot \hat{I}_{\text{LOOP}} \tag{4.77}$$

it follows that

$$[\hat{Z}]_{\text{LOOP}} = [\hat{C}]^t[z][\hat{C}] \tag{4.78}$$

Rewriting Eq. (4.78) in a partitioned form

$$\begin{bmatrix} \hat{Z}_A & \hat{Z}_B \\ \hline \hat{Z}_C & \hat{Z}_D \end{bmatrix} = \begin{bmatrix} U_b & 0 \\ C_b^t & U_l \end{bmatrix} \cdot \begin{bmatrix} z_{bb} & z_{bl} \\ z_{lb} & z_{ll} \end{bmatrix} \begin{bmatrix} U_b & C_b \\ 0 & U_l \end{bmatrix} \tag{4.79}$$

$$\hat{Z}_D = [C_b]^t[z_{bb}][C_b] + [z_{lb}][C_b] + [C_b]^t[z_{bl}] + [z_{ll}] \tag{4.80}$$

We have also

$$[Z_{LOOP}] = [C^t][z][C] \tag{4.81}$$

$$= [C_b^t \, U_l] \begin{bmatrix} z_{bb} & z_{bl} \\ z_{lb} & z_{ll} \end{bmatrix} \begin{bmatrix} C_b \\ U_l \end{bmatrix} \tag{4.82}$$

Then

$$Z_{LOOP} = C_b^t[z_{bb}][c_b] + [z_{lb}][c_b] + C_b^t[z_{bl}] + z_{ll} \tag{4.83}$$

Comparing Eqs. (4.80) and (4.83)

$$Z_{LOOP} = \hat{Z}_D$$

The loop admittance matrix, if required, is then

$$[Y_{LOOP}] = [\hat{Z}_D]^{-1}$$

It is left to the student to prove as an exercise that

$$Z_{BR} = A_b Z_{BUS} A_b^t$$

and

$$Z_{BUS} = K^t Z_{BR} K$$

Summary of Nonsingular Transformations

$$[Y_{BR}] = [y_{bb}] + [B_l^t][Y_{lb}] + [Y_{bl}][B_l] + [B_l^t][Y_{ll}][B_l]$$

where $[y] = \begin{bmatrix} y_{bb} & y_{bl} \\ y_{lb} & y_{ll} \end{bmatrix}$ for the primitive network

$$[Z_{LOOP}] = [C_b^t][z_{bb}][C_b] + [z_{lb}][c_b] + [C_b^t][z_{bl}] + [z_{ll}]$$

where $[Z] = \begin{bmatrix} z_{bb} & z_{bl} \\ z_{lb} & z_{ll} \end{bmatrix}$ for the primitive network

$$[Z_{BR}] = [Y_{BR}]^{-1}$$

$$[Y_{LOOP}] = [Z_{LOOP}]^{-1}$$

$$[Z_{BR}] = [A_b][Z_{BUS}][A_b^t]$$

$$[Z_{BUS}] = [k^t] \cdot [Z_{BR}] \cdot [k]$$

4.3 BUS ADMITTANCE MATRIX BY DIRECT INSPECTION

Bus admittance matrix can be obtained for any network, if there are no mutual impedances between elements, by direct inspection of the network. This is explained by taking an example.

Consider the three bus power system as shown in Fig. 4.3.

The equivalent circuit is shown in Fig. 4.4. The generator is represented by a voltage source in series with the impedance. The three transmission lines are replaced by their "π equivalents."

The equivalent circuit is further simplified as in Fig. 4.5 combining the shunt admittance wherever feasible.

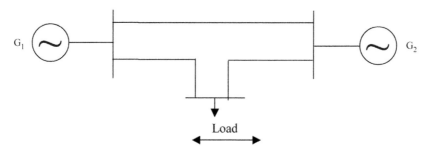

FIGURE 4.3

Three Bus power system.

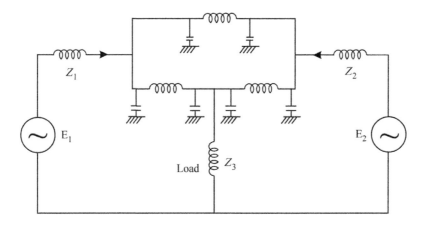

FIGURE 4.4

Equivalent circuit for network in Fig. 4.3.

The three nodes are at voltage V_1, V_2, and V_3, respectively, above the ground. The Kirchoff's nodal current equations are written as follows:

At Node 1:

$$I_1 = I_7 + I_8 + I_4$$
$$I_1 = (V_1 - V_2)Y_7 + (V_1 - V_3)Y_8 + V_1 Y_4 \tag{4.84}$$

At Node 2:

$$I_2 = I_5 + I_9 - I_7$$
$$= V_2 Y_5 + (V_2 - V_3)Y_9 - (V_1 - V_2)Y_7 \tag{4.85}$$

At Node 3:

$$I_3 = I_8 + I_9 - I_6$$
$$= (V_1 - V_3)Y_8 + (V_2 - V_3)Y_9 + V_3 Y_6 \tag{4.86}$$

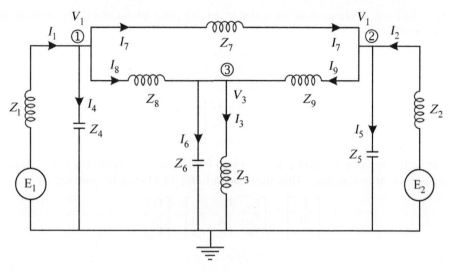

FIGURE 4.5

Simplified equivalent circuit.

Rearranging the terms, the equations will become

$$I_1 = V_1(Y_4 + Y_7 + Y_8) - V_2 Y_7 - V_3 Y_8 \qquad (4.87a)$$

$$I_2 = -V_1 Y_7 + V_2(Y_5 + Y_7 + Y_9) - V_3 Y_9 \qquad (4.87b)$$

$$I_3 = -V_1 Y_8 - V_2 Y_9 + V_3(Y_6 + Y_8 + Y_9) \qquad (4.87c)$$

The last of the above equations may be rewritten as

$$-I_3 = -V_1 Y_8 - V_2 Y_9 + V_3(Y_6 + Y_8 + Y_9) \qquad (4.88)$$

Thus we get the matrix relationship from the above

$$\begin{bmatrix} I_1 \\ I_2 \\ -I_3 \end{bmatrix} = \begin{bmatrix} (Y_4 + Y_7 + Y_8) & -Y_7 & -Y_8 \\ -Y_7 & (Y_5 + Y_7 + Y_9) & -Y_9 \\ -Y_8 & -Y_9 & (Y_6 + Y_8 + Y_9) \end{bmatrix} \cdot \begin{bmatrix} V_1 \\ V_2 \\ V_3 \end{bmatrix} \qquad (4.89)$$

It may be recognized that the diagonal terms in the admittance matrix at each of the nodes are the sum of the admittances of the branches incident to the node. The off-diagonal terms are the negative of these admittances branch-wise incident on the node. Thus the diagonal element is the negative sum of the off-diagonal elements. The matrix can be written easily by direct inspection of the network.

The diagonal elements are denoted by

$$\left.\begin{aligned} Y_{11} &= Y_4 + Y_7 + Y_8 \\ Y_{22} &= Y_5 + Y_7 + Y_9 \\ Y_{33} &= Y_6 + Y_8 + Y_9 \end{aligned}\right\} \qquad (4.90)$$

They are called self-admittances of the nodes or driving point admittances. The off-diagonal elements are denoted by

$$
\left.\begin{aligned}
Y_{12} &= -Y_7 \\
Y_{13} &= -Y_8 \\
Y_{21} &= -Y_7 \\
Y_{23} &= -Y_8 \\
Y_{31} &= -Y_8 \\
Y_{32} &= -Y_9
\end{aligned}\right\}
\tag{4.91}
$$

using double suffix denoting the nodes across which the admittances exist. They are called mutual admittances or transfer admittances. Thus the relation in Eq. (4.91) can be rewritten as

$$
\begin{bmatrix} I_1 \\ I_2 \\ -I_3 \end{bmatrix} = \begin{bmatrix} Y_{11} & Y_{12} & Y_{13} \\ Y_{21} & Y_{22} & Y_{23} \\ Y_{31} & Y_{32} & Y_{33} \end{bmatrix} \cdot \begin{bmatrix} V_1 \\ V_2 \\ V_3 \end{bmatrix}
\tag{4.92}
$$

$$
\bar{I}_{\text{BUS}} = [Y_{\text{BUS}}] \cdot \bar{V}_{\text{BUS}}
\tag{4.93}
$$

In power systems, each node is called a bus. Thus, if there are n independent buses, the general expression for the source current towards the node i is given by

$$
I_i = \sum_{j=1}^{n} Y_{ij} V_j; \quad i \neq j
\tag{4.94}
$$

WORKED EXAMPLES

E.4.1. **Form the Y_{BUS} by using singular transformation for the network shown in Fig. E.4.1 including the generator buses.**

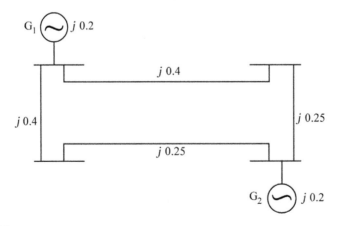

FIGURE E.4.1

Network for E4.1.

Solution:

The given network is represented in admittance form (Fig. E.4.2).

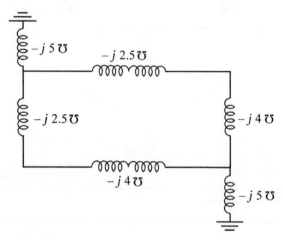

FIGURE E.4.2

Network in admittance form.

The oriented graph is shown in Fig. E.4.3.

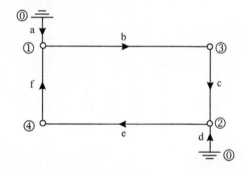

FIGURE E.4.3

Oriented graph for network in E.4.2.

The above graph can be converted into the following form for convenience (Fig. E.4.4) the element-node incidence matrix is given by

$$\hat{A} = \begin{array}{c|ccccc} e \setminus n & 0 & 1 & 2 & 3 & 4 \\ \hline a & +1 & -1 & 0 & 0 & 0 \\ b & 0 & +1 & 0 & -1 & 0 \\ c & 0 & 0 & -1 & +1 & 0 \\ d & +1 & 0 & -1 & 0 & 0 \\ e & 0 & 0 & +1 & 0 & -1 \\ f & 0 & -1 & 0 & 0 & +1 \end{array}$$

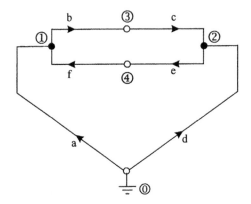

FIGURE E.4.4

Oriented graph in another mode.

Bus incidence matrix is obtained by deleting the column corresponding to the reference bus.

$$
A = \begin{array}{c|cccc}
e \backslash b & 1 & 2 & 3 & 4 \\
\hline
a & -1 & 0 & 0 & 0 \\
b & +1 & 0 & -1 & 0 \\
c & 0 & -1 & +1 & 0 \\
d & 0 & -1 & 0 & 0 \\
e & 0 & +1 & 0 & -1 \\
f & -1 & 0 & 0 & +1
\end{array}
$$

$$
A^t = \begin{array}{c|cccccc}
b \backslash e & a & b & c & d & e & f \\
\hline
1 & -1 & 1 & 0 & 0 & 0 & -1 \\
2 & 0 & 0 & -1 & -1 & 1 & 0 \\
3 & 0 & -1 & 1 & 0 & 0 & 0 \\
4 & 0 & 0 & 0 & 0 & -1 & 1
\end{array}
$$

The bus admittance matrix

$$Y_{BUS} = [A]^t[y][A]$$

$$
[y][A] = \begin{array}{c|cccccc}
 & a & b & c & d & e & f \\
\hline
a & y_a & 0 & 0 & 0 & 0 & 0 \\
b & 0 & y_b & 0 & 0 & 0 & 0 \\
c & 0 & 0 & y_c & 0 & 0 & 0 \\
d & 0 & 0 & 0 & y_d & 0 & 0 \\
e & 0 & 0 & 0 & 0 & y_e & 0 \\
f & 0 & 0 & 0 & 0 & 0 & y_f
\end{array}
\begin{array}{cccc}
(1) & (2) & (3) & (4) \\
-1 & 0 & 0 & 0 \\
1 & 0 & -1 & 0 \\
0 & -1 & 1 & 0 \\
0 & -1 & 0 & 0 \\
0 & 1 & 0 & -1 \\
-1 & 0 & 0 & 1
\end{array}
$$

$$
= \begin{bmatrix}
5 & 0 & 0 & 0 \\
-2.5 & 0 & 2.5 & 0 \\
0 & 4 & -4 & 0 \\
0 & 5 & 0 & 0 \\
0 & -4 & 0 & 4 \\
2.5 & 0 & 0 & -2.5
\end{bmatrix}
$$

$$Y_{BUS} = [A]^t[y][A] = \begin{array}{c} \\ (1) \\ (2) \\ (3) \\ (4) \end{array} \begin{array}{cccccc} a & b & c & d & e & f \\ \end{array} \begin{bmatrix} -1 & 1 & 0 & 0 & 0 & -1 \\ 0 & 0 & -1 & -1 & 1 & 0 \\ 0 & -1 & 1 & 0 & 0 & 0 \\ 0 & 0 & 0 & 0 & -1 & 1 \end{bmatrix} \begin{bmatrix} 2 & 0 & - & 0 \\ -2.5 & 0 & 2.5 & 0 \\ 0 & 4 & -4 & 0 \\ 0 & 5 & 0 & 0 \\ 0 & -4 & 0 & 4 \\ 2.5 & 0 & 0 & -2.5 \end{bmatrix}$$

whence,

$$Y_{BUS} = \begin{bmatrix} -10 & 0 & 2.5 & 2.5 \\ 0 & -13 & 4 & 4 \\ 2.5 & 4 & -6.5 & 0 \\ 2.5 & 4 & 0 & -6.5 \end{bmatrix}$$

E.4.2. **Find the Y_{BUS} using singular transformation for the system shown in** Fig. E.4.5.

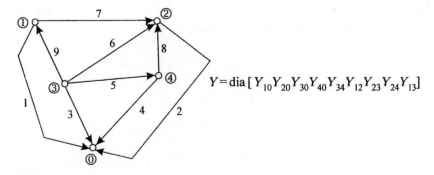

$$Y = dia\,[\,Y_{10}\,Y_{20}\,Y_{30}\,Y_{40}\,Y_{34}\,Y_{12}\,Y_{23}\,Y_{24}\,Y_{13}]$$

FIGURE E.4.5

Graph for system in E.4.2.

Solution:

The graph may be redrawn for convenient as follows (see Fig. E.4.6.):

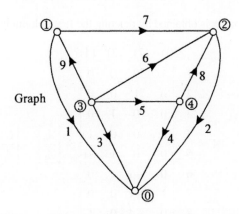

FIGURE E.4.6

Graph for E.4.5 redrawn.

A tree and a co-tree are identified as shown in Fig. E.4.7.

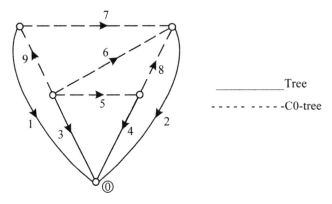

_____ Tree

- - - - - - - - - - - -C0-tree

FIGURE E.4.7

Trees and co-trees for E.4.2.

The element mode incidence matrix $\hat{A}$ is given by

$$\hat{A} = $$

	(0)	(1)	(2)	(3)	(4)
1	−1	1	0	0	0
2	−1	0	1	0	0
3	−1	0	0	1	0
4	−1	0	0	0	1
5	0	0	0	1	−1
6	0	0	−1	1	0
7	0	1	−1	0	0
8	0	0	−1	0	1
9	0	−1	0	1	0

The bus incidence matrix is obtained by deleting the first column taking (0) node as reference.

$$A = \begin{array}{c} 1 \\ 2 \\ 3 \\ 4 \\ 5 \\ 6 \\ 7 \\ 8 \\ 9 \end{array}
\begin{matrix} (1) & (2) & (3) & (4) \end{matrix}
\begin{bmatrix} 1 & 0 & 0 & 0 \\ 0 & 1 & 0 & 0 \\ 0 & 0 & 1 & 0 \\ 0 & 0 & 0 & 1 \\ 0 & 0 & 1 & -1 \\ 0 & -1 & 1 & 0 \\ 1 & -1 & 0 & 0 \\ 0 & -1 & 0 & 1 \\ -1 & 0 & 1 & 0 \end{bmatrix} = \begin{bmatrix} A_b \\ \hline A_l \end{bmatrix} = \begin{bmatrix} U \\ \hline A_l \end{bmatrix}$$

Given

$$[y] = \begin{bmatrix}
y_{10} & 0 & 0 & 0 & 0 & 0 & 0 & 0 & 0 \\
0 & y_{20} & 0 & 0 & 0 & 0 & 0 & 0 & 0 \\
0 & 0 & y_{30} & 0 & 0 & 0 & 0 & 0 & 0 \\
0 & 0 & 0 & y_{40} & 0 & 0 & 0 & 0 & 0 \\
0 & 0 & 0 & 0 & y_{34} & 0 & 0 & 0 & 0 \\
0 & 0 & 0 & 0 & 0 & y_{23} & 0 & 0 & 0 \\
0 & 0 & 0 & 0 & 0 & 0 & y_{12} & 0 & 0 \\
0 & 0 & 0 & 0 & 0 & 0 & 0 & y_{24} & 0 \\
0 & 0 & 0 & 0 & 0 & 0 & 0 & 0 & y_{3}
\end{bmatrix}$$

$$Y_{\text{BUS}} = A^{t}[y]A$$

$$[y][A] = \begin{bmatrix}
y_{10} & 0 & 0 & 0 & 0 & 0 & 0 & 0 & 0 \\
0 & y_{20} & 0 & 0 & 0 & 0 & 0 & 0 & 0 \\
0 & 0 & y_{30} & 0 & 0 & 0 & 0 & 0 & 0 \\
0 & 0 & 0 & y_{40} & 0 & 0 & 0 & 0 & 0 \\
0 & 0 & 0 & 0 & y_{34} & 0 & 0 & 0 & 0 \\
0 & 0 & 0 & 0 & 0 & y_{23} & 0 & 0 & 0 \\
0 & 0 & 0 & 0 & 0 & 0 & y_{12} & 0 & 0 \\
0 & 0 & 0 & 0 & 0 & 0 & 0 & y_{24} & 0 \\
0 & 0 & 0 & 0 & 0 & 0 & 0 & 0 & y_{3}
\end{bmatrix}
\begin{bmatrix}
1 & 0 & 0 & 0 \\
0 & 1 & 0 & 0 \\
0 & 0 & 1 & 0 \\
0 & 0 & 0 & 1 \\
0 & 0 & 1 & -1 \\
0 & -1 & 1 & 0 \\
1 & -1 & 0 & 0 \\
0 & -1 & 0 & 1 \\
-1 & 0 & 1 & 0
\end{bmatrix}$$

$$= \begin{bmatrix}
y_{10} & 0 & 0 & 0 \\
0 & y_{20} & 0 & 0 \\
0 & 0 & y_{30} & 0 \\
0 & 0 & 0 & y_{40} \\
0 & 0 & y_{34} & -y_{34} \\
0 & -y_{23} & y_{23} & 0 \\
y_{12} & -y_{12} & 0 & 0 \\
0 & -y_{24} & 0 & y_{24} \\
-y_{13} & 0 & y_{13} & 0
\end{bmatrix}$$

$$[A_{t}][y][A] = \begin{array}{c} \\ (1) \\ (2) \\ (3) \\ (4) \end{array}
\begin{array}{ccccccccc}
1 & 2 & 3 & 4 & 5 & 6 & 7 & 8 & 9 \\
\end{array}
\begin{bmatrix}
1 & 0 & 0 & 0 & 0 & 0 & 1 & 0 & -1 \\
0 & 1 & 0 & 0 & 0 & -1 & -1 & -1 & 0 \\
0 & 0 & 1 & 0 & 1 & 1 & 0 & 0 & 1 \\
0 & 0 & 0 & 1 & -1 & 0 & 0 & 1 & 0
\end{bmatrix}
\begin{bmatrix}
y_{10} & 0 & 0 & 0 \\
0 & y_{20} & 0 & 0 \\
0 & 0 & y_{30} & 0 \\
0 & 0 & 0 & y_{40} \\
0 & 0 & y_{34} & -y_{34} \\
0 & -y_{23} & y_{23} & 0 \\
y_{12} & -y_{12} & 0 & 0 \\
0 & -y_{24} & 0 & y_{24} \\
-y_{13} & 0 & y_{13} & 0
\end{bmatrix}$$

$$Y_{BUS} = \begin{bmatrix} (y_{10} + y_{12} + y_{13}) & -y_{12} & -y_{13} & 0 \\ -y_{12} & (y_{20} + y_{12} + y_{23} + y_{24}) & -y_{23} & -y_{24} \\ -y_{13} & -y_{23} & (y_{30} + y_{13} + y_{23} + y_{34}) & -y_{34} \\ 0 & -y_{24} & -y_{34} & (y_{40} + y_{34} + y_{24}) \end{bmatrix}$$

E.4.3. Derive an expression for Z_{loop} **for the oriented graph shown in** Fig. E.4.8.

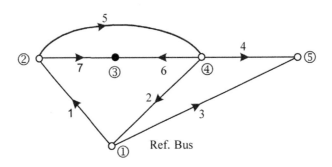

FIGURE E.4.8

Oriented graph for E4.3.

Solution:

Consider the tree and co-tree identified in Fig. E.4.9.

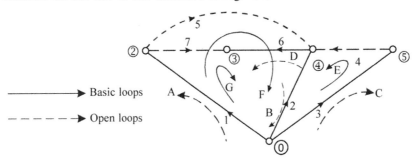

FIGURE E.4.9

Basic loops and open loops for E.4.3.

The augmented loop incidence matrix $\hat{C}$ is obtained as shown in the figure.

$$\hat{C} = \begin{array}{|c|c|c|c|c|c|c|c|} \hline e & 1 & A & B & C & D & E & F & G \\ \hline 1 & 1 & & & & & 0 & 1 & 1 \\ \hline 2 & & 1 & & & & 1 & 1 & 1 \\ \hline 3 & & & 1 & & & 1 & 0 & 0 \\ \hline 4 & & & & 1 & & 0 & 0 & -1 \\ \hline 5 & & & & & & 1 & & \\ \hline 6 & & & & & & & 1 & \\ \hline 7 & & & & & & & & 1 \\ \hline \end{array} = \begin{bmatrix} U_b & C_b \\ 0 & U_l \end{bmatrix}$$

The basic loop incidence matrix

e	1	E	F	G
1	0	1	1	
2	1	1	1	
3	1	0	0	
6	0	0	−1	
4	1			
5		1		
7			1	

$$C = \begin{bmatrix} C_b \\ \hline 1 \end{bmatrix}$$

$$Z_{\text{LOOP}} = [C^t][Z][C]$$

$$= [C_b^t [U_1]] \begin{bmatrix} Z_{bb} & Z_{bl} \\ Z_{lb} & Z_{ll} \end{bmatrix} \begin{bmatrix} C_b \\ U_l \end{bmatrix}$$

$$[Z_{\text{LOOP}}] = C_b^t[Z_{bb}]C_b + [Z_{bb}]C_b + C_b^t[Z_{bl}] + [Z_{ll}]$$

Note: It is not necessary to form the augmented loop incidence matrix for this problem only loop incidence matrix suffices].

E.4.4. **For the system shown in** Fig. E.4.10, **obtain** Y_{BUS} **by inspection method. Take bus (1) as reference. The impedance marked are in p.u.**

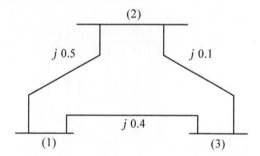

FIGURE E.4.10

Network for E.4.4.

Solution:

$$Y_{\text{BUS}} = \begin{matrix} (2) \\ (3) \end{matrix} \begin{bmatrix} \dfrac{1}{j0.5} + \dfrac{1}{j0.1} & -\dfrac{1}{j0.1} \\ -\dfrac{1}{j0.1} & \dfrac{1}{j0.1} + \dfrac{1}{j0.4} \end{bmatrix} \begin{matrix} (2) \\ (3) \end{matrix} \begin{matrix} (2) \qquad (3) \\ \begin{bmatrix} -j12 & +j10 \\ j10 & -j12.5 \end{bmatrix} \end{matrix}$$

E.4.5. Consider the linear graph shown below which represents a **4 bus transmission system** with all the shunt admittance lumped together. Each line has a series impedance of $(0.02 + j0.08)$ and half-line charging admittance of $j0.02$. Compute the Y_{BUS} by singular transformation. Compute the Y_{BUS} also by inspection.

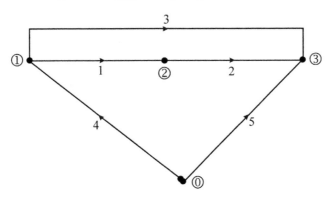

FIGURE E.4.11

Graph for E4.5.

Solution:

The half-line charging admittances are all connected to ground. Taking this ground as reference and eliminating it, the bus incidence matrix is given by

$$
A = \begin{array}{c|cccc}
 & (0) & (1) & (2) & (3) \\
\hline
1 & 0 & 1 & -1 & 0 \\
\hline
2 & 0 & 0 & 1 & -1 \\
\hline
3 & 0 & 1 & 0 & -1 \\
\hline
4 & 1 & -1 & 0 & 0 \\
\hline
5 & 1 & 0 & 0 & -1
\end{array} \, a
$$

incidence matrix is given by the transport of the bus

$$
A^t = \begin{array}{c|ccccc}
 & 1 & 2 & 3 & 4 & 5 \\
\hline
0 & 0 & 0 & 0 & 1 & 1 \\
\hline
1 & 1 & 0 & 1 & -1 & 0 \\
\hline
2 & -1 & 1 & 0 & 0 & 0 \\
\hline
3 & 0 & -1 & -1 & 0 & -1
\end{array}
$$

The primitive admittance matrix [y] is shown below.

$$[y] = \begin{bmatrix} y_1 & 0 & 0 & 0 & 0 \\ 0 & y_2 & 0 & 0 & 0 \\ 0 & 0 & y_3 & 0 & 0 \\ 0 & 0 & 0 & y_4 & 0 \\ 0 & 0 & 0 & 0 & y_5 \end{bmatrix}$$

The admittance of all the branches are the same, i.e.,

$$y_1 = y_2 = y_3 = y_4 = y_5 = \frac{1}{Z} = \frac{1}{0.02 + j0.08}$$

$$[y] = \begin{bmatrix} 2.94 - j11.75 & 0 & 0 & 0 & 0 \\ 0 & 2.94 - j11.75 & 0 & 0 & 0 \\ 0 & 0 & 2.94 - j11.75 & 0 & 0 \\ 0 & 0 & 0 & 2.94 - j11.75 & 0 \\ 0 & 0 & 0 & 0 & 2.94 - j11.75 \end{bmatrix}$$

$$[y][A] = \begin{bmatrix} 2.94 - j11.75 & 0 & 0 & 0 & 0 \\ 0 & 2.94 - j11.75 & 0 & 0 & 0 \\ 0 & 0 & 2.94 - j11.75 & 0 & 0 \\ 0 & 0 & 0 & 2.94 - j11.75 & 0 \\ 0 & 0 & 0 & 0 & 2.94 - j11.75 \end{bmatrix} \cdot \begin{bmatrix} 0 & 1 & -1 & 0 \\ 0 & 0 & 1 & -1 \\ 0 & 1 & 0 & -1 \\ 1 & -1 & 0 & 0 \\ 1 & 0 & 0 & -1 \end{bmatrix}$$

$$= \begin{bmatrix} 0 + j0 & 2.94 - j11.75 & -2.94 + j11.75 & 0 + j0 \\ 0 + j0 & 0 + j0 & 2.94 - j11.75 & -2.94 + j11.75 \\ 0 + j0 & 2.94 - j11.75 & 0 + j0 & -2.94 + j11.75 \\ 2.94 - j11.75 & -2.94 + j11.75 & 0 + j0 & 0 + j0 \\ 2.94 - j11.75 & 0 + j0 & 0 + j0 & -2.94 + j11.75 \end{bmatrix}$$

$$Y_{BUS} = [A^t][y][A] \begin{bmatrix} 0 & 0 & 0 & 1 & 1 \\ 1 & 0 & 1 & -1 & 0 \\ -1 & 1 & 0 & 0 & 0 \\ 0 & -1 & -1 & 0 & -1 \end{bmatrix} \cdot \begin{bmatrix} 0 + j0 & 2.94 - j11.75 & -2.94 + j11.75 & 0 + j0 \\ 0 + j0 & 0 + j0 & 2.94 - j11.75 & -2.94 + j11.75 \\ 0 + j0 & 2.94 - j11.75 & 0 + j0 & -2.94 + j11.75 \\ 2.94 - j11.75 & -2.94 + j11.75 & 0 + j0 & 0 + j0 \\ 2.94 - j11.75 & 0 + j0 & 0 + j0 & -2.94 + j11.75 \end{bmatrix}$$

$$Y_{BUS} = \begin{bmatrix} 5.88 - j23.5 & -2.94 + j11.75 & 0 + j0 & -2.94 + j11.75 \\ -2.94 + j11.75 & 8.82 - j35.25 & -2.94 + j11.75 & -2.94 + j11.75 \\ 0 + j0 & -2.94 + j11.75 & 5.88 - j23.5 & -2.94 + j11.75 \\ -2.94 + j11.75 & -2.94 + j11.75 & -2.94 + j11.75 & 8.82 - j35.7 \end{bmatrix}$$

Solution by inspection including line charging admittances:

$$y_{00} = y_{01} + y_{03} + y_{0\frac{1}{2}} + y_{0\ 3/2}$$

$$y_{00} = [2.94 - j11.75 + j0.02 + j0.02 + 2.94 - j11.75]$$

$$y_{00} = [5.88 - j23.46]$$

$$y_{00} = y_{22} = 5.88 - j23.46$$

$$y_{11} = y_{10} + y_{12} + y_{13} + y_{10/2} + y_{12/2} + y_{13/2}$$

$$y_{33} = [y_{30} + y_{31} + y_{32} + y_{30/2} + y_{31/2} + y_{32/2}]$$

$$y_{11} = y_{33} = 3(2.94 - j11.75) + 3(j0.02) = 8.82 - j35.25 + j0.006$$

$$y_{11} = y_{33} = [8.82 - j35.19]$$

the off-diagonal elements are

$$y_{01} = y_{10} = -y_{01} = -2.94 + j11.75$$
$$y_{12} = y_{21} = (-y_{02}) = 0$$
$$y_{03} = y_{30} = (-y_{03}) = -2.94 + j11.75$$
$$y_{2b} = y_{32} = (-y_{23}) = (-2.94 + j11.75)$$
$$y_{31} = y_{13} = (-y_{13}) = (-2.94 + j11.75)$$

$$Y_{\text{BUS}} = \begin{bmatrix} 5.88 - j23.46 & -2.94 + j11.75 & 0 & -2.94 + j11.75 \\ -2.94 + j11.75 & 8.82 - j35.19 & -2.94 + j11.75 & -2.94 + j11.75 \\ 0 & -2.94 + j11.75 & 5.88 - j23.46 & -2.94 + j11.75 \\ -2.94 + j11.75 & -2.94 + j11.75 & -2.94 + j11.75 & 8.82 - j35.19 \end{bmatrix}$$

The slight changes in the imaginary part of the diagonal elements are due to the line charging capacitances which are not neglected here.

E.4.6. **A power system consists of 4 buses. Generators are connected at buses 1 and 3 reactances of which are $j0.2$ and $j0.1$, respectively. The transmission lines are connected between buses 1-2, 1-4, 2-3, and 3-4 and have reactances $j0.25$, $j0.5$, $j0.4$, and $j0.1$, respectively. Find the bus admittance matrix (1) by direct inspection (2) using bus incidence matrix and admittance matrix.**

Solution:

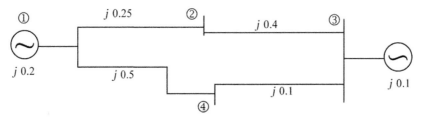

FIGURE E.4.12

Network for E.4.6.

Taking bus (1) as reference, the graph is drawn as shown in Fig. E.4.13.

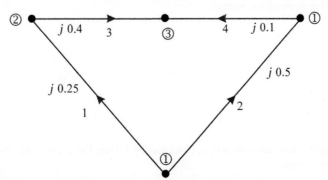

FIGURE E.4.13

Graph for E4.6.

Only the network reactances are considered. Generator reactances are not considered.
By direct inspection:

$$Y_{BUS} =$$

	(1)	(2)	(3)	(4)
1	$\dfrac{1}{j0.25}+\dfrac{1}{j0.5}$	$-\dfrac{1}{j0.25}$	0	$-\dfrac{1}{j0.5}$
2	$-\dfrac{1}{j0.25}$	$\dfrac{1}{j0.4}+\dfrac{1}{j0.25}$	$-\dfrac{1}{j0.4}$	0
3	0	$-\dfrac{1}{j0.4}$	$\dfrac{1}{j0.4}+\dfrac{1}{j0.1}$	$-\dfrac{1}{j0.1}$
4	$-\dfrac{1}{j0.5}$	0	$-\dfrac{1}{j0.1}$	$\dfrac{1}{j0.1}+\dfrac{1}{j0.5}$

This reduces to

$$Y_{BUS} =$$

	(1)	(2)	(3)	(4)
1	$-j6.0$	$j4.0$	0	$j2.0$
2	$j4.0$	$-j6.5$	$j2.5$	0
3	0	$j2.5$	$-j12.5$	$j10$
4	$+j2$	0	$j10$	$-j12$

Deleting the reference bus (1)

$$Y_{BUS} =$$

	(2)	(3)	(4)
(2)	$-j6.5$	$j2.5$	0
(3)	$j2.5$	$-j12.5$	$j10$
(4)	0	$j10$	$-j12.0$

By singular transformation
The primitive impedance matrix

$$[z] = \begin{array}{c} \\ 1 \\ 2 \\ 3 \\ 4 \end{array} \begin{array}{cccc} 1 & 2 & 3 & 4 \\ \begin{bmatrix} j0.25 & 0 & 0 & 0 \\ 0 & j0.5 & 0 & 0 \\ 0 & 0 & j0.4 & 0 \\ 0 & 0 & 0 & j0.1 \end{bmatrix} \end{array}$$

The primitive admittance matrix is obtained by taking the reciprocals of z elements since there are no matrices.

$$[y] = \begin{array}{c} \\ 1 \\ 2 \\ 3 \\ 4 \end{array} \begin{array}{cccc} 1 & 2 & 3 & 4 \\ \begin{bmatrix} -j4 & & & \\ & j2 & & \\ & & -j2.5 & \\ & & & -j10 \end{bmatrix} \end{array}$$

The bus incidence matrix is from the graph:

$$A = \begin{array}{c} \\ 1 \\ 2 \\ 3 \\ 4 \end{array} \begin{array}{ccc} (2) & (3) & (4) \\ \begin{array}{|c|c|c|} \hline -1 & 0 & 0 \\ \hline 0 & 0 & -1 \\ \hline +1 & -1 & 0 \\ \hline 0 & -1 & +1 \\ \hline \end{array} \end{array}$$

and

$$y \cdot A = \begin{bmatrix} j4 & 0 & 0 \\ 0 & 0 & j2 \\ -j2.5 & j2.5 & 0 \\ 0 & j10 & -j10 \end{bmatrix}$$

$$A^t y A = \begin{bmatrix} -1 & 0 & 1 & 0 \\ 0 & 0 & -1 & -1 \\ 0 & -1 & 0 & 1 \end{bmatrix} \cdot \begin{bmatrix} j4 & 0 & 0 \\ 0 & 0 & j2 \\ -j2.5 & j2.5 & 0 \\ 0 & j10 & -j10 \end{bmatrix} = \begin{bmatrix} -j6.5 & j2.5 & 0 \\ j2.5 & -j12.5 & j10 \\ 0 & j10 & -j12.0 \end{bmatrix}$$

E.4.7. For the system shown in Fig. E.4.14 form Y_{BUS}.

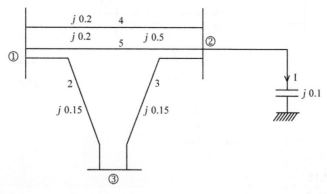

FIGURE E.4.14

Network for E.4.7.

Solution:

Solution is obtained using singular transformation

The primitive admittance matrix is obtained by inverting the primitive impedance as

$$
[y] = \begin{array}{c} \\ 1 \\ 2 \\ \\ \\ \end{array}
\begin{array}{c}
\begin{array}{ccccc} 1 & 2 & 3 & 4 & 5 \end{array} \\
\begin{bmatrix}
10 & 0 & 0 & 0 & 0 \\
0 & 6.66 & 0 & 0 & 0 \\
0 & 0 & 6.66 & 0 & 0 \\
0 & 0 & 0 & -0.952 & 2.381 \\
0 & 0 & 0 & 2.831 & -0.952
\end{bmatrix}
\end{array}
$$

From the graph shown in Fig. E.4.15, the element-node incidence matrix is given by

$$
\bar{A} =
\begin{array}{c|cccc}
e\backslash node & 0 & 1 & 2 & 3 \\
\hline
1 & -1 & 0 & +1 & 0 \\
2 & 0 & +1 & -1 & 0 \\
3 & 0 & +1 & -1 & 0 \\
4 & 0 & -1 & 0 & +1 \\
5 & 0 & 0 & -1 & +1
\end{array}
$$

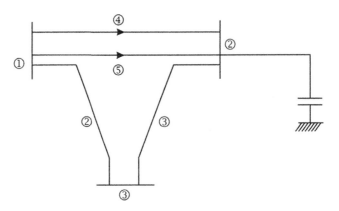

FIGURE E.4.15

Graph for E4.7.

Taking bus zero as reference and eliminating its column, the bus incidence matrix A is given by

$$
A = \begin{matrix} & (1) & (2) & (3) \\ 1 \\ 2 \\ 3 \\ 4 \\ 5 \end{matrix}
\begin{bmatrix}
0 & +1 & 0 \\
+1 & -1 & 0 \\
+1 & -1 & 0 \\
-1 & 0 & +1 \\
0 & -1 & +1
\end{bmatrix}
$$

$$
[y] = \begin{matrix} 1 \\ 2 \end{matrix}
\begin{bmatrix}
10 & 0 & 0 & 0 & 0 \\
0 & 6.66 & 0 & 0 & 0 \\
0 & 0 & 6.66 & 0 & 0 \\
0 & 0 & 0 & -0.952 & 2.381 \\
0 & 0 & 0 & 2.831 & -0.952
\end{bmatrix}
\cdot
\begin{bmatrix}
0 & 1 & 0 \\
1 & -1 & 0 \\
1 & -1 & 0 \\
-1 & 0 & 1 \\
0 & -1 & 1
\end{bmatrix}
$$

The bus admittance matrix Y_{BUS} is obtained from

$$
Y_{BUS} = A^t y A
$$

$$
= \begin{bmatrix}
0 & +1 & +1 & -1 & 0 \\
+1 & -1 & -1 & 0 & -1 \\
0 & 0 & 0 & +1 & +1
\end{bmatrix}
\cdot
\begin{bmatrix}
0 & 10 & 0 \\
-6.66 & 0 & 6.66 \\
0 & -6.66 & 6.66 \\
1.42 & -1.42 & 0 \\
1.42 & -1.42 & 0
\end{bmatrix}
$$

$$
= \begin{bmatrix}
9.518 & -2.858 & -6.66 \\
-2.8580 & 19.51 & -6.66 \\
-6.66 & -6.66 & 13.32
\end{bmatrix}
$$

E.4.8. Consider a six element network with admittances y_1, y_2, and y_3 for branches and y_4, y_5, and y_6 for links corresponding to a tree and co-tree. There are no mutual admittances. Obtain Y_{BR}, by nonsingular transformation.

Tree and co-tree for the given graph from Fig. E.4.16.

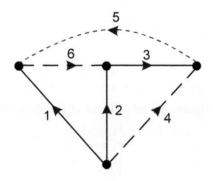

FIGURE E.4.16

Network graph with branches and links for E.4.8.

Solution:

The augmented cut-set incidence matrix is

$$
\tilde{B} = \begin{array}{c} \\ 1 \\ 2 \\ 3 \\ 4 \\ 5 \\ 6 \end{array}
\begin{array}{c} \begin{array}{cccccc} 1 & 2 & 3 & 4 & 5 & 6 \end{array} \\
\left[\begin{array}{ccc|ccc}
1 & 0 & 0 & 0 & 0 & 0 \\
0 & 1 & 0 & 0 & 0 & 0 \\
0 & 0 & 1 & 0 & 0 & 0 \\ \hline
0 & 1 & 1 & 1 & 0 & 0 \\
1 & -1 & -1 & 0 & 1 & 0 \\
-1 & 1 & 0 & 0 & 0 & 1
\end{array} \right] \end{array}
= \left[\begin{array}{c|c} U_b & 0 \\ \hline B_l & U_l \end{array} \right]
$$

$$Y_{BR} = [y_{bb}] + [B_l^t] + [y_{lb}] + [y_{bl}] + [B_l] + [B_l^t][y_{ll}] + [B_l]$$

Since, there are no mutuals

$$Y_{BR} = [y_{bb}] + [B_l^t] + [y_{ll}] + [B_l]$$

$$
B_1 = \begin{bmatrix} 0 & 1 & 1 \\ 1 & -1 & -1 \\ -1 & 1 & 0 \end{bmatrix}; \quad
[B_l^t] = \begin{bmatrix} 0 & 1 & -1 \\ 1 & -1 & 1 \\ 1 & -1 & 0 \end{bmatrix}; \quad
y_{ll} = \begin{bmatrix} y_4 & 0 & 0 \\ 0 & y_5 & 0 \\ 0 & 0 & y_6 \end{bmatrix}
$$

$$
[B_l^t][y_{ll}] = \begin{bmatrix} 0 & 1 & -1 \\ 1 & -1 & 1 \\ 1 & -1 & 0 \end{bmatrix} \cdot
\begin{bmatrix} y_4 & 0 & 0 \\ 0 & y_5 & 0 \\ 0 & 0 & y_6 \end{bmatrix} =
\begin{bmatrix} 0 & y_5 & -y_6 \\ y_4 & -y_5 & y_6 \\ y_4 & -y_5 & 0 \end{bmatrix}
$$

$$[B_l^t][y_{ll}][B_l] = \begin{bmatrix} 0 & y_5 & -y_6 \\ y_4 & -y_5 & y_6 \\ y_4 & -y_5 & 0 \end{bmatrix} \begin{bmatrix} 0 & 1 & 1 \\ 1 & -1 & -1 \\ -1 & 1 & 0 \end{bmatrix} = \begin{bmatrix} y_5 + y_6 & -y_5 - y_6 & -y_5 \\ -y_5 - y_6 & y_4 + y_5 + y_6 & +(y_4 + y_5) \\ -y_5 & y_4 + y_5 & y_4 + y_5 \end{bmatrix}$$

$$Y_{BR} = \begin{bmatrix} y_1 & 0 & 0 \\ 0 & y_2 & 0 \\ 0 & 0 & y_3 \end{bmatrix} + \begin{bmatrix} y_5 + y_6 & -y_5 - y_6 & -y_5 \\ -y_5 - y_6 & y_4 + y_5 + y_6 & +(y_4 + y_5) \\ -y_5 & y_4 + y_5 & y_4 + y_5 \end{bmatrix}$$

$$= \begin{bmatrix} y_1 + y_5 + y_6 & -y_5 - y_6 & -y_5 \\ -y_5 - y_6 & y_2 + y_5 + y_6 + y_4 & y_5 + y_4 \\ -y_5 & y_4 + y_5 & y_3 + y_4 + y_5 \end{bmatrix}$$

E.4.9. **For the graph given in figure, the admittance of the elements is**

$$Y_1 = -j\,2.0$$
$$Y_2 = -j\,2.0$$
$$Y_3 = -j\,3.0$$
$$Y_4 = -j\,2.0$$
$$Y_5 = -j\,4.0$$
$$Y_6 = -j\,3.0$$

Obtain A, B, and K matrices

Determine $Y_{BUS} = [A^t][y][A]$

Obtain $Y_{BR} = [B]^t[y][B]$

Determine $Z_{BR} = [Y_{BR}]^{-1}$

Calculate $Z_{BUS} = [k^t][Z_{BR}][k]$

Show that $Y_{BUS} \cdot Z_{BUS} = [I]$

Solution:

The required incidence matrices are

$$
\begin{array}{c}
\quad\quad\quad (1)\ \ (2)\ \ (3) \\
A = \begin{array}{c} 1 \\ 2 \\ 3 \\ 4 \\ 5 \\ 6 \end{array}
\begin{bmatrix}
-1 & 0 & 0 \\
0 & -1 & 0 \\
0 & 1 & -1 \\
0 & 0 & -1 \\
-1 & 0 & 1 \\
1 & 1 & 0
\end{bmatrix}
\end{array}
$$

$$Y_{BUS} = A^t[y]A$$

where $|y|$ is primitive admittance matrix

$$A^t|y| = \begin{bmatrix} -1 & 0 & 0 & 0 & -1 & 1 \\ 0 & -1 & 1 & 0 & 0 & -1 \\ 0 & 0 & -1 & -1 & 1 & 0 \end{bmatrix} \cdot \begin{bmatrix} y_1 & 0 & 0 & 0 & 0 & 0 \\ 0 & y_2 & 0 & 0 & 0 & 0 \\ 0 & 0 & y_3 & 0 & 0 & 0 \\ 0 & 0 & 0 & y_4 & 0 & 0 \\ 0 & 0 & 0 & 0 & y_5 & 0 \\ 0 & 0 & 0 & 0 & 0 & y_6 \end{bmatrix}$$

where $y_1, y_2, \ldots, y_6$ are branch admittance of given network.

$$= \begin{bmatrix} -y_1 & 0 & 0 & 0 & -y_5 & y_6 \\ 0 & -y_2 & y_3 & 0 & 0 & -y_6 \\ 0 & 0 & -y_3 & -y_4 & y_5 & 0 \end{bmatrix} \cdot \begin{bmatrix} -1 & 0 & 0 \\ 0 & -1 & 0 \\ 0 & 1 & -1 \\ 0 & 0 & -1 \\ -1 & 0 & 1 \\ 1 & -1 & 0 \end{bmatrix}$$

$$Y_{BUS} = \begin{bmatrix} y_1 + y_5 + y_6 & -y_6 & -y_5 \\ -y_6 & y_2 + y_3 + y_6 & -y_3 \\ -y_5 & -y_3 & y_3 + y_4 + y_5 \end{bmatrix}$$

Substituting the values $= \begin{bmatrix} 9 & -3 & 4 \\ -3 & 8 & -3 \\ -4 & -3 & 9 \end{bmatrix}$

$$Y_{BR} = B^t[y]B$$

$$B = \begin{bmatrix} 1 & 0 & 0 \\ 0 & 1 & 0 \\ 0 & 0 & 1 \\ 1 & -1 & -1 \\ -1 & 1 & 0 \end{bmatrix}$$

where "y" is primitive admittance matrix.

$$B^T[y] = \begin{bmatrix} 1 & 0 & 0 & 0 & 1 & -1 \\ 0 & 1 & 0 & 1 & -1 & 1 \\ 0 & 0 & 1 & 1 & -1 & 0 \end{bmatrix} \cdot \begin{bmatrix} y_1 & 0 & 0 & 0 & 0 & 0 \\ 0 & y_2 & 0 & 0 & 0 & 0 \\ 0 & 0 & y_3 & 0 & 0 & 0 \\ 0 & 0 & 0 & y_4 & 0 & 0 \\ 0 & 0 & 0 & 0 & y_5 & 0 \\ 0 & 0 & 0 & 0 & 0 & y_6 \end{bmatrix}$$

where $y_1, y_2, \ldots, y_5$ are the branch admittance of given network.

$$B^T[y] = \begin{bmatrix} y_1 & 0 & 0 & 0 & y_5 & -y_6 \\ 0 & y_2 & 0 & y_4 & -y_5 & y_6 \\ 0 & 0 & y_3 & y_4 & -y_5 & 0 \end{bmatrix}$$

$$B^T[y]B = \begin{bmatrix} y_1 & 0 & 0 & 0 & y_5 & -y_6 \\ 0 & y_2 & 0 & y_4 & -y_5 & y_6 \\ 0 & 0 & y_3 & y_4 & -y_5 & 0 \end{bmatrix} \cdot \begin{bmatrix} 1 & 0 & 0 \\ 0 & 1 & 0 \\ 0 & 0 & 1 \\ 0 & 1 & 1 \\ 0 & -1 & -1 \\ -1 & 1 & 0 \end{bmatrix}$$

$$Y_{BR} = \begin{bmatrix} y_1 + y_5 + y_6 & -y_5 - y_6 & -y_5 \\ -y_5 - y_6 & y_2 + y_4 + y_5 + y_6 & y_4 + y \\ -y_5 & y_4 + y_5 & y_3 + y_4 + y_5 \end{bmatrix}$$

$$Z_{BUS} = K^t(Z_{BR})K$$

$$Z_{BR} = [Y_{BR}]^{-1} = \begin{bmatrix} y_1 + y_4 + y_5 & -y_4 - y_5 & -y_4 \\ -y_4 - y_5 & y_2 + y_4 + y_5 & y_4 \\ -y_4 & y_4 & y_3 + y_4 \end{bmatrix}^{-1}$$

After substitution of y values

$$Z_{BR} = \begin{bmatrix} 9 & -7 & -4 \\ -7 & 11 & 6 \\ -4 & 6 & 9 \end{bmatrix}^{-1}$$

$$Z_{BR} = \begin{bmatrix} 0.2203 & 0.1364 & 0.007 \\ 0.1364 & 0.2273 & -0.0909 \\ 0.007 & -0.0909 & 0.1748 \end{bmatrix}$$

$$Z_{BUS} = K^t(Z_{BUS})K$$

$$K = \begin{bmatrix} -1 & 0 & 0 \\ 0 & -1 & -1 \\ 0 & 0 & -1 \end{bmatrix}$$

$$Z_{BR} = \begin{bmatrix} 0.2203 & 0.1364 & 0.007 \\ 0.1364 & 0.2273 & -0.0909 \\ 0.007 & -0.0909 & 0.1748 \end{bmatrix}$$

$$K^t[Z_{BR}] = \begin{bmatrix} -1 & 0 & 0 \\ 0 & -1 & 0 \\ 0 & -1 & -1 \end{bmatrix} \cdot \begin{bmatrix} 0.2203 & 0.1364 & 0.007 \\ 0.1364 & 0.2273 & -0.0909 \\ 0.007 & -0.0909 & 0.1748 \end{bmatrix}$$

$$= \begin{bmatrix} 0.2203 & 0.1364 & 0.007 \\ 0.1364 & 0.2273 & -0.09 \\ 0.1434 & 0.1364 & 0.0839 \end{bmatrix}$$

$$K^t[Z_{BR}]K = \begin{bmatrix} 0.2203 & 0.1364 & 0.007 \\ 0.1364 & 0.2273 & -0.09 \\ 0.1434 & 0.1364 & 0.0839 \end{bmatrix} \cdot \begin{bmatrix} -1 & 0 & 0 \\ 0 & -1 & -1 \\ 0 & 0 & -1 \end{bmatrix}$$

$$Z_{BUS} = \begin{bmatrix} 0.2203 & 0.1364 & 0.1434 \\ 0.1364 & 0.2273 & 0.1364 \\ 0.1434 & 0.1364 & 0.2203 \end{bmatrix}$$

Hence

$$Z_{BUS} \cdot Y_{BUS} = \begin{bmatrix} 0.2203 & 0.1364 & 0.1434 \\ 0.1364 & 0.2273 & 0.1364 \\ 0.1434 & 0.1364 & 0.2203 \end{bmatrix} \begin{bmatrix} 9 & -3 & -4 \\ -3 & 8 & -3 \\ -4 & -3 & 9 \end{bmatrix}$$

$$= \begin{bmatrix} 1 & 0 & 0 \\ 0 & 1 & 0 \\ 0 & 0 & 1 \end{bmatrix} = \textbf{[I] Unit Matrix}.$$

QUESTIONS

4.1. Derive the bus admittance matrix by singular transformation

4.2. Prove that $Z_{BUS} = K^t Z_{BR} K$

4.3. Explain how do you form Y_{BUS} by direct inspection with a suitable example

4.4. Derive the expression for bus admittance matrix Y_{BUS} in terms of primitive admittance matrix and bus incidence matrix

4.5. Derive the expression for the loop impedance matrix Z_{LOOP} using singular transformation in terms of primitive impedance matrix Z and the basic loop incidence matrix C

4.6. Show that $Z_{LOOP} = C^t [z] C$

4.7. Show that $Y_{BR} = B^t [y] B$ where $[y]$ is the primitive admittance matrix and B is the basic cut-set matrix

4.8. Prove that $Z_{BR} = A_B Z_{BUS} A_B^T$ with usual notation

4.9. Prove that $Y_{BR} = K Y_{BUS} K^t$ with usual notation

4.10. What is the significance of nonsingular transformation?

PROBLEMS

P.4.1. Determine Z_{LOOP} for the following network (Fig. P.4.1) using basic loop incidence matrix.

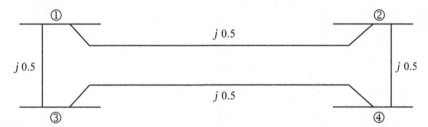

FIGURE P.4.1

Network for P.4.1.

P.4.2. Compute the bus admittance matrix for the power shown in Fig. P.4.2 by (1) direct inspection method and (2) by using singular transformation.

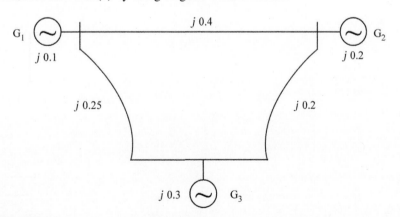

FIGURE P.4.2

System connection for P.4.2.

BUILDING OF NETWORK MATRICES 5

5.1 INTRODUCTION

In this chapter, methods for obtaining various network matrices are presented. These methods basically depend upon incidence matrices. *A*, *B*, *C*, *K* for singular and nonsingular transformation, respectively. Thus the procedure for obtaining *Y* or *Z* matrices in any frame of reference requires matrix transformations involving inversions and multiplications. This could be a very laborious and time-consuming process for large systems involving hundreds of nodes. It is possible to build the Z_{BUS} by using an algorithm where in systematically element by element is considered for addition and build the complete network directly from the element parameters. Such an algorithm would be very convenient for various manipulations that may be needed while the system is in operation such as addition of lines, removal of lines, and change in parameters.

The basic equation that governs the performance of a network is

$$\overline{V}_{BUS} = [Z_{BUS}] \cdot \overline{I}_{BUS}$$

5.2 PARTIAL NETWORK

In order to build the network element by element, a partial network is considered. At the beginning to start with, the building up of the network and its Z_{BUS} or Y_{BUS} model, a single element 1 is considered. Further, this element having two terminals connected to two nodes say (1) and (2) will have one of the terminals as reference or ground. Thus if node (1) is the reference then the element will have its own self-impedance as Z_{BUS}. When we connect any other element 2 to this element 1, then it may be either a branch or a link. The branch is connected in series with the existing node either (1) or (2) giving rise to a third node (3). On the contrary, a link is connected across the terminals (1) and (2) parallel to element 1. This is shown in Fig. 5.1.

In this case, no new bus is formed. The element 1 with nodes (1) and (2) is called the partial network that is already existing before the branch or link 2 is connected to the element. We shall use the notations (*a*) and (*b*) for the nodes of the element added either as a branch or as a link. The terminals of the already existing network will be called (*x*) and (*y*). Thus as element by element is added to an existing network, the network already in existence is called the partial network, to which, in step that follows a branch or a link is added. Thus generalizing the process consider m buses or nodes already contained in the partial network in which (*x*) and (*y*) are any buses (Fig. 5.2).

Recalling Eq. (5.1)

Power Systems Analysis. DOI: http://dx.doi.org/10.1016/B978-0-08-101111-9.00005-7

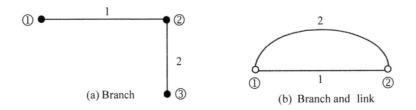

FIGURE 5.1

Branches and links.

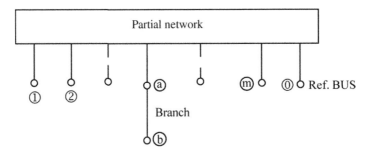

Addition of a branch

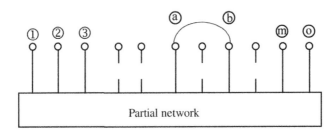

FIGURE 5.2

Addition of branch or link to partial network.

$$\overline{V}_{BUS} = [Z_{BUS}]\overline{I}_{BUS}$$

in the partial network, $[Z_{BUS}]$ will be of $[m \times m]$ dimension while $\overline{V}_{BUS}$ and $\overline{I}_{BUS}$ will be of $(m \times 1)$ dimension.

The voltage and currents are indicated in Fig. 5.3.

The performance equation (5.1) for the partial network is represented in the matrix form:

$$\begin{bmatrix} V_1 \\ V_2 \\ \vdots \\ V_m \end{bmatrix} = \begin{bmatrix} Z_{11} & Z_{12} & \dots & Z_{1m} \\ Z_{21} & Z_{22} & \dots & Z_{2m} \\ \vdots & \vdots & \vdots & \vdots \\ Z_{m1} & Z_{m2} & \dots & Z_{mm} \end{bmatrix} \begin{bmatrix} I_1 \\ I_2 \\ \vdots \\ I_m \end{bmatrix} \tag{5.1}$$

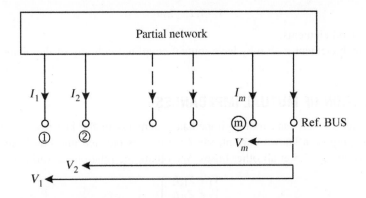

FIGURE 5.3

Partial network.

5.3 **ADDITION OF A BRANCH**

Consider an element $a-b$ added to the node (a) existing in the partial network. An additional node (b) is created as in Fig. 5.4.

The performance equation will be

$$
\begin{bmatrix} V_1 \\ V_2 \\ \vdots \\ V_a \\ \vdots \\ V_m \\ \hline V_b \end{bmatrix} =
\begin{bmatrix} Z_{11} & Z_{12} & \cdots & Z_{1a} & \cdots & Z_{1m} & Z_{1b} \\ Z_{21} & Z_{22} & \cdots & Z_{2a} & \cdots & Z_{2m} & Z_{2b} \\ \vdots & \vdots & \vdots & \vdots & \vdots & \vdots & \vdots \\ Z_{a1} & Z_{a2} & \cdots & Z_{aa} & \cdots & Z_{am} & Z_{ab} \\ \vdots & \vdots & \vdots & \vdots & \vdots & \vdots & \vdots \\ Z_{m1} & Z_{m2} & \cdots & Z_{ma} & \cdots & Z_{mm} & Z_{mb} \\ \hline Z_{b1} & Z_{b2} & \cdots & Z_{ba} & \cdots & Z_{bm} & Z_{bb} \end{bmatrix}
\begin{bmatrix} I_1 \\ I_2 \\ \vdots \\ I_a \\ \vdots \\ I_m \\ \hline I_b \end{bmatrix}
\tag{5.2}
$$

The last row and the last column in the Z-matrix are due to the added node b

$$ Z_{bi} = Z_{ib} $$

where

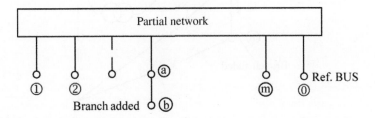

FIGURE 5.4

Addition of a branch.

$$i = 1, 2, \ldots, m$$

for all passive bilateral elements.

The added branch element $a-b$ may have mutual coupling with any of the elements of the partial network.

5.3.1 CALCULATION OF MUTUAL IMPEDANCES

It is required to find the self- and mutual impedance elements of the last row and last column of Eq. (5.2). For this purpose a known current, say $I = 1$ p.u., is injected into bus k and the voltage is measured as shown in Fig. 5.5 at all other buses. We obtain the relations as follows:

$$\left.\begin{aligned}
V_1 &= Z_{1k}I_k \\
V_2 &= Z_{2k}I_k \\
&\;\;\vdots \\
V_a &= Z_{ak}I_k \\
&\;\;\vdots \\
V_m &= Z_{mk}I_k \\
&\;\;\vdots \\
V_b &= Z_{bk}I_k
\end{aligned}\right\} \tag{5.3}$$

Since I_k is selected as 1 p.u. and all other bus currents are zero.
Z_{bk} can be known setting $I_k = 1.0$, from the measured value of V_b.

$$\left.\begin{aligned}
V_1 &= Z_{1k} \\
V_2 &= Z_{2k} \\
&\;\;\vdots \\
V_a &= Z_{ak} \\
&\;\;\vdots \\
V_m &= Z_{mk} \\
&\;\;\vdots \\
V_b &= Z_{bk}
\end{aligned}\right\} \tag{5.3a}$$

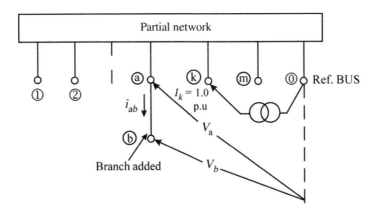

FIGURE 5.5

Partial network with branch added (calculations of mutual impedances).

We have that

$$V_b = V_a - e_{ab} \tag{5.4}$$

Also

$$\begin{bmatrix} i_{ab} \\ i_{xy} \end{bmatrix} = \begin{bmatrix} y_{ab-ab} & y_{ab-xy} \\ y_{xy-ab} & y_{xy-xy} \end{bmatrix} \begin{bmatrix} e_{ab} \\ e_{xy} \end{bmatrix} \tag{5.5}$$

where

y_{ab-ab} = self-admittance of added branch $a-b$
$\bar{y}_{ab-xy}$ = mutual admittance between added branch ab and the elements $x-y$ of partial network
$\bar{y}_{xy-ab}$ = transpose of y_{ab-xy}
y_{xy-xy} = primitive admittance of the partial network
i_{ab} = current in element $a-b$
e_{ab} = voltage across the element $a-b$.

It is clear from Fig. 5.5 that

$$i_{ab} = 0 \tag{5.6}$$

But, e_{ab} is not zero, since it may be mutually coupled to some elements in the partial network. Also,

$$e_{xy} = V_x - V_y \tag{5.7}$$

where V_x and V_y are the voltages at the buses x and y in the partial network.
The current in $a-b$

$$i_{ab} = y_{ab-ab}e_{ab} + \bar{y}_{ab-xy}\bar{e}_{xy} = 0 \tag{5.8}$$

From Eq. (5.6)

$$y_{ab-ab}e_{ab} + \bar{y}_{ab-xy}\bar{e}_{xy} = 0$$

$$e_{ab} = \frac{-\bar{y}_{ab-xy}\bar{e}_{xy}}{y_{ab-ab}} \tag{5.9}$$

substituting Eq. (5.7)

$$e_{ab} = \frac{-\bar{y}_{ab-xy}(\bar{V}_x - \bar{V}_y)}{y_{ab-ab}} \tag{5.10}$$

From Eq. (5.4)

$$V_b = V_a + \frac{\bar{y}_{ab-xy}(\bar{V}_x - \bar{V}_y)}{y_{ab-ab}} \tag{5.11}$$

Using Eq. (5.3), a general expression for the mutual impedance Z_{bi} between the added branch and other elements becomes

$$Z_{bi} = Z_{ai} + \frac{\bar{y}_{ab-xy}(\bar{Z}_{xi} - \bar{Z}_{yi})}{y_{ab-ab}} \tag{5.12}$$

$$i = 1,2,\ldots,m; i \neq b$$

5.3.2 CALCULATION OF SELF-IMPEDANCE OF ADDED BRANCH Z_{AB}

In order to calculate the self-impedance Z_{bb} once again, unit current $I_b = 1$ p.u. will be injected into bus b and the voltages at all the buses will be measured. Since all other currents are zero.

$$\left.\begin{array}{l} V_1 = Z_{1b}I_b \\ V_2 = Z_{2b}I_b \\ \vdots \\ V_a = Z_{ab}I_b \\ \vdots \\ V_m = Z_{mb}I_b \\ V_b = Z_{bb}I_b \end{array}\right\} \tag{5.13}$$

The voltage across the elements of the partial network is given in Eq. (5.7). The currents are given in Eq. (5.5).

Also,

$$i_{ab} = -I_b = -1 \tag{5.14}$$

From Eq. (5.8)

$$i_{ab} = y_{ab-ab}e_{ab} + \bar{y}_{ab-xy}\bar{e}_{xy} = -1$$

But

$$\bar{e}_{xy} = V_x - V_y$$

Therefore

$$-1 = y_{ab-ab}e_{ab} + y_{eb-xy}(V_x - V_y)$$

Hence

$$e_{ab} = \left[\frac{-1 + \bar{y}_{ab-xy}(\overline{V}_x - \overline{V}_y)}{y_{ab-ab}}\right] \tag{5.15}$$

Note:

$$\begin{aligned} \overline{V}_x - \overline{V}_y &= Z_{xb}I_b - Z_{yb}\,I_b \\ &= (Z_{xb} - Z_{yb})I_b \\ &= (Z_{xb} - Z_{yb}) \end{aligned} \tag{5.16}$$

substituting from Eq. (5.16) into Eq. (5.15)

$$e_{ab} = \frac{-[1 + \bar{y}_{ab-xy}(Z_{xb} - Z_{yb})]}{y_{ab-ab}} \tag{5.16a}$$

From Eq. (5.4), $V_b = V_a - e_{ab}$
Therefore

$$V_b = V_a + \frac{[1 + \bar{y}_{ab-xy}(Z_{xb} - Z_{yb})]}{y_{ab-ab}}$$

where y_{ab-ab} is the self-admittance of branch added $a-b$ and $\bar{y}_{ab-xy}$ is the mutual admittance vector between $a-b$ and $x-y$.

$$V_b = Z_{bb} \, I_b,$$

$$V_a = Z_{ab} \, I_b,$$

and

$$I_b = 1 \text{ p.u.}$$

Hence

$$Z_{bb} = Z_{ab} + \frac{1 + \bar{y}_{ab-xy}(\overline{Z}_{xb} - \overline{Z}_{yb})}{y_{ab-ab}} \qquad (5.17)$$

5.3.3 SPECIAL CASES

If there is no mutual coupling from Eq. (5.12) with

$$Z_{ab-ab} = \frac{1}{y_{ab-ab}}$$

and since

$$\bar{y}_{ab-xy} = \overline{0}$$

$$\boxed{Z_{bi} = Z_{ai};} \quad i = 1, 2, \ldots\ldots\ldots\ldots, m; \ i \neq b$$

From Eq. (5.17)

$$\boxed{Z_{bb} = \ Z_{ab} + z_{ab\text{-}ab}}$$

If there is no mutual coupling and a is the reference bus, Eq. (5.12) further reduces to $Z_{ai} = 0$ $i = 1, 2, \ldots, m; \ i \neq b$

$$Z_{bi} = 0$$

From Eq. (5.17)

$$Z_{ab} = 0$$

and

$$Z_{bb} = Z_{ab-ab}$$

5.4 ADDITION OF A LINK

Consider the partial network shown in Fig. 5.6.

Consider a link connected between a and b as shown.

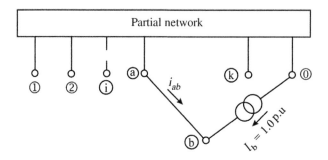

FIGURE 5.6

Partial network with branch added (calculations of self-impedance).

The procedure for building up Z_{BUS} for the addition of a branch is already developed. Now, the same method will be used to develop an algorithm for the addition of a link. Consider a fictitious node l between a and b. Imagine an voltage source V_1 in series with it between l and b as shown in Fig. 5.7.

Voltage V_1 is such that the current through the link ab, i.e., $i_{ab} = 0$
e_{ab} = voltage across the link $a-b$
V_1 = source voltage across $1-b = e_{lb}$.

Thus we may consider that a branch $a-1$ is added at the node (a) since the current through the link is made zero by introducing a source voltage V_1.

Now consider the performance equation

$$\overline{E}_{BUS} = [Z_{BUS}] \cdot \overline{I}_{BUS} \tag{5.18}$$

once again the partial network with the link added

$$
\begin{bmatrix} V_1 \\ V_2 \\ \cdots \\ V_i \\ \cdots \\ V_a \\ \cdots \\ V_m \\ \hline V_l \end{bmatrix}
=
\begin{array}{c}
1 \\ 2 \\ \\ i \\ \\ a \\ \\ m \\ l
\end{array}
\begin{bmatrix}
Z_{11} & Z_{12} & \cdots & Z_{1i} & \cdots & Z_{1a} & \cdots & Z_{1m} & Z_{1l} \\
Z_{21} & Z_{22} & \cdots & Z_{2i} & \cdots & Z_{2a} & \cdots & Z_{2m} & Z_{2l} \\
\cdots & \cdots & \cdots & \cdots & \cdots & \cdots & \cdots & \cdots & \cdots \\
Z_{i1} & Z_{i2} & \cdots & Z_{ii} & \cdots & Z_{ia} & \cdots & Z_{im} & Z_{il} \\
\cdots & \cdots & \cdots & \cdots & \cdots & \cdots & \cdots & \cdots & \cdots \\
Z_{a1} & Z_{a2} & \cdots & Z_{i2} & \cdots & Z_{aa} & \cdots & Z_{am} & Z_{al} \\
\cdots & \cdots & \cdots & \cdots & \cdots & \cdots & \cdots & \cdots & \cdots \\
Z_{m1} & Z_{m2} & \cdots & Z_{mi} & \cdots & Z_{ma} & \cdots & Z_{mm} & Z_{ml} \\
Z_{l1} & Z_{l2} & \cdots & Z_{li} & \cdots & Z_{la} & \cdots & Z_{lm} & Z_{ll}
\end{bmatrix}
\cdot
\begin{bmatrix} I_1 \\ I_2 \\ \cdots \\ I_i \\ \cdots \\ I_a \\ \cdots \\ I_m \\ I_l \end{bmatrix}
\tag{5.19}
$$

Also, the last row and the last column in Z-matrix are due to the added fictitious node l.

$$v_1 = V_1 - V_b \tag{5.20}$$

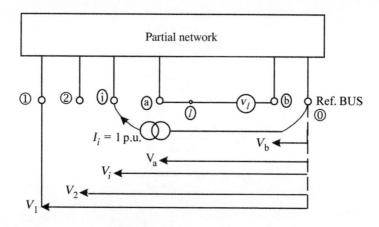

FIGURE 5.7

Addition of a link (calculation of mutual impedance).

5.4.1 CALCULATION OF MUTUAL IMPEDANCES

The element Z_{li}, in general, can be determined by injecting a current at the ith bus and measuring the voltage at the node i with respect to bus b. Since all other bus currents are zero, we obtain from the above consideration.

$$V_1 = Z_{li}\, I_i; \quad i = 1, 2, \ldots, m \tag{5.21}$$

and

$$V_1 = Z_{li}\, I_i \tag{5.22}$$

Letting $I_i = 1.0$ p.u. Z_{li} can be seen as v_1 which is the same as v_1.
But

$$v_l = V_a - V_b - e_{al} \tag{5.23}$$

It is already stated that the current through the added link $i_{ab} = 0$.

Treating $a-l$ as a branch, current in this element in terms of primitive admittances and the voltages across the elements

$$i_{al} = y_{al-al} \cdot e_{al} + \overline{y}_{al-xy} \cdot e_{xy} \tag{5.24}$$

where $\overline{y}_{al-xy}$ are the mutual admittances of any element $x-y$ in the partial network with respect to al and e_{xy} is the voltage across the element $x-y$ in the partial network.
But

$$i_{al} = i_{ab} = 0$$

Hence, Eq. (5.23) gives

$$e_{al} = \left(\frac{-\overline{y}_{al-xy} \cdot \overline{e}_{xy}}{y_{al-al}} \right) \tag{5.25}$$

Note that

$$y_{al-al} = y_{ab-ab} \tag{5.26}$$

and

$$\bar{y}_{al-xy} = \bar{y}_{ab-xy} \tag{5.27}$$

Therefore

$$e_{al} = \left(\frac{-\bar{y}_{ab-xy} \cdot \bar{e}_{xy}}{y_{ab-ab}} \right) \tag{5.28}$$

$$v_l = v_a - v_b + \frac{\bar{y}_{ab-xy} \bar{e}_{xy}}{y_{ab-ab}}$$

i.e.,

$$Z_{li}I_l = Z_{ai}I_l - Z_{bi}I_l + \frac{\bar{y}_{ab-xy}\bar{v}_x - \bar{v}_y}{y_{ab-ab}}$$

since

$$I_i = 1.0 \text{ p.u.}$$

$$i = 1, \dots, m$$

$$i \neq l$$

Also

$$\bar{e}_{xy} = \bar{v}_x - \bar{v}_y \tag{5.29}$$

$$Z_{li} = Z_{ai} - Z_{bi} + \frac{\bar{y}_{ab-xy}(Z_{xi} - Z_{yi})I_i}{y_{ab-ab}}$$

Thus using Eq. (5.20) and putting $I_i = 1.0$ p.u.

$$Z_{li} = Z_{ai} - Z_{bi} + \left(\frac{\bar{y}_{ab-xy}(Z_{xi} - Z_{yi})}{y_{ab-ab}} \right) \tag{5.30}$$

$$i = 1, 2, \dots, m$$

$$i \neq l$$

In this way, all the mutual impedance in the last row and last column of Eq. (5.19) can be calculated.

5.4.2 COMPUTATION OF SELF-IMPEDANCE

Now, the value of Z_{ll}, the self-impedance in Eq. (5.30) remains to be computed. For this purpose, as in the case of a branch, a unit current is injected at bus l and the voltage with respect to bus b is measured at bus l. Since all other bus currents are zero (Fig. 5.8).

$$V_k = Z_{kl} I_l; \quad k = 1, 2 \dots, m \tag{5.31}$$

and

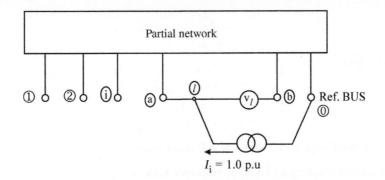

FIGURE 5.8

Addition of a link (calculation of self-impedance).

$$v_l = Z_{ll} \, I_l \qquad (5.32)$$

But

$$I_l = 1 \text{ p.u.} = - \, i_{al} \qquad (5.33)$$

The current i_{al} in terms of the primitive admittances and voltages across the elements

$$\begin{aligned} i_{al} \quad &= y_{al-al} \, e_{al} + \overline{y}_{al-xy} \overline{e}_{xy} \\ &= -1 \end{aligned} \qquad (5.34)$$

Again, as

$$\overline{y}_{al-xy} = y_{ab-xy} \quad \text{and} \qquad (5.35)$$

$$y_{al-al} = y_{ab-ab} \qquad (5.36)$$

Then, from Eq. (5.34), $-1 = \overline{y}_{al-al} e_{al} + \overline{y}_{ab-xy} \overline{e}_{xy}$

$$e_{al} = \frac{-(1 + \overline{y}_{ab-xy} \overline{e}_{xy})}{y_{ab-ab}} \qquad (5.37)$$

Substituting

$$\begin{aligned} e_{xy} \quad &= V_x - V_y \\ &= \overline{Z}_{xi} I_i - \overline{Z}_{yi} \cdot I_i \\ &= Z_{xi} - Z_{yi} (\text{since } I_i = 1.0 \text{ p.u.}) \end{aligned}$$

$$Z_{ll} = Z_{al} - Z_{bl} + \frac{1 + \overline{y}_{ab-xy} (\overline{Z}_{xi} - \overline{Z}_{yi})}{y_{ab-ab}} \qquad (5.38)$$

Case (i): no mutual impedance

If there is no mutual coupling between the added link and the other elements in the partial network

$$\overline{y}_{ab-xy} \text{ are all zero}$$

$$\frac{1}{y_{ab-ab}} = z_{ab-ab} \tag{5.39}$$

Hence we obtain

$$Z_{li} = Z_{ai} - Z_{bi}; \quad i = 1, 2, \ldots, m \tag{5.40}$$

$$i \neq 1$$

$$Z_{ll} = Z_{al} - Z_{bl} + z_{ab-ab} \tag{5.41}$$

Case (ii): No mutual impedance and a is reference node

If there is no mutual coupling and a is the reference node
Also

$$\left.\begin{array}{l} Z_{ai} = 0 \\ Z_{li} = - Z_{bi} \\ Z_{ll} = - Z_{bl} + z_{ab-ab} \end{array}\right\} \tag{5.42)(5.43}$$

Thus all the elements introduced in the performance equation of the network with the link added and node l created are determined.

It is required now to eliminate the node l.

For this, we short circuit the series voltage source v_l, which does not exist in reality from Eq. (5.19)

$$\overline{V}_{\text{BUS}} = [Z_{\text{BUS}}]\,\overline{I}_{\text{BUS}} + \overline{Z}_{il}I_l \tag{5.44}$$

$$V_l = \overline{Z}_{lj} \cdot \overline{I}_{\text{BUS}} + Z_{ll}I_l; \quad j = 1, 2, \ldots, m$$

$$= 0 \text{ (since the source is short circuited)} \tag{5.45}$$

Solving for I_l from Eq. (5.45)

$$I_l = \frac{- Z_{lj} \cdot \overline{I}_{\text{BUS}}}{Z_{ll}} \tag{5.46}$$

Substituting in Eq. (5.44)

$$V_{\text{BUS}} = [Z_{\text{BUS}}] \cdot \overline{I}_{\text{BUS}} - \frac{Z_{il} \cdot Z_{lj}}{Z_{ll}}\overline{I}_{\text{BUS}}$$

$$= \left[Z_{\text{BUS}} - \frac{Z_{il} \cdot Z_{lj}}{Z_{ll}}\right] I_{\text{BUS}} \tag{5.47}$$

This is the performance equation for the partial network including the link $a-b$ incorporated. From Eq. (5.47), we obtain

$$Z_{\text{BUS}}(\text{modified}) = \left[Z_{\text{BUS}}(\text{before addition of link}) - \frac{\overline{Z}_{il}\overline{Z}_{lj}}{Z_{ll}}\right]$$

and for any element

$$Z_{ij}(\text{modified}) = \left[Z_{ij}(\text{before addition of link} - \frac{Z_{il}Z_{lj}}{Z_{ll}}\right] \tag{5.48}$$

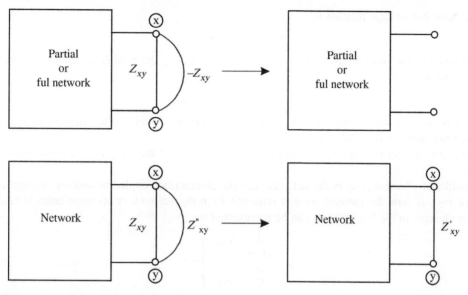

FIGURE 5.9

Removal or change in impedance of an element.

5.4.3 REMOVAL OF ELEMENTS OR CHANGES IN ELEMENT

Consider the removal of an element from a network. The modified impedance can be obtained by adding in parallel with the element a link, whose impedance is equal to the negative of the impedance to be removed.

In a similar manner, if the impedance of the element is changed, then the modified impedance matrix obtained by adding a link in parallel with the element such that the equivalent impedance of the two elements is the desired value (Fig. 5.9).

$$Z_{xy} \text{ changed to } Z_{xy}^{|}$$

$$\frac{1}{Z_{xy}^{|}} = \frac{1}{Z_{xy}} + \frac{1}{Z_{xy}^{||}}$$

However, the above are applicable only when there is no mutual coupling between the element to be moved or changed with any element or elements of partial network.

5.5 REMOVAL OR CHANGE IN IMPEDANCE OF ELEMENTS WITH MUTUAL IMPEDANCE

Changes in the configuration of the network elements introduce changes in the bus currents of the original network. In order to study the effect of addition of a link consider the partial network shown in Fig. 5.7.

The basic bus voltage relation is

$$\overline{V}_{BUS} = [Z_{BUS}]\,\overline{I}_{BUS}$$

with changes in the bus currents denoted by the vector $\Delta\overline{I}_{BUS}$ the modified voltage performance relation will become

$$\overline{V}'_{BUS} = [Z'_{BUS}](\overline{I}_{BUS} + \Delta\overline{I}_{BUS}) \tag{5.49}$$

$\overline{V}_{BUS}$ is the new bus voltage vector. It is desired now to calculate the impedances Z_{ij}^1 of the modified impedance matrix $[\overline{Z}'_{BUS}]$.

The usual method is to inject a known current (say 1 p.u.) at the bus j and measure the voltage at bus i.

Consider an element $p-q$ in the network. Let the element be coupled to another element in the network $r-s$. If now the element $p-q$ is removed from the network or its impedance is changed then the changes in the bus currents can be represented by

$$\left.\begin{array}{l} \Delta I_p = \Delta i_{pq} \\ \Delta I_q = -\Delta i_{pq} \\ \Delta I_r = \Delta i_{rs} \\ \Delta I_s = -\Delta i_{rs} \end{array}\right\} \tag{5.50}$$

Inject a current of 1 p.u. at any jth bus

$$I_j = 1.0$$
$$I_k = 0[k = 1, 2, \ldots, n;\ \ k \neq j]$$

Then

$$V_i^1 = \sum_{k=1}^{n} Z_{ik}(I_k + \Delta I_k)$$

$i = 1, 2, \ldots, n$. With the index k introduced, equation may be understood from (Fig. 5.10)

$$\left.\begin{array}{l} \Delta I_k = \Delta i_{pq};\ k = p \\ \Delta I_k = -\Delta i_{pq};\ k = q \\ \Delta I_k = \Delta i_{rs};\ k = r \\ \Delta I_k = -\Delta i_{rs};\ k = s \end{array}\right\} \tag{5.50a}$$

From equation

$$\begin{aligned} V_i^1 &= Z_{ij}^1 \cdot 1 + Z_{ij}\Delta i_{pq} - Z_{iq}\Delta i_{pq} + Z_{ir}\Delta i_{rs} - Z_{is}\Delta i_{rs} \\ &= Z_{ij} + (Z_{ip} - Z_{iq})\Delta i_{pq} + (Z_{ir} - Z_{is})\Delta i_{rs} \end{aligned} \tag{5.51}$$

If α, β are used as the subscripts for the elements of both $p-q$ and $r-s$, then

$$V_i^1 = Z_{ij} + (Z_{i\alpha} - Z_{i\beta})\Delta\overline{i}_{\alpha\beta};\ \ i = 1, 2, 3, \ldots, n \tag{5.52}$$

From the performance equation of the primitive network:

$$\Delta i_{\alpha\beta} = ([y_{sm}] - [y_{sm}]^{-1})v_{\gamma\delta} \tag{5.53}$$

where $[y_{sm}]$ and $[y_{sm}]^{-1}$ are the square submatrices of the original and modified primitive admittance matrices. Consider as an example, the system in Fig. 5.11A, y matrix is shown. If y_2 element is removed, the y_1 matrix is shown in Fig. 5.11B.

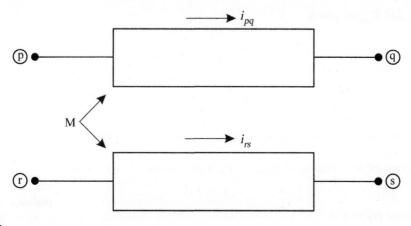

FIGURE 5.10

Network elements for measurement of change in impedance.

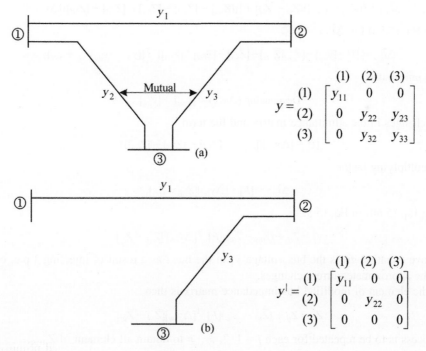

FIGURE 5.11

Computation of change in impedance. Network (A) and element removed (B).

Then, Y_{sm} and Y'_{sm} are given:

$$Y_{sm} = \begin{bmatrix} y_{22} & y_{23} \\ y_{32} & y_{33} \end{bmatrix}$$

and

$$Y'_{sm} = \begin{bmatrix} y_{22} & 0 \\ 0 & 0 \end{bmatrix}$$

Thus the rows and columns of the submatrices Y_{sm} and Y'_{sm} correspond to the network elements $p-q$ and $r-s$ (Fig. 5.10).

The subscripts of the elements of $(|y_{sm}| - |y'_{sm}|)$ are $\alpha\beta$ and $\gamma\delta$. We know that

$$v^|_{\gamma\delta} = v^|_{\gamma} - v^|_{\delta} \tag{5.54}$$

substituting from Eq. (5.52)
for

$$\bar{v}^|_{\gamma} \text{ and } \bar{v}^|_{\delta}$$

$$v^|_{\gamma\delta} = \bar{Z}_{\gamma j} - \bar{Z}_{\delta j} + ([Z_{\gamma\alpha}] - [Z_{\gamma\alpha}] - [Z_{\delta\alpha}] + [Z_{\gamma\beta}])\Delta\bar{i}_{\alpha\beta} \tag{5.55}$$

Substituting from Eq. (5.55) for $v^|_{\gamma\delta}$ into Eq. (5.53)

$$\Delta\bar{i}_{\alpha\beta} = (|y_{sm}| - |y'_{sm}|)(\bar{Z}_{\gamma j} - \bar{Z}_{\delta j}) + [([Z_{\gamma\alpha}] - [Z_{\gamma\alpha}] - [Z_{\delta\alpha}] - [Z_{\gamma\beta}] + [Z_{\delta\beta}])]\Delta\bar{i}_{\alpha\beta} \tag{5.56}$$

Solving Eq. (5.56) for $\Delta\bar{i}_{\alpha\beta}$

$$\Delta\bar{i}_{\alpha\beta} = \{U - [y_{sm}] - [y'_{sm}][\bar{Z}_{\gamma\alpha}] - [Z_{\delta\alpha}] - [Z_{\delta\alpha}] + [Z_{\delta\beta}]\}^{-1}\{(y_{sm} - y^|_{sm})(z_{rj} - z_{\delta j})\} \tag{5.57}$$

where U is unit matrix

$$\text{Designating } (\Delta y_{sm}) = [y_{sm}] - [y^|_{sm}] \tag{5.58}$$

giving the changes in the admittance matrix and the term

$$\{U - [\Delta y_{sm}]([Z_{\gamma\alpha}] - [Z_{\delta\alpha}] - [Z_{\gamma\beta}] - [Z_{\delta\beta}])\} \tag{5.59}$$

by F, the multiplying factor

$$\Delta\bar{i}_{\alpha\beta} = [F]^{-1}[\Delta y_{sm}][\bar{Z}_{\gamma j} - \bar{Z}_{\delta j}] \tag{5.60}$$

substituting Eq. (5.60) in Eq. (5.52)

$$V^|_i = Z_{ij} + (\bar{Z}_{i\alpha} - \bar{Z}_{i\beta})[F]^{-1}[\Delta y_{sm}]^|[Z_{\gamma j} - Z_{\delta j}] \tag{5.61}$$

The above equation gives the bus voltage $V^|_i$ at the bus i as a result of injecting 1 p.u. current at bus j and the approximate current changes.

The ij the element of modified bus impedance matrix is then

$$Z^|_{ij} = Z_{ij} + (\bar{Z}_{i\alpha} - \bar{Z}_{i\beta})[F]^{-1}[\Delta y_{sm}][\bar{Z}_{\gamma i} - \bar{Z}_{\delta j}]$$

The process is to be repeated for each $j = 1, 2, \ldots, n$ to obtain all element of $Z^|_{BUS}$.

WORKED EXAMPLES

E.5.1. **A transmission line exists between buses 1 and 2 with per unit impedance 0.4. Another line of impedance 0.2 p.u. is connected in parallel with it making it a double-circuit line with mutual impedance of 0.1 p.u. Obtain by building algorithm method the impedance of the two-circuit system.**

Solution:

Consider the system with one line

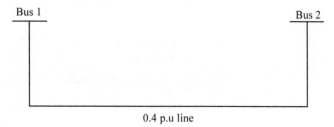

0.4 p.u line

FIGURE E.5.1

Single line system for E.5.1.

Taking bus (1) as reference, the Z_{BUS} is obtained as

$$
Z_{BUS} =
\begin{array}{c|c|c|}
 & (1) & (2) \\
\hline
(1) & 0 & 0 \\
\hline
(2) & 0 & 0.4 \\
\end{array}
=
\begin{array}{c|c|}
 & (2) \\
\hline
(2) & 0.4 \\
\end{array}
$$

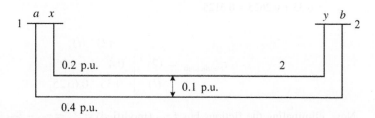

FIGURE E.5.2

Double line system for E.5.2.

Now consider the addition of the second line in parallel with it.

The addition of the second line is equivalent to the addition of a link. The augmented impedance matrix with the fictitious node *l* introduced.

$$
\begin{array}{cc}
 & (2) \quad (l) \\
Z_{zl} = \begin{array}{c}(2) \\ (l)\end{array} & \begin{bmatrix} 0.4 & z_{2l} \\ z_{l2} & z_{ll} \end{bmatrix}
\end{array}
$$

$$
\begin{array}{cc}
 & 1-2\,(1) \quad 1-2\,(2) \\
Z_{ab-xy} = \begin{array}{c}1-2(2) \\ 1-2(l)\end{array} & \begin{bmatrix} 0.4 & 0.1 \\ 0.1 & 0.2 \end{bmatrix}
\end{array}
$$

$$
a = 1; \quad x = 1
$$
$$
b = 2; \quad y = 2
$$

$$
z_{li} = z_{ai} - z_{bi} + \frac{y_{ab-xy}(z_{xi} - z_{yi})}{y_{ab-ab}}
$$

$$
y_{ab-xy} = \left[z_{ab-xy}\right]^{-1} \left[\begin{pmatrix} 0.2 & -0.1 \\ -0.1 & 0.4 \end{pmatrix}\right] \cdot \frac{1}{(0.08 - 0.01)} = \begin{array}{c} \\ 1-2(1) \\ 1-2(2)\end{array} \begin{array}{cc} 1-2\,(1) \quad 1-2\,(2) \\ \begin{bmatrix} 2.857 & -1.4286 \\ -1.4286 & 5.7143 \end{bmatrix}\end{array}
$$

$$
z_{li} = z_{ai} - z_{bi} + \frac{\bar{y}_{ab-xy}(\bar{z}_{xi} - \bar{z}_{yi})}{y_{ab-ab}}
$$

$$
z_{21} = z_{12} = 0 - 0.4 + \frac{(-1.4286)(0 - 0.2)}{5.7143} = -0.35
$$

$$
z_{ll} = z_{al} - a_{bl} + \frac{[1 + \bar{y}_{ab-xy}(\bar{z}_{xl} - \bar{z}_{yl})]}{y_{ab-ab}} = 0 - (-0.35) + \frac{[1 + (-1.4286)(0 - (-0.35))]}{5.7143}
$$

$$
= 0.35 + 0.2625 = 0.6125
$$

$$
\begin{array}{cc}
 & (2) \quad (l) \\
z_{\text{augmented}} = \begin{array}{c}(2) \\ (l)\end{array} & \begin{bmatrix} 0.4 & -0.35 \\ -0.35 & 0.6125 \end{bmatrix}
\end{array}
$$

Now, eliminating the fictions bus l z_{22} (modified) $= z_{22}^l = z_{22} - \frac{z_{2l} \cdot z_{l2}}{z_{ll}}$

$$
= 0.4 - \frac{(-0.35)(-0.35)}{0.6125} = 0.4 - 0.2 = 0.2
$$

$$
\begin{array}{cc}
 & (2) \\
Z_{\text{BUS}} = \begin{array}{c}(2)\end{array} & \boxed{2}
\end{array}
$$

E.5.2. **The double-circuit line in the problem E.4.1 is further extended by the addition of a transmission line from bus (1). The new line by virtue of its proximity to the existing lines has a mutual impedance of 0.05 p.u. and a self-impedance of 0.3 p.u. Obtain the bus impedance matrix by using the building algorithm.**

Solution:

Consider the extended system

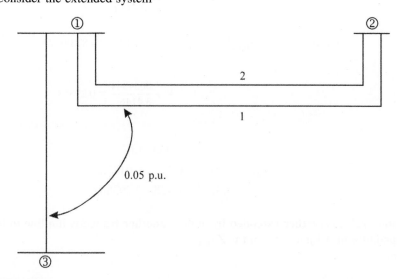

FIGURE E.5.3

Double line system with the addition of branch.

Now $a = 1$ and $b = 3$

Also a is the reference bus

$$z_{bi} = z_{ai} + \frac{\bar{y}_{ab-xy}(\bar{z}_{xi} - \bar{z}_{yi})}{y_{ab-ab}}$$

The primitive impedance matrix z_{ab-xy} is given by

$$
z_{ab-xy} =
\begin{array}{c}
 \\
1-2(1) \\
1-2(2) \\
1-3
\end{array}
\begin{array}{ccc}
1-2(1) & 1-2(2) & 1-3 \\
\begin{bmatrix} 0.4 & 0.1 & 0.05 \\ 0.1 & 0.2 & 0 \\ 0.05 & 0 & 0.3 \end{bmatrix}
\end{array}
$$

$$
[y_{ab-xy}] = [z_{ab-xy}]^{-1} =
\begin{array}{c}
 \\
1-2(1) \\
1-2(2) \\
1-3
\end{array}
\begin{array}{ccc}
1-2(1) & 1-2(2) & 1-3 \\
\begin{bmatrix} 2.9208 & -1.4634 & -0.4878 \\ -1.4634 & 5.7317 & 0.2439 \\ -0.4878 & 0.2439 & 3.4146 \end{bmatrix}
\end{array}
$$

Setting $b = 3$; $i = 2$; $a = 1$

$$z_{32} = z_{12} + \frac{[y_{1-31-2(1)}y_{1-31-2(2)}]\begin{bmatrix} z_{12} - z_{22} \\ z_{12} - z_{22} \end{bmatrix}}{y_{13-13}} = \frac{0 + [(-0.4878)(0.2439)]\begin{bmatrix} 0 - 0.2 \\ 0 - 0.2 \end{bmatrix}}{3.4146}$$

$$= \frac{0.04828}{3.4146} = 0.014286$$

$$z_{33} = z_{13} + \frac{1 + [y_{1-31-2(1)}y_{1-31-2(2)}]\begin{bmatrix} z_{13} - z_{23} \\ z_{13} - z_{23} \end{bmatrix}}{y_{13-13}} = 0 + \frac{1 + [(-0.4878)(0.2439)]\begin{bmatrix} -0.014286 \\ -0.014286 \end{bmatrix}}{3.4146}$$

$$= \frac{1.0034844}{3.4146} = 0.29388$$

$$Z_{BUS} = \begin{array}{c} \\ (2) \\ (3) \end{array} \begin{array}{cc} (2) & (3) \\ \begin{bmatrix} 0.2 & 0.01428 \\ 0.01428 & 0.29388 \end{bmatrix} \end{array}$$

E.5.3. The system E.5.2 is further extended by adding another transmission line to bus 3 with self-impedance of 0.3 p.u. Obtain the Z_{BUS}.

Solution:

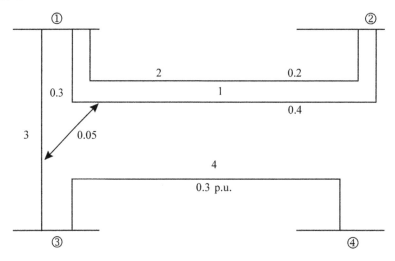

FIGURE E.5.4

Another branch added to system in Fig. E.5.3.

Consider the system shown above with the line 4 added to the previous system. This is the case of the addition of a branch. Bus (3) is not the reference bus.

$$a = 3$$
$$b = 4$$

There is no mutual coupling.

$$z_{bi} = z_{ai}; \quad i = 1, 2, \ldots, m; \quad i \neq b$$
$$z_{bb} = z_{ab} + z_{ab-ab}$$

setting $i = 1, 2,$ and 3, respectively, we can compute.

$$z_{4i} = z_{41} = 0 \ (\text{ref. Node}) = z_{31}$$
$$z_{42} = z_{32} = 0 = 0.01428$$
$$z_{4i} = z_{33} = 0.29288$$
$$z_{4i} = z_{31} + z_{44-44} = 0.29288 + 0.3 = 0.59288$$

$$
Z_{\text{BUS}} = \begin{array}{c} (2) \\ (3) \\ (4) \end{array}
\begin{array}{ccc} (2) & (3) & (4) \end{array}
\left[\begin{array}{ccc}
0.2 & 0.1428 & 0.1428 \\
0.01428 & 0.29288 & 0.29288 \\
0.01428 & 0.29288 & 0.59288
\end{array} \right]
$$

E.5.4. **The system in E.5.3 is further extended and the radial system is converted into a ring system joining bus (2) to bus (4) for reliability of supply. Obtain the Z_{BUS}.**
   The self-impedance of element 5 is 0.1 p.u
Solution:

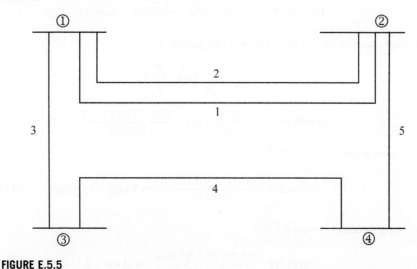

FIGURE E.5.5

Addition of a link to system in Fig. E.5.4.

The ring system is shown in the figure, now let $a = 2$ and $b = 4$ addition of the line 5 is addition of a link to the existing system. Hence initially a fictitious node l is created. However, there is no mutual impedance bus (2) is not a reference node.

$$z_{li} = z_{ai} - z_{bi}$$

$$z_{ll} = z_{al} - z_{bl} + z_{ab-ab}$$

with

$$i = 2: z_{21} = z_{22} - z_{42} = z_{12} = 0.2 - 0.01428 = 0.18572$$

with

$$i = 3: z_{31} = z_{23} - z_{43} = z_{13} = 0.01428 - 0.29288 = -0.27852$$

setting

$$i = 4: z_{41} = z_{14} = z_{24} - z_{44} = 0.1428 - 0.59288 = -0.5786$$

$$
\begin{aligned}
z \text{ (augmented)} \quad &= z_{ll} + z_{al} - z_{bl} + z_{ab-ab} \\
&= z_{21} - z_{41} + 0.1 = 0.18572 + 0.5786 + 0.1 = 0.86432
\end{aligned}
$$

$$
\begin{array}{c c c c c}
 & (2) & (3) & (4) & (l) \\
(2) & 0.2 & 0.01428 & 0.01428 & 0.18572 \\
(3) & 0.01428 & 0.29288 & 0.29288 & -0.27852 \\
(4) & 0.01428 & 0.29288 & 0.59288 & -0.5786 \\
(l) & 0.18572 & -0.27582 & -0.5786 & 0.86432
\end{array}
$$

The expression below fictitious node l introduced is

$$
Z_{2l} = \begin{array}{c} (2) \\ (l) \end{array}
\begin{bmatrix}
0.4 & Z_{2l} \\
Z_{l2} & Z_{ll}
\end{bmatrix}
\begin{array}{c} (2) \quad (l) \end{array}
$$

$$z_{22}(\text{modified}) = z_{22} - \frac{z_{2l}z_{l2}}{z_{ll}} = 0.2 - \frac{(0.18572)(0.18572)}{0.86432} = 0.16$$

$$
\begin{aligned}
z_{23}(\text{modified}) \quad &= z_{23} - \frac{z_{2l}\, z_{l3}}{z_{ll}} \\
&= 0.01428 - \frac{(0.18572)(-0.27852)}{0.86432} = 0.01428 + 0.059467 = 0.0741267
\end{aligned}
$$

$$
\begin{aligned}
z_{24}(\text{modified}) \quad &= z_{33} - \frac{z_{2l}\, z_{l4}}{z_{ll}} \\
&= 0.01428 - \frac{(0.18572)(-0.5786)}{0.86432} = 0.01428 + 0.1243261 = 0.1386
\end{aligned}
$$

$$
\begin{aligned}
z_{33}(\text{modified}) \quad &= z_{33} = \frac{z_{3l}\, z_{l3}}{z_{ll}} \\
&= 0.59288 - \frac{(-0.27582)(-0.27852)}{0.86432} = 0.59288 - 0.0897507 = 0.50313
\end{aligned}
$$

$$z_{34}(\text{modified}) \quad = z_{43}(\text{modified})$$

$$= 0.2928 - \frac{(-0.27852)(-0.5786)}{0.86432} = 0.29288 - 0.186449 = 0.106431$$

$$z_{44}(\text{modified}) \quad = z_{44} - \frac{z_{4l}\, z_{l4}}{z_{ll}}$$

$$= 0.59288 - \frac{(-0.5786)(-0.57861)}{0.86432} = 0.59288 - 0.38733 = 0.2055$$

The Z_{BUS} for the entire ring system is obtained as

$$Z_{\text{BUS}} = \begin{array}{c} \\ (2) \\ (3) \\ (4) \end{array} \begin{array}{ccc} (2) & (3) & (4) \\ \left[\begin{array}{ccc} 0.16 & 0.0741267 & 0.1386 \\ 0.0741267 & 0.50313 & 0.106431 \\ 0.1386 & 0.106431 & 0.2055 \end{array} \right] \end{array}$$

E.5.5. **Compute the bus impedance matrix for the system shown in the figure by adding element by element. Take bus (2) as reference bus.**

Solution:

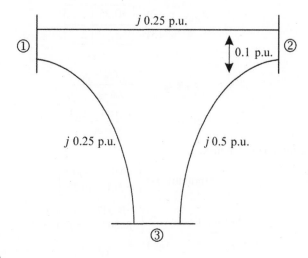

FIGURE E.5.6

Network for E.5.5.

Step 1: Taking bus (1) as reference bus

$$Z_{\text{BUS}} = \begin{array}{c} \\ (2) \end{array} \begin{array}{c} (2) \\ \boxed{j0.25} \end{array}$$

Step 2: Ass line joining buses (2) and (3). This is the addition of a branch with mutuals.

$$a = (2); b = (3)$$

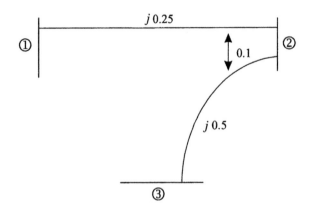

FIGURE E.5.7

Line added to buses 2 and 3.

$$Z_{\text{BUS}} = \begin{array}{c} \\ (2) \\ (3) \end{array} \begin{array}{cc} (2) & (3) \\ \hline \boxed{\begin{array}{c|c} j0.25 & z_{23} \\ \hline z_{32} & z_{33} \end{array}} \end{array}$$

$$z_{bi} = z_{ai} + \frac{\bar{y}_{ab-xy}(\bar{z}_{xi} - \bar{z}_{yi})}{y_{ab-ab}}$$

$$z_{32} = z_{22} + \frac{y_{23-12}(z_{12} - z_{22})}{y_{23-23}}$$

The primitive impedance matrix:

$$z \text{ (primitive)} = \begin{bmatrix} j0.5 & -j0.1 \\ -j0.1 & j0.25 \end{bmatrix}$$

$$y_{ab-xy} = [z]^{-1}_{primitive} = \begin{bmatrix} -j4.347 & j0.869 \\ j0.869 & -j2.1739 \end{bmatrix}$$

Hence

$$z_{32} = j0.25 + \frac{j0.869(0 - j0.25)}{-j2.1739} = j0.25 + j0.099 = j0.349$$

$$z_{33} = j0.349 + \frac{1 + 0.869(0 - j0.349)}{-j2.1739} = j0.349 + j0.5 = j0.946$$

$$
Z_{\text{BUS}} =
\begin{array}{c}
 \\
(2) \\
(3)
\end{array}
\begin{array}{cc}
(2) & (3) \\
\hline
j0.25 & j0.349 \\
\hline
j0.349 & j0.946 \\
\hline
\end{array}
$$

Step 3: Add the live joining (1) and (3) buses. This is the addition of a link to the existing system without mutual impedance.
A fictitious bus l is created.

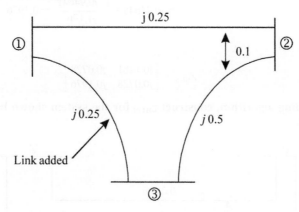

j 0.25

0.1

j 0.25

j 0.5

① ② ③

Link added

FIGURE E.5.8

Link added to network in Fig. E.5.7.

$$
\begin{aligned}
z_{li} &= -z_{bi} \\
z_{ll} &= -z_{bi} + z_{ab-ab} \\
z_{12} &= -z_{32} = -j0.349 \\
z_{13} &= -z_{33} = -j0.9464 \\
z_{ll} &= -z_{31} + z_{13-13} \\
&= +j0.9464 + j0.25 \\
&= j1.196
\end{aligned}
$$

The augmented impedance matrix

$$
Z_{\text{BUS}} =
\begin{array}{c}
(2) \\
(3) \\
(l)
\end{array}
\begin{array}{ccc}
(2) & (3) & (4) \\
\hline
j0.25 & j0.349 & -j0.349 \\
\hline
j0.349 & j0.946 & -j0.9464 \\
\hline
-j0.349 & -j0.9464 & j1.196 \\
\hline
\end{array}
$$

The factious node (l) is now eliminated.

$$
\begin{aligned}
z_{22} \text{ (modified)} &= z_{22} - \frac{z_{2l}\,z_{l2}}{z_{ll}} \\
&= j0.25 - \frac{(-j0.349)(-j0.349)}{j1.196} = 0.1481
\end{aligned}
$$

$$z_{23}(\text{modified}) = z_{32}(\text{modified}) \quad = z_{23} - \frac{z_{2l}\,z_{l3}}{z_{ll}}$$

$$= j0.349 - \frac{(-j0.349)(-j0.9464)}{j1.196} = 0.072834$$

$$z_{33}(\text{modified}) \quad = z_{33} - \frac{z_{3l}\,z_{l3}}{z_{ll}}$$

$$= j0.94645 - \frac{(-j0.9464)^2}{j1.196} = 0.1976$$

Hence

$$Z_{BUS} = \begin{bmatrix} j0.1481 & j0.0728 \\ j0.0728 & j0.1976 \end{bmatrix}$$

E.5.6. **Using the building algorithm, construct z_{BUS} for the system shown below. Choose 4 as reference BUS.**

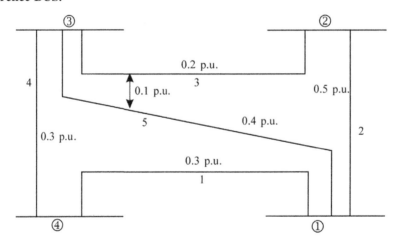

FIGURE E.5.9

Network for E.5.6. Please also correct the example number from E.4.6 to E.5.6.

Solution:

Step 1: Start with element (1) which is a branch $a = 4$ to $b = 1$. The elements of the bus impedance matrix for the partial network containing the single branch are as follows.

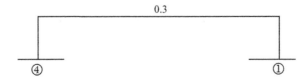

FIGURE E.5.10

Single line element with bus 4 as reference.

Taking bus (4) as reference bus

$$Z_{BUS} = \begin{array}{c} \\ (4) \\ (1) \end{array} \begin{array}{cc} (4) & (1) \\ \hline \boxed{\begin{array}{c|c} 0 & 0 \\ \hline 0 & 0.3 \end{array}} \end{array}$$

Since node 4 chosen as reference. The elements of the first row and column are zero and need not be written thus

$$Z_{BUS} = \begin{array}{c} \\ (1) \end{array} \begin{array}{c} (1) \\ \hline \boxed{0.3} \end{array}$$

Step 2: Add element (2) which is a branch $a = 1$ to $b = 2$. This adds a new bus.

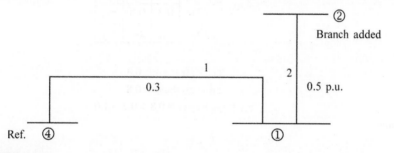

FIGURE E.5.11

Element 2 added to network in Fig. E.5.10.

$$Z_{BUS} = \begin{array}{c} \\ (1) \\ (2) \end{array} \begin{array}{cc} (1) & (2) \\ \hline \boxed{\begin{array}{c|c} 0.3 & z_{12} \\ \hline z_{21} & z_{22} \end{array}} \end{array}$$

$$z_{12} = z_{21} = z_{11} = 0.3$$
$$z_{22} = z_{12} + z_{1212} = 0.3 + 0.5 = 0.8$$

$$Z_{BUS} = \begin{array}{c} \\ (1) \\ (2) \end{array} \begin{array}{cc} (1) & (2) \\ \hline \boxed{\begin{array}{c|c} 0.3 & 0.3 \\ \hline 0.3 & 0.8 \end{array}} \end{array}$$

Step 3: Add element (3) which is a branch $a = 2$ to $b = 3$. This adds a new BUS, the BUS impedance matrix.

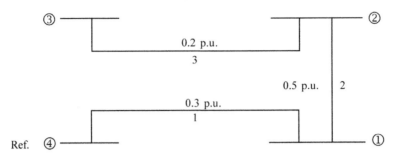

FIGURE E.5.12

Another element 3 added as branch to network in Fig. E.5.11.

$$
Z_{BUS} = \begin{array}{c} \\ (1) \\ (2) \\ (3) \end{array}
\begin{array}{ccc} (1) & (2) & (3) \\ \hline
0.3 & 0.3 & z_{13} \\
0.3 & 0.8 & z_{23} \\
z_{31} & z_{23} & z_{33} \end{array}
$$

$$z_{13} = z_{31} = z_{21} = 0.3$$
$$z_{32} = z_{23} + z_{22} = 0.8$$
$$z_{33} = z_{23} + z_{2323} = 0.8 + 0.2 = 1.0$$

Hence

$$
Z_{BUS} = \begin{array}{c} \\ (1) \\ (2) \\ (3) \end{array}
\begin{array}{ccc} (1) & (2) & (3) \\ \hline
0.3 & 0.3 & 0.3 \\
0.3 & 0.8 & 0.8 \\
0.3 & 0.8 & 1.0 \end{array}
$$

Step 4: Add element (4) which is a link $a = 4$; $b = 3$. The augmented impedance matrix with the fictitious node 1 is shown below.

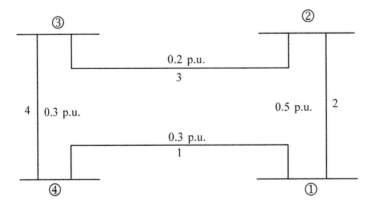

FIGURE E.5.13

Element 4 added to network in Fig. E.5.12 as link.

	(1)	(2)	(3)	(l)
(1)	0.3	0.3	0.3	z_{1l}
(2)	0.3	0.8	0.8	z_{2l}
(3)	0.3	0.8	1.0	z_{3l}
(l)	z_{l1}	z_{l2}	z_{l3}	z_{ll}

$$z_{1l} = z_{l1} - z_{31} = -0.3$$

$$z_{2l} = z_{12} - z_{32} = -0.8$$

$$z_{3l} = z_{13} - z_{33} = -1.0$$

$$z_{1l} = -z_{31} + z_{4343} = -(-1) + 0.3 = 1.3$$

The augmented matrix is

	(1)	(2)	(3)	(l)
(1)	0.3	0.3	0.3	−0.3
(2)	0.3	0.8	0.8	−0.8
(3)	0.3	0.8	1.0	−1.0
(l)	−0.3	−0.8	−1.0	1.3

To eliminate the lth row and column

$$z_{11}^1 = z_{11} - \frac{(z_{1l})(z_{l1})}{z_{11}} = 0.3 - \frac{(-0.3)(-0.3)}{1.3} = 0.230769$$

$$z_{21}^1 = z_{12}^1 = z_{12} - \frac{(z_{1l})(z_{21})}{z_{11}} = 0.3 - \frac{(-0.3)(-0.8)}{1.3} = 0.1153$$

$$z_{31}^1 = z_{13}^1 = z_{13} - \frac{(z_{1l})(z_{13})}{z_{11}} = 0.3 - \frac{(-0.3)(-1.0)}{1.3} = 0.06923$$

$$z_{22}^1 = z_{22} - \frac{(z_{2l})(z_{12})}{z_{11}} = 0.8 - \frac{(-0.8)(-0.8)}{1.3} = 0.30769$$

$$z_{23}^1 = z_{32}^1 = z_{32} - \frac{(z_{3l})(z_{12})}{z_{11}} = 0.8 - \frac{(-1.0)(-0.8)}{1.3} = 0.18461$$

$$z_{33}^1 = z_{33} - \frac{(z_{3l})(z_{13})}{z_{11}} = 1.0 - \frac{(-1.0)(-1.0)}{1.3} = 0.230769$$

and, thus,

		(1)	(2)	(3)
$Z_{BUS} =$	(1)	0.230769	0.1153	0.06923
	(2)	0.1153	0.30769	0.18461
	(3)	0.06923	0.18461	0.230769

Step 5: Add element (5) which is a link $a = 3$ to $b = 1$, mutually coupled with element (4). The augmented impedance matrix with the fictitious node l is shown below.

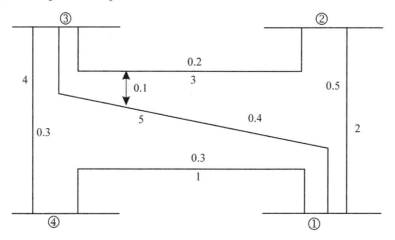

FIGURE E.5.14

Element 5 added as a link to network in Fig. E.5.13.

	(1)	(2)	(3)	(*l*)
(1)	0.230769	0.1153	0.06923	z_{1l}
(2)	0.1153	0.30769	0.18461	z_{2l}
(3)	0.06923	0.18461	0.230769	z_{3l}
(*l*)	z_{l1}	z_{12}	z_{13}	z_{1l}

$$z_{1l} = z_{l1} = z_{31} - z_{11} + \frac{y_{31\ 23}(z_{21} - z_{31})}{y_{31\ 31}}$$

$$z_{l2} = z_{2l} = z_{32} - z_{12} + \frac{y_{31\ 23}(z_{22} - z_{32})}{y_{31\ 31}}$$

$$z_{3l} = z_{l3} = z_{33} - z_{13} + \frac{y_{31\ 23}(z_{23} - z_{33})}{y_{31\ 31}}$$

$$z_{ll} = z_{31} - z_{11} + \frac{1 + y_{31\ 23}(z_{21} - z_{31})}{y_{31\ 31}}$$

Invert the primitive impedance matrix of the partial network to obtain the primitive admittance matrix.

$$[z_{xy\,xy}] = \begin{matrix} & \begin{matrix} 4-1(1) & 1-2(2) & 2-3(3) & 4-3(4) & 3-1(5) \end{matrix} \\ \begin{matrix} 4-1(1) \\ 1-2(2) \\ 2-3(3) \\ 4-3(4) \\ 3-1(5) \end{matrix} & \begin{bmatrix} 0.3 & 0 & 0 & 0 & 0 \\ 0 & 0.5 & 0 & 0 & 0 \\ 0 & 0 & 0.2 & 0 & 0.1 \\ 0 & 0 & 0 & 0.3 & 0 \\ 0 & 0 & 0.1 & 0 & 0.4 \end{bmatrix} \end{matrix}$$

$$[z_{xy\,xy}]^{-1} = [y_{xy\,xy}] = \begin{matrix} & \begin{matrix} 4-1(1) & 1-2(2) & 2-3(3) & 4-3(4) & 3-1(5) \end{matrix} \\ \begin{matrix} 4-1(1) \\ 1-2(2) \\ 2-3(3) \\ 4-3(4) \\ 3-1(5) \end{matrix} & \begin{bmatrix} 3.33 & 0 & 0 & 0 & 0 \\ 0 & 0.2 & 0 & 0 & 0 \\ 0 & 0 & 5.714 & 0 & -1.428 \\ 0 & 0 & 0 & 3.33 & 0 \\ 0 & 0 & -1.428 & 0 & 2.8571 \end{bmatrix} \end{matrix}$$

$$z_{11} = z_{12} = 0.06923 - 0.230769 + \frac{(-1.428)(0.1153 - 0.06923)}{2.8571} = -0.18456$$

$$z_{12} = z_{21} = 0.18461 - 0.1153 + \frac{(-1.428)(0.30769 - 0.18461)}{2.8571} = 0.00779$$

$$z_{31} = z_{13} = 0.230769 - 0.06923 + \frac{(-1.428)(0.18461 - 0.230769)}{2.8571} = 0.1846$$

$$z_{ll} = 0.1846 - (-0.1845) + \frac{1 + (-1.428)(0.00779 - 0.1846)}{2.8571} = 0.8075$$

	(1)	(2)	(3)	(l)
(1)	0.230769	0.1153	0.06923	−0.18456
(2)	0.1153	0.30769	0.18461	0.00779
(3)	0.06923	0.18461	0.230769	0.186
(l)	−0.18456	0.00779	0.1846	0.8075

To eliminate lth row and column:

$$z_{21}^1 = z_{11} - \frac{(z_{1l})(z_{l1})}{z_{11}} = 0.230769 - \frac{(-0.18456)(-0.18456)}{0.8075} = 0.18858$$

$$z_{12}^1 = z_{21}^1 = z_{12} = -\frac{(z_{1l})(z_{l2})}{z_{11}} = 0.1153 - \frac{(-0.18456)(0.00779)}{0.8075} = 0.11708$$

$$z_{22}^1 = z_{22} = -\frac{(z_{2l})(z_{l2})}{z_{11}} = 0.30769 - \frac{(-0.00779)(0.00779)}{0.8075} = 0.30752$$

$$z^1_{13} = z^1_{31} = z_{13} = -\frac{(z_{1l})(z_{13})}{z_{11}} = 0.06923 - \frac{(-0.18456)(0.1846)}{0.8075} = 0.11142$$

$$z^1_{33} = z_{33} - \frac{(z_{3l})(z_{13})}{z_{11}} = 0.230769 - \frac{(-0.1846)(0.1846)}{0.8075} = 0.18857$$

$$z^1_{32} = z^1_{23} = z_{23} - \frac{(z_{2l})(z_{31})}{z_{11}} = 0.18461 - \frac{(0.00779)(0.1846)}{0.8075} = 0.18283$$

and

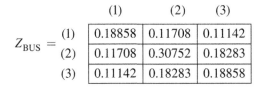

		(1)	(2)	(3)
$Z_{BUS} =$	(1)	0.18858	0.11708	0.11142
	(2)	0.11708	0.30752	0.18283
	(3)	0.11142	0.18283	0.18858

E.5.7. Given the network shown in Fig. E.5.15.
Its z_{BUS} is as follows.

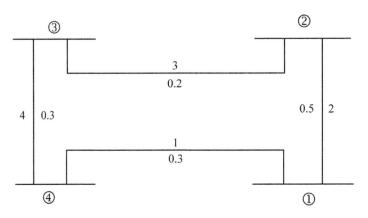

FIGURE E.5.15

Network for E.5.7.

		(1)	(2)	(3)
$Z_{BUS} =$	(1)	0.230769	0.1153	0.0623
	(2)	0.1153	0.30769	0.18461
	(3)	0.06923	0.18461	0.230769

If the line 4 is removed determine the z_{BUS} for the changed network.

Solution:

Add an element parallel to the element 4 having an impedance equal to impedance of element 4 with negative sign.

$$\frac{1}{Z_{\text{new}}} = \frac{1}{Z_{\text{added}}} + \frac{1}{Z_{\text{existing}}} = -\frac{1}{0.3} + \frac{1}{0.3} = 0$$

This amount to addition of a link.

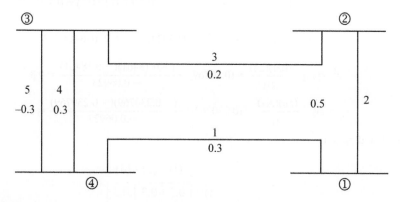

FIGURE E.5.16

Addition if a link with equal negative value.

	(1)	(2)	(3)	(*l*)
(1)	0.230769	0.1153	0.0623	z_{1l}
Z_{BUS} = (2)	0.1153	0.30769	0.18461	z_{2l}
(3)	0.06923	0.18461	0.230769	z_{3l}
(*l*)	z_{l1}	z_{l2}	z_{l3}	z_{ll}

where

$$z_{1l} = z_{l1} = -z_{31} = -0.06923$$

$$z_{2l} = z_{l2} = -z_{32} = -0.18461$$

$$z_{3l} = z_{l3} = -z_{33} = -0.230769$$

$$z_{ll} = -z_{31} + z_{4343}$$
$$= (-0.230769) + (-0.3) = -0.06923$$

The augmented z_{BUS} is then

	(1)	(2)	(3)	(*l*)
(1)	0.230769	0.1153	0.0623	−0.06923
Z_{BUS} = (2)	0.1153	0.30769	0.18461	−0.18461
(3)	0.06923	0.18461	0.230769	−0.230769
(*l*)	−0.06923	−0.18461	−0.230769	−0.06923

Eliminating the fictitious node l

$$z_{32}^1 = z_{23}^1 = z_{23} - \frac{(z_{2l})(z_{13})}{z_{11}} = 0.18461 - \frac{(0.18461)(-0.2307669)}{-0.06923} = 0.8$$

$$z_{11}^1 = z_{11} - \frac{(z_{2l})(z_{11})}{z_{11}} = (0.230769) - (-0.06923) = 0.3$$

$$z_{21}^1 = z_{12}^1 = z_{12} - \frac{(z_{2l})(z_{21})}{z_{11}} = (0.1153) - (-0.18461) = 0.3$$

$$z_{31}^1 = z_{13}^1 = z_{13} - \frac{(z_{2l})(z_{31})}{z_{11}} = (0.06923) - (-0.230769) = 0.3$$

$$z_{22}^1 = z_{22} - \frac{(z_{2l})(z_{2l})}{z_{11}} = (0.30769) - \frac{(-0.18461)(-0.18461)}{-0.06923} = 0.8$$

$$z_{33}^1 = z_{33} - \frac{(z_{3l})(z_{13})}{z_{11}} = (0.230769) - \frac{(-0.230769)(-0.230769)}{-0.06923} = 1.0$$

The modified z_{BUS} is

$$
Z_{BUS} =
\begin{array}{c}
\quad\ (1)\ \ (2)\ \ (3) \\
\begin{array}{c}
(1) \\
(2) \\
(3)
\end{array}
\begin{array}{|c|c|c|}
\hline
0.3 & 0.3 & 0.3 \\
\hline
0.3 & 0.8 & 0.8 \\
\hline
0.3 & 0.8 & 1.0 \\
\hline
\end{array}
\end{array}
$$

E.5.8. **Consider the system in** Fig. E.5.17.
Obtain z_{BUS} by using building algorithm.
Solution:
Bus (1) is chosen as reference. Consider element 1 (between bus (1) and (2))

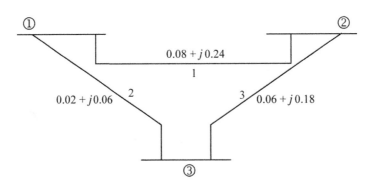

FIGURE E.5.17

Network for E.5.8.

$$Z_{\text{BUS}} = \begin{array}{c} \quad\quad (2) \\ (2) \quad \boxed{0.08 + j0.24} \end{array}$$

Step 1: Add element 2 (which is between bus (1) and (3)).

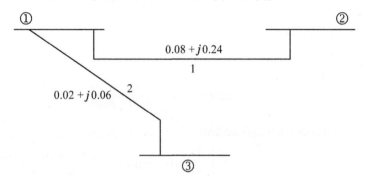

FIGURE E.5.18

Application of building algorithm with a branch added.

This is addition of a branch. A new bus (3) is created. There is no mutual impedance.

$$Z_{\text{BUS}} = \begin{array}{c} \\ (2) \\ (3) \end{array} \begin{array}{|c|c|} \hline (2) & (3) \\ \hline 0.08 + j0.24 & 0.0 + j0.0 \\ \hline 0.0 + j0.0 & 0.02 + j0.06 \\ \hline \end{array}$$

Step 2: Add element 3 which is between buses (2) and (3).

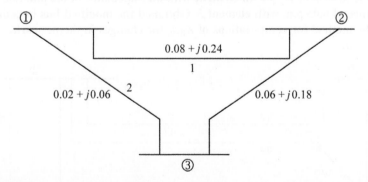

FIGURE E.5.19

Application of building algorithm with link added.

A link is added. Fictitious node l is introduced.

	(2)	(3)	(*l*)
(2)	$0.08 + j0.24$	$0.0 + j0.0$	$0.08 + j0.24$
(3)	$0.0 + j0.0$	$0.02 + j0.06$	$-(0.02 + j0.06)$
(*l*)	$0.08 + j0.24$	$-(0.02 + j0.006)$	$0.16 + j0.48$

$Z_{BUS} =$ (with row labels (2), (3), (*l*))

Eliminating the fictitious node *l*

$$z_{22}(\text{modified}) = z_{22} - \frac{z_{2l}\, z_{l2}}{z_{ll}}$$

$$= (0.08 + j0.24) - \frac{(0.08 + j0.24)^2}{0.49 + j0.48} = 0.04 + j0.12$$

$$z_{23}(\text{modified}) = z_{32}(\text{modified})$$

$$= \left[z_{23} - \frac{z_{2l}\, z_{l3}}{z_{ll}} \right] = 0.0 + j0.0 + \frac{(0.8 + j0.24)(0.02 + j0.06)}{0.16 + j0.48}$$

$$= 0.01 + j0.03$$

$$z_{33}(\text{modified}) = 0.0175 + j0.0526$$

The z_{BUS} matrix is thus

$$Z_{BUS} = \begin{array}{c} (2) \\ (3) \end{array} \begin{bmatrix} 0.04 + j0.12 & 0.01 + j0.03 \\ 0.01 + j0.03 & 0.0175 + j0.0526 \end{bmatrix}$$

with column headers (2) and (3)

E.5.9. **Given the system of E.5.4. An element with an impedance of 0.2 p.u. and mutual impedance of 0.05 p.u. with element 5. Obtained the modified bus impedance method using the method for computations of Z_{BUS} for changes in the network.**

Solution:

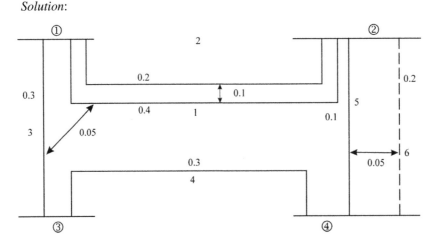

FIGURE E.5.20

Network for E.5.9.

Element added is a link
$a = 2; b = 4$. A fictitious node l is created.

	(2)	(3)	(4)	l
(2)	0.16	0.0741267	0.1386	Z_{2l}
(3)	0.0741267	0.50313	0.106431	Z_{3l}
(4)	0.1386	0.106437	0.2055	Z_{4l}
(l)	Z_{l2}	Z_{l3}	Z_{l4}	Z_{ll}

$$Z_{li} = Z_{ai} - Z_{bi} + \frac{\bar{y}_{ab-xy}(Z_{xi} - Z_{yi})}{y_{ab-ab}}$$

and

$$Z_{ll} = Z_{ai} - Z_{bi} + \frac{1 + \bar{y}_{ab-xy}(Z_{xi} - Z_{yi})}{y_{ab-ab}}$$

The primitive impedance matrix is

	$1-2(1)$	$1-2(2)$	$1-3$	$3-4$	$2-4(1)$	$2-4(2)$
	0.4	0.1	0.05	0	0	0
$1-2(2)$	0.1	0.2	0	0	0	0
$[z] = \quad 1-3$	0.05	0	0.3	0	0	0
$3-4$	0	0	0	0	0	0
$2-4(1)$	0	0	0	0.3	0.1	0.05
$2-4(2)$	0	0	0	0	0.05	0.2

The added element 6 is coupled to only one element, i.e., element 5. It is sufficient to invert the submatrix for the coupled element.

	$2-4(1)$	$2-4(2)$
$Z_{ab-xy} = 2-4(1)$	0.1	0.05
$2-4(2)$	0.05	0.2

$$y_{ab-xy} = \begin{bmatrix} 11.4285 & -2.857 \\ -2.857 & 5.7143 \end{bmatrix}$$

$$Z_{21} = Z_{12} = Z_{22} - Z_{42} + \frac{(-2.857)(Z_{xi} - Z_{yi})}{11.43}$$

$$= 0.16 - 0.1386 + \frac{(-2.857)(0 - 16 - 0.1386)}{5.7143} = 0.01070$$

$$Z_{3l} = Z_{l3} = 0.741267 - 0.10643 + \frac{(-2.857)(0.741 - 0.10643)}{5.7143} = -0.1614$$

$$Z_{l4} = Z_{41} = Z_{24} - Z_{44} + \frac{\bar{y}_{ab-xy}(Z_{xi} - Z_{yi})}{y_{ab-ab}}$$

$$= 0.1386 - 0.2055 + \frac{(-2.857)(0.1386 - 0.2055)}{5.7143} = -0.03345$$

$$Z_{ll} = 0.0107 - (-0.03345) + \frac{1 + (-2.857)(0.0107 - (-0.03345))}{5.7143} = 0.19707$$

The augmented matrix is then

	(2)	(3)	(4)	(l)
(2)	0.1600	0.0741	0.1386	0.0107
(3)	0.0741	0.5031	0.1064	-0.01614
(4)	0.1386	0.1064	0.2055	-0.03345
(l)	0.0107	-0.01614	-0.03345	0.19707

$$Z_{22}(\text{modified}) = 0.16 - \frac{(0.0107)^2}{0.19707} = 0.1594$$

$$Z_{23}(\text{modified}) = Z_{32}(\text{modified})$$
$$= 0.0741 \frac{0.01070 \times (-0.01614)}{0.19707} = 0.7497$$

$$Z_{24}(\text{modified}) = Z_{42}(\text{modified})$$
$$= 0.1386 - \frac{0.01070 \times (-0.03345)}{0.19707} = 0.1404$$

$$Z_{33}(\text{modified}) = 0.50313 - \frac{(-0.01614)^2}{0.19707} = 0.5018$$

$$Z_{43}(\text{modified}) = Z_{34}(\text{modified})$$
$$= 0.106431 - \frac{(-0.03345)(-0.01614)}{0.19707} = 0.10369$$

$$Z_{44}(\text{modified}) = 0.2055 - \frac{(-0.0334)^2}{0.19707} = 0.1998$$

Hence the Z_{BUS} is obtained by

$$Z_{\text{BUS}} = \begin{array}{c} (2) \\ (3) \\ (4) \end{array} \begin{bmatrix} (2) & (3) & (4) \\ 0.1594 & 0.07497 & 0.1404 \\ 0.07497 & 0.5018 & 0.10369 \\ 0.1404 & 0.10369 & 0.1998 \end{bmatrix}$$

E.5.10. **Consider the problem E.5.9. If the element 6 is now removed obtain the Z_{BUS}.**
  Solution:

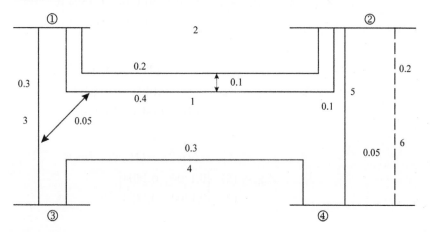

FIGURE E.5.21

Network for E.5.10.

$$Z'_{ij} = Z_{ij} + (\bar{Z}_{i\alpha} - \bar{Z}_{i\beta})[F]^{-1}[\Delta y_i]\begin{bmatrix} \bar{Z} & -Z \\ \gamma & \delta i \end{bmatrix}$$

where

$$\alpha = 2, \beta = 4; i = 1, 2, \ldots, n$$

and

$$\gamma = 2; \delta = 4$$

the original primitive admittance matrix

$$[y_{sm}] = \begin{matrix} & 2-4(1) & 2-4(2) \\ 2-4(1) & \\ 2-4(2) & \end{matrix} \begin{bmatrix} 11.4285 & -2.857 \\ -2.857 & 5.7143 \end{bmatrix}$$

The modified primitive admittance matrix

$$[y_{sm}]' = \begin{matrix} & 1-2(1) & 1-2(2) \\ 1-2(1) & \\ 1-2(2) & \end{matrix} \begin{bmatrix} +\dfrac{1}{0.1} & 0 \\ 0 & 0 \end{bmatrix} = \begin{bmatrix} 10 & 0 \\ 0 & 0 \end{bmatrix}$$

$$\{[y_{sm}][y_{sm}]'\} = \begin{bmatrix} 11.4285 & -2.857 \\ -2.857 & 5.7143 \end{bmatrix} = \begin{bmatrix} 10 & 0 \\ 0 & 0 \end{bmatrix} = \begin{bmatrix} 1.4285 & -2.857 \\ -2.857 & 5.7143 \end{bmatrix} = \Delta y_{sm}$$

Computing term by term

$$Z_{\gamma\alpha} - Z_{\delta\alpha} - Z_{\gamma\beta} - Z_{\delta\beta} = \begin{array}{c} \\ (2) \\ (2) \end{array} \begin{matrix} (2) & (2) \\ \begin{bmatrix} 0.1594 & 0.1594 \\ 0.1594 & 0.1594 \end{bmatrix} \end{matrix}$$

$$Z_{\delta\alpha} = Z_{24} = Z_{42} = \begin{array}{c} \\ (4) \\ (4) \end{array} \begin{matrix} (2) & (2) \\ \begin{bmatrix} 0.1404 & 0.1404 \\ 0.1404 & 0.1404 \end{bmatrix} \end{matrix}$$

$$Z_{\gamma\alpha} = \begin{array}{c} \\ (2) \\ (2) \end{array} \begin{matrix} (4) & (4) \\ \begin{bmatrix} 0.1404 & 0.1404 \\ 0.1404 & 0.1404 \end{bmatrix} \end{matrix}$$

$$Z_{\delta\beta} = \begin{array}{c} \\ (4) \\ (4) \end{array} \begin{matrix} (4) & (4) \\ \begin{bmatrix} 0.1998 & 0.1998 \\ 0.1998 & 0.1998 \end{bmatrix} \end{matrix}$$

$$Z_{\gamma\alpha} - Z_{\delta\alpha} - Z_{\gamma\beta} + Z_{\delta\beta} = \begin{bmatrix} 0.1594 & 0.1594 \\ 0.1594 & 0.1594 \end{bmatrix} - \begin{bmatrix} 0.1404 & 0.1404 \\ 0.1404 & 0.1404 \end{bmatrix}$$
$$- \begin{bmatrix} 0.1404 & 0.1404 \\ 0.1404 & 0.1404 \end{bmatrix} + \begin{bmatrix} 0.1998 & 0.1998 \\ 0.1998 & 0.1998 \end{bmatrix}$$
$$= \begin{bmatrix} 0.0784 & 0.0784 \\ 0.0784 & 0.0784 \end{bmatrix}$$

$$[\Delta y_{sm}][Z_{\gamma\alpha} - Z_{\delta\alpha} - Z_{\gamma\beta} - Z_{\delta\beta}] = \begin{bmatrix} 1.4285 & -2.857 \\ -2.857 & 5.7143 \end{bmatrix}\begin{bmatrix} 0.0784 & 0.784 \\ 0.0784 & 0.0784 \end{bmatrix} = \begin{bmatrix} -0.1120 & -0.1120 \\ 0.2240 & 0.2240 \end{bmatrix}$$

$$F = U - \Delta y_{sm}(Z_{\gamma\alpha} - Z_{\delta\alpha} - Z_{\gamma\beta} + Z_{\delta\beta}) = \begin{bmatrix} 1 & 0 \\ 0 & 1 \end{bmatrix} - \begin{bmatrix} -0.1120 & -0.1120 \\ 0.2240 & 0.2240 \end{bmatrix} = \begin{bmatrix} 1.112 & 0.112 \\ -0.224 & 0.7760 \end{bmatrix}$$

$$[F]^{-1} = \begin{bmatrix} 0.87387 & -0.12612 \\ 0.25225 & 1.25225 \end{bmatrix}$$

$$[F]^{-1}\Delta y_{sm} = \begin{bmatrix} 0.87387 & -0.12612 \\ 0.25225 & 1.25225 \end{bmatrix}\begin{bmatrix} 1.4285 & -2.8571 \\ -2.8571 & 5.7143 \end{bmatrix} = \begin{bmatrix} 1.6086 & -3.2173 \\ -3.2173 & 6.435 \end{bmatrix}$$

The elements of modified Z_{BUS} are then given by

$$Z_{ij}^{\mid} = Z_{ij} + (\overline{Z}_{i\alpha} - \overline{Z}_{i\beta})[F]^{-1}[\Delta y_{sm}][\overline{Z}_{\gamma i} - \overline{Z}_{\delta i}]$$

For $i = 2; j = 2$

$$Z_{22}^{|} = Z_{22} + ([Z_{22}Z_{22}] - [Z_{24}Z_{24}][M]^{-1}\{\Delta y_{sm}\})\left(\begin{bmatrix} Z_{22} \\ Z_{22} \end{bmatrix} - \begin{bmatrix} Z_{42} \\ Z_{42} \end{bmatrix}\right)$$

$$= 0.1594 + ([0.1594 \ 0.1594] - [0.1404 \ 0.1404])$$

$$\begin{bmatrix} 1.6086 & -3.2173 \\ -3.2173 & 6.435 \end{bmatrix} \left(\begin{pmatrix} 0.1594 \\ 0.1594 \end{pmatrix} - \begin{pmatrix} 0.1404 \\ 0.1404 \end{pmatrix}\right)$$

$$= 0.1594 + [0.019 \ 0.019]\begin{bmatrix} 1.6086 & -3.2173 \\ -3.2173 & 6.435 \end{bmatrix}\begin{bmatrix} 0.019 \\ 0.019 \end{bmatrix}$$

$$= 0.1594 + [-0.03056 \ 0.0611]\begin{bmatrix} 0.019 \\ 0.019 \end{bmatrix} = 0.1594 + 0.00058026 = 0.15998 = 0.16$$

Let $i = 2; j = 3$

$$Z_{23}^{|} = 0.07497 + [0.019 \ 0.019]\begin{bmatrix} 1.6086 & -3.2173 \\ -3.2173 & 6.435 \end{bmatrix}\begin{bmatrix} -0.03146 \\ -0.03146 \end{bmatrix}$$

$$= 0.07497 + [-0.03056 \ 0.0611]\begin{bmatrix} -0.03146 \\ -0.03146 \end{bmatrix} = 0.07497 - 0.00096075 = 0.074040$$

Let $i = 2; j = 4$

$$Z_{24}^{|} = 0.1404 + [-0.03056 \ 0.0611]\begin{bmatrix} -0.0594 \\ -0.0594 \end{bmatrix} = 0.1404 - 0.001814 = 0.13858$$

Let $i = 3; j = 3$

$$Z_{33}^{|} = 0.5018 + [-0.03146 \ -0.03146]\begin{bmatrix} -1.6086 & -3.2173 \\ -3.2173 & 6.435 \end{bmatrix}\begin{bmatrix} -0.03146 \\ -0.03146 \end{bmatrix}$$

$$= 0.5018 + [0.0506 \ -0.101228]\begin{bmatrix} -0.03146 \\ -0.03146 \end{bmatrix} = 0.5018 + 0.001592 = 0.503392$$

For $i = 3; j = 4$

$$Z_{34}^{|} = 0.103691 + [0.05060 \ -0.101228]\begin{bmatrix} -0.0594 \\ -0.0594 \end{bmatrix}$$

$$= 0.103691 + 0.003006 = 0.106696$$

Similarly for $I = j =$

$$Z_{44}^{|} = 0.1998 + [-0.0594 \ -0.0594]\begin{bmatrix} 1.6086 & -3.2173 \\ -3.2173 & 6.435 \end{bmatrix}\begin{bmatrix} -0.0594 \\ -0.0594 \end{bmatrix}$$

$$= 0.1998 + [0.09556 \ -0.19113]\begin{bmatrix} -0.0594 \\ -0.0594 \end{bmatrix} = 0.205476$$

Hence the Z_{BUS} with the element 6 removed will be

		(2)	(3)	(4)
$Z_{BUS} =$	(2)	0.15998	0.074040	0.13858
	(3)	0.07404	0.503392	0.106696
	(4)	0.13858	0.106696	0.20576

PROBLEMS

P.5.1. Form the bus impedance matrix for the system shown in Fig. P.5.1, the line data is given below.

Element Numbers	Bus Code	Self-Impedance (p.u.)
1	(2)–(3)	0.6
2	(1)—(3)	0.5
3	(1)—(2)	0.4

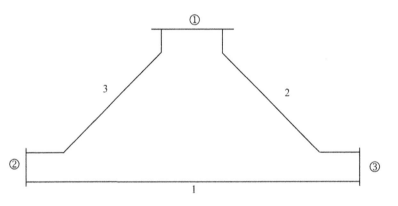

FIGURE P.5.1

Network for P.5.1.

P.5.2. Obtain the modified Z_{BUS} if a line 4 is added parallel to line 1 across the busses (2) and (3) with a self-impedance of 0.5 p.u.

P.5.3. In the problem P.5.2, if the added line element 4 has a mutual impedance with respect to line element 1 of 0.1 p.u., how will the Z_{BUS} matrix change?

QUESTIONS

5.1. Starting from Z_{BUS} for a partial network, describe step by step how will you obtain the Z_{BUS} for a modified network when a new line is to be added to a bus in the existing network?

5.2. Starting from Z_{BUS} for a partial network, describe step by step how will you obtain the Z_{BUS} for a modified network when a new line is to be added between two buses of the existing network?

5.3. What are the advantages of Z_{BUS} building algorithm?

5.4. Describe the procedure for the modification of Z_{BUS} when a line is added or removed which has no mutual impedance.

5.5. Describe the procedure for the modification of Z_{BUS} when a line with mutual impedance is added or removed.

5.6. Derive the necessary expressions for the building up of Z_{BUS} when (1) new element is added and (2) new element is added between two existing buses. Assume mutual coupling between the added element and the elements in the partial network.

5.7. Write short notes on removal of a link in Z_{BUS} with no mutual coupling between the element deleted and the other elements in the network.

5.8. Derive an expression for adding a link to a network with mutual inductance.

5.9. Derive an expression for adding a branch element between two buses in the Z_{BUS} building algorithm.

5.10. Explain the modifications necessary in the Z_{BUS} when a mutually coupled element is removed or its impedance is changed.

5.11. Develop the equation for modifying the elements of a bus impedance matrix when it is coupled to other elements in the network, adding the element not creating a new bus.

SYMMETRICAL COMPONENTS

Three-phase systems are accepted as the standard system for generation, transmission, and utilization of the bulk of electric power generated world over. The above holds good even when some of the transmission lines are replaced by d-c links. When the three-phase system becomes unbalanced while in operation, analysis becomes difficult. Dr. C.L. Fortesque proposed in 1918 at a meeting of the American Institute of Electrical Engineers through a paper titled "Method of Symmetrical Coordinates Applied to the Solution of Polyphase Networks", a very useful method for analyzing unbalanced three-phase networks.

Faults of various types such as line-to-ground, line-to-line, three-phase short circuits with different fault impedances, etc. create unbalances. Breaking down of line conductors is also another source for unbalances in Power Systems Operation. The symmetrical Coordinates proposed by Fortesque are known more commonly as symmetrical components or sequence components.

An unbalanced system of n phasors can be resolved into n systems of balanced phasors. These subsystems of balanced phasors are called symmetrical components. With reference to three-phase systems, the following balanced set of three components are identified and defined (Fig. 6.1).

1. Set of three phasors equal in magnitude, displaced from each other by 120° in phase and having the same phase sequence as the original phasors constitute positive sequence components. They are denoted by the suffix 1.

2. Set of three phasors equal in magnitude, displaced from each other by 120° in phase, and having a phase sequence opposite to that of the original phasors *constitute the negative sequence components. They are denoted by the suffix 2.*

3. Set of three phasors equal in magnitude and all in phase (with no mutual phase displacement) constitute zero sequence components. They are denoted by the suffix 0. Denoting the phases as R, Y, and B, V_R, V_Y, and V_B are the unbalanced phase voltages. These voltages are expressed in terms of the sequence components V_{R1}, V_{Y1}, V_{B1}, V_{R2}, V_{Y2}, V_{B2}, and V_{R0}, V_{Y0}, V_{B0} as follows:

$$V_R = V_{R1} + V_{R2} + V_{R0} \qquad (6.1)$$

$$V_Y = V_{Y1} + V_{Y2} + V_{Y0} \qquad (6.2)$$

$$V_B = V_{B1} + V_{B2} + V_{B0} \qquad (6.3)$$

Power Systems Analysis. DOI: http://dx.doi.org/10.1016/B978-0-08-101111-9.00006-9

113

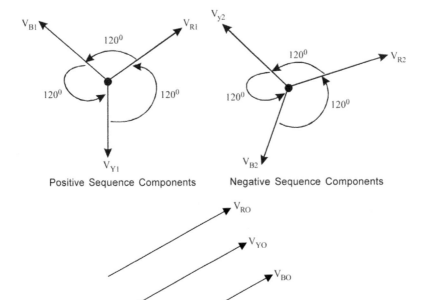

Positive Sequence Components   Negative Sequence Components

Zero Sequence Components

FIGURE 6.1

Sequence components.

6.1 THE OPERATOR "*a*"

In view of the phase displacement of 120°, an operator "*a*" is used to indicate the phase displacement, just as j operator is used to denote 90° phase displacement.

$$a = 1 \angle 120° = -0.5 + j0.866$$
$$a^2 = 1 \angle 240° = -0.5 - j0.866$$
$$a^3 = 1 \angle 360° = 1 + j0$$

so that $1 + a + a^2 = 0 + j0$

The operator is represented graphically in Fig. 6.2.

Note that

$$a = 1 \angle 120° = 1 \cdot e^{j2\pi/3}$$

$$a^2 = 1 \angle 240° = 1 \cdot e^{j4\pi/3}$$

$$a^3 = 1 \angle 360° = 1 \cdot e^{j6\pi/3} = 1 \cdot e^{j2\pi}$$

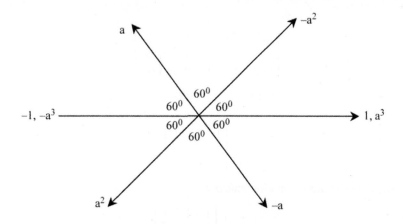

FIGURE 6.2

Operator "*a*."

6.2 SYMMETRICAL COMPONENTS OF UNSYMMETRICAL PHASES

With the introduction of the operator "*a*," it is possible to redefine the relationship between unbalanced phasors of voltages and currents in terms of the symmetrical components or sequence components as they are known otherwise. We can write the sequence phasors with the operator as follows:

$$\left.\begin{array}{l} V_{R1} = V_{R1} \\ V_{R2} = V_{R2} \\ V_{R0} = V_{R0} \\ V_{Y1} = a^2 V_{R1} \\ V_{Y2} = a V_{R2} \\ V_{Y0} = V_{R0} \end{array}\right\} \quad \left.\begin{array}{l} V_{B1} = a V_{R1} \\ V_{B1} = a^2 V_{R2} \\ V_{B0} = V_{R0} \end{array}\right\} \tag{6.4}$$

The voltage and current phasors for a three-phase unbalanced system are then represented by

$$\left.\begin{array}{l} V_R = V_{R1} + V_{R2} + V_{R0} \\ V_Y = a^2 V_{R1} + a V_{R2} + V_{R0} \\ V_B = a V_{R1} + a^2 V_{R2} + V_{R0} \end{array}\right\} \tag{6.5}$$

$$\left.\begin{array}{l} I_R = I_{R1} + I_{R2} + I_{R0} \\ I_Y = a^2 I_{R1} + a I_{R2} + I_{R0} \\ I_B = a I_{R1} + a^2 I_{R2} + I_{R0} \end{array}\right\} \tag{6.6}$$

The above equations can be put in matrix form considering zero sequence relation as the first for convenience.

$$\begin{bmatrix} V_R \\ V_Y \\ V_B \end{bmatrix} = \begin{bmatrix} 1 & 1 & 1 \\ 1 & a^2 & a \\ 1 & a & a^2 \end{bmatrix} \begin{bmatrix} V_{R0} \\ V_{R1} \\ V_{R2} \end{bmatrix} \tag{6.7}$$

and

$$
\begin{bmatrix} I_R \\ I_Y \\ I_B \end{bmatrix} = \begin{bmatrix} 1 & 1 & 1 \\ 1 & a^2 & a \\ 1 & a & a^2 \end{bmatrix} \begin{bmatrix} I_{R0} \\ I_{R1} \\ I_{R2} \end{bmatrix}
\tag{6.8}
$$

Eqs. (6.7) and (6.8) relate the sequence components to the phase components through the transformation matrix

$$
C = \begin{bmatrix} 1 & 1 & 1 \\ 1 & a^2 & a \\ 1 & a & a^2 \end{bmatrix}
\tag{6.9}
$$

consider the inverse of the transformation matrix C

$$
C^{-1} = \frac{1}{3} \begin{bmatrix} 1 & 1 & 1 \\ 1 & a & a^2 \\ 1 & a^2 & a \end{bmatrix}
\tag{6.10}
$$

Then the sequence components can be obtained from the phase values as

$$
\begin{bmatrix} V_{R0} \\ V_{R1} \\ V_{R2} \end{bmatrix} = \frac{1}{3} \begin{bmatrix} 1 & 1 & 1 \\ 1 & a & a^2 \\ 1 & a^2 & a \end{bmatrix} \begin{bmatrix} V_R \\ V_Y \\ V_B \end{bmatrix}
\tag{6.11}
$$

and

$$
\begin{bmatrix} I_{R0} \\ I_{R1} \\ I_{R2} \end{bmatrix} = \frac{1}{3} \begin{bmatrix} 1 & 1 & 1 \\ 1 & a & a^2 \\ 1 & a^2 & a \end{bmatrix} \begin{bmatrix} I_R \\ I_Y \\ I_B \end{bmatrix}
\tag{6.12}
$$

6.3 POWER IN SEQUENCE COMPONENTS

The total complex power flowing into a three-phase circuit through the lines R, Y, B is

$$
\begin{aligned}
S = P + jQ &= \overline{V}\,\overline{I}^* \\
&= \overline{V}_R \overline{I}_R^* + \overline{V}_Y \overline{I}_Y^* + \overline{V}_Z \overline{I}_Z^*
\end{aligned}
\tag{6.13}
$$

written in matrix notation

$$
S = [V_R \quad V_Y \quad V_B] \begin{bmatrix} I_R \\ I_Y \\ I_B \end{bmatrix}
\tag{6.14}
$$

$$
= \begin{bmatrix} V_R \\ V_Y \\ V_B \end{bmatrix} \begin{bmatrix} I_R \\ I_Y \\ I_B \end{bmatrix}^*
\tag{6.15}
$$

Also

$$\begin{bmatrix} V_R \\ V_Y \\ V_B \end{bmatrix} = C \begin{bmatrix} V_{R0} \\ V_{R1} \\ V_{R2} \end{bmatrix} \tag{6.16}$$

$$\begin{bmatrix} I_R \\ I_Y \\ I_B \end{bmatrix}^* = C^* \begin{bmatrix} I_{R0} \\ I_{R1} \\ I_{R2} \end{bmatrix}^* \tag{6.17}$$

$$\begin{bmatrix} V_R \\ V_Y \\ V_B \end{bmatrix}^t = \begin{bmatrix} V_{R0} \\ V_{R1} \\ V_{R2} \end{bmatrix}^t C^t \tag{6.18}$$

From Eq. (6.14)

$$S = \begin{bmatrix} V_{R0} & V_{R1} & V_{R2} \end{bmatrix} \begin{bmatrix} 1 & 1 & 1 \\ 1 & a^2 & a \\ 1 & a & a^2 \end{bmatrix} \begin{bmatrix} 1 & 1 & 1 \\ 1 & a & a^2 \\ 1 & a^2 & a \end{bmatrix} \begin{bmatrix} I_{R0} \\ I_{R1} \\ I_{R2} \end{bmatrix}^* \tag{6.19}$$

Note that $C^t C^* = 3U$

$$S = 3 \begin{bmatrix} V_{R0} & V_{R1} & V_{R2} \end{bmatrix} \begin{bmatrix} I_{R0} \\ I_{R1} \\ I_{R2} \end{bmatrix}^* \tag{6.20}$$

Power in phase components is three times the power in sequence components.

The disadvantage with these symmetrical components is that the transformation matrix C is not power invariant or is not orthogonal or unitary.

6.4 UNITARY TRANSFORMATION FOR POWER INVARIANCE

It is more convenient to define "C" as a unitary matrix so that the transformation becomes power invariant.

That is power in phase components = power in sequence components. Defining a transformation matrix T which is unitary, such that,

$$T = \begin{bmatrix} \dfrac{1}{\sqrt{3}} \end{bmatrix} \begin{bmatrix} 1 & 1 & 1 \\ 1 & a^2 & a \\ 1 & a & a^2 \end{bmatrix} \tag{6.21}$$

$$\begin{bmatrix} V_R \\ V_Y \\ V_B \end{bmatrix} = \begin{bmatrix} \dfrac{1}{\sqrt{3}} \end{bmatrix} \begin{bmatrix} 1 & 1 & 1 \\ 1 & a^2 & a \\ 1 & a & a^2 \end{bmatrix} \begin{bmatrix} V_{R0} \\ V_{R1} \\ V_{R2} \end{bmatrix} \tag{6.22}$$

and

$$\begin{bmatrix} I_R \\ I_Y \\ I_B \end{bmatrix} = \begin{bmatrix} \dfrac{1}{\sqrt{3}} \end{bmatrix} \begin{bmatrix} 1 & 1 & 1 \\ 1 & a^2 & a \\ 1 & a & a^2 \end{bmatrix} \begin{bmatrix} I_{R0} \\ I_{R1} \\ I_{R2} \end{bmatrix} \tag{6.23}$$

so that

$$T^{-1} = \left[\sqrt{3}\right] \begin{bmatrix} 1 & 1 & 1 \\ 1 & a & a^2 \\ 1 & a^2 & a \end{bmatrix} \tag{6.24}$$

$$\begin{bmatrix} V_{R0} \\ V_{R1} \\ V_{R2} \end{bmatrix} = \left[\frac{\sqrt{3}}{3}\right] \begin{bmatrix} 1 & 1 & 1 \\ 1 & a & a^2 \\ 1 & a^2 & a \end{bmatrix} \begin{bmatrix} V_R \\ V_Y \\ V_B \end{bmatrix} \tag{6.25}$$

and

$$\begin{bmatrix} I_{R0} \\ I_{R1} \\ I_{R2} \end{bmatrix} = \left[\frac{\sqrt{3}}{3}\right] \begin{bmatrix} 1 & 1 & 1 \\ 1 & a & a^2 \\ 1 & a^2 & a \end{bmatrix} \begin{bmatrix} I_R \\ I_Y \\ I_B \end{bmatrix} \tag{6.26}$$

$$S = P + jQ = VI^* \tag{6.27}$$

$$= \begin{bmatrix} V_R & V_Y & V_B \end{bmatrix} \begin{bmatrix} I_R \\ I_Y \\ I_B \end{bmatrix}^* \tag{6.28}$$

$$= \begin{bmatrix} V_R \\ V_Y \\ V_B \end{bmatrix}^t \begin{bmatrix} I_R \\ I_Y \\ I_B \end{bmatrix}^* \tag{6.29}$$

$$\begin{bmatrix} V_R \\ V_Y \\ V_B \end{bmatrix} = \left[\frac{1}{\sqrt{3}}\right] \begin{bmatrix} 1 & 1 & 1 \\ 1 & a^2 & a \\ 1 & a & a^2 \end{bmatrix} \begin{bmatrix} V_{R0} \\ V_{R1} \\ V_{R2} \end{bmatrix} \tag{6.30}$$

$$\begin{bmatrix} I_R \\ I_Y \\ I_B \end{bmatrix}^* = \left[\frac{1}{\sqrt{3}}\right] \begin{bmatrix} 1 & 1 & 1 \\ 1 & a^2 & a \\ 1 & a & a^2 \end{bmatrix} \begin{bmatrix} I_{R0} \\ I_{R1} \\ I_{R2} \end{bmatrix} \tag{6.31}$$

$$\begin{bmatrix} V_R \\ V_Y \\ V_B \end{bmatrix}^t = \begin{bmatrix} V_{R0} \\ V_{R1} \\ V_{R2} \end{bmatrix} \left[\frac{1}{\sqrt{3}}\right] \begin{bmatrix} 1 & 1 & 1 \\ 1 & a^2 & a \\ 1 & a & a^2 \end{bmatrix} \tag{6.32}$$

$$S = \begin{bmatrix} V_{R0} & V_{R1} & V_{R2} \end{bmatrix} \left[\frac{1}{\sqrt{3}}\right] \begin{bmatrix} 1 & 1 & 1 \\ 1 & a^2 & a \\ 1 & a & a^2 \end{bmatrix} \left[\frac{1}{\sqrt{3}}\right] \begin{bmatrix} 1 & 1 & 1 \\ 1 & a & a^2 \\ 1 & a^2 & a \end{bmatrix} \begin{bmatrix} I_{R0} \\ I_{R1} \\ I_{R2} \end{bmatrix} \tag{6.33}$$

$$\begin{bmatrix} V_{R0} & V_{R1} & V_{R2} \end{bmatrix} \left[\frac{1}{3}\right] \begin{bmatrix} 1 & 1 & 1 \\ 1 & a^2 & a \\ 1 & a & a^2 \end{bmatrix} \begin{bmatrix} 1 & 1 & 1 \\ 1 & a & a^2 \\ 1 & a^2 & a \end{bmatrix} \begin{bmatrix} I_{R0} \\ I_{R1} \\ I_{R2} \end{bmatrix} \tag{6.34}$$

$$\begin{bmatrix} V_{R0} & V_{R1} & V_{R2} \end{bmatrix} \begin{bmatrix} \dfrac{1}{3} \end{bmatrix} [3] \begin{bmatrix} I_{R0} \\ I_{R1} \\ I_{R2} \end{bmatrix} \tag{6.35}$$

$$S = \begin{bmatrix} V_{R0} & V_{R1} & V_{R2} \end{bmatrix} \begin{bmatrix} I_{R0} \\ I_{R1} \\ I_{R2} \end{bmatrix} \tag{6.36}$$

Thus with the unitary transformation matrix

$$T = \begin{bmatrix} \dfrac{1}{\sqrt{3}} \end{bmatrix} \begin{bmatrix} 1 & 1 & 1 \\ 1 & a^2 & a \\ 1 & a & a^2 \end{bmatrix} \tag{6.37}$$

we obtain power invariant transformation with sequence components.

THREE-PHASE NETWORKS

Three-phase networks are generally balanced. Any problem in this case can be solved on a single-phase basis. In case of unbalanced systems, the actual network can be solved in terms of either actual phase quantities or using sequence components.

7.1 THREE-PHASE NETWORK ELEMENT REPRESENTATION

A three-phase network element connected across the terminals a and b is represented in impedance form as shown in Fig. 7.1.

The voltages across the element $a-b$ for phases R, Y, and B are, respectively

$$\left.\begin{array}{l} v_{ab}^{R} = V_{a}^{R} - V_{b}^{R} \\ v_{ab}^{Y} = V_{a}^{Y} - V_{b}^{Y} \\ v_{ab}^{B} = V_{a}^{B} - V_{b}^{B} \end{array}\right\} \tag{7.1}$$

$e_{ab}^{R}, e_{ab}^{Y}, e_{ab}^{B}$ are the source voltages in series with phases R, Y, and B.

$i_{ab}^{R}, i_{ab}^{Y}, i_{ab}^{B}$ are the currents through the element $a-b$ for phases R, Y, and B, respectively.

Fig. 7.2 shows the same element in the admittance form.

$j_{ab}^{R}, j_{ab}^{Y}, j_{ab}^{B}$ are the source currents in parallel with phases R, Y, and B, respectively, of the element $a-b$.

$$\left.\begin{array}{l} v_{ab}^{R} = V_{a}^{R} - V_{b}^{R} \\ v_{ab}^{V} = V_{a}^{Y} - V_{b}^{Y} \\ v_{ab}^{B} = V_{a}^{B} - V_{b}^{B} \end{array}\right\} \tag{7.2}$$

The network element in impedance form represented in Fig. 7.1 can be put in matrix form as follows:

$$\begin{bmatrix} v_{ab}^{R} \\ v_{ab}^{Y} \\ v_{ab}^{B} \end{bmatrix} + \begin{bmatrix} e_{ab}^{R} \\ e_{ab}^{Y} \\ e_{ab}^{B} \end{bmatrix} = \begin{bmatrix} z_{ab}^{RR} & z_{ab}^{RY} & z_{ab}^{RB} \\ z_{ab}^{YR} & z_{ab}^{YY} & z_{ab}^{YB} \\ z_{ab}^{BR} & z_{ab}^{BY} & z_{ab}^{BB} \end{bmatrix} \cdot \begin{bmatrix} i_{ab}^{R} \\ i_{ab}^{Y} \\ i_{ab}^{B} \end{bmatrix} \tag{7.3}$$

where z_{ab}^{RR} is the self-impedance of phase "R" of the three-phase impedance element connected between nodes a and b.

z_{ab}^{RY} is the mutual impedance of phases R and Y of the three-phase impedance element connected between nodes a and b, etc.

Eq. (7.3) can be written more concisely as follows:

Power Systems Analysis. DOI: http://dx.doi.org/10.1016/B978-0-08-101111-9.00007-0
Copyright © 2017 BSP Books Pvt. Ltd. Published by Elsevier Ltd. All rights reserved.

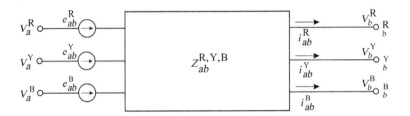

FIGURE 7.1

Three-phase element in impedance form.

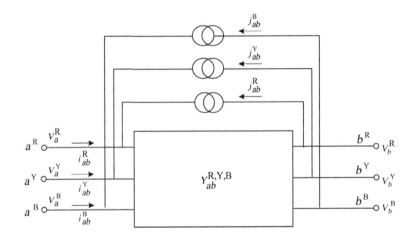

FIGURE 7.2

Three-phase element in admittance form.

$$v_{ab}^{R,Y,B} + e_{ab}^{R,Y,B} = z_{ab}^{R,Y,B} i_{ab}^{R,Y,B} \tag{7.4}$$

In a similar manner, in admittance form Fig. 7.2 can be represented as a matrix relation as follows:

$$\begin{bmatrix} i_{ab}^{R} \\ i_{ab}^{Y} \\ i_{ab}^{B} \end{bmatrix} + \begin{bmatrix} j_{ab}^{R} \\ j_{ab}^{Y} \\ j_{ab}^{B} \end{bmatrix} = \begin{bmatrix} y_{ab}^{RR} & y_{ab}^{RY} & y_{ab}^{RB} \\ y_{ab}^{YR} & y_{ab}^{YY} & y_{ab}^{YB} \\ y_{ab}^{BR} & y_{ab}^{BY} & y_{ab}^{BB} \end{bmatrix} \cdot \begin{bmatrix} v_{ab}^{R} \\ v_{ab}^{Y} \\ v_{ab}^{B} \end{bmatrix} \tag{7.5}$$

Eq. (7.5) can be represented more consciously as follows:

$$i_{ab}^{R,Y,B} + j_{ab}^{R,Y,B} = Y_{ab}^{R,Y,B} \cdot v_{ab}^{R,Y,B} \tag{7.6}$$

It is to be noted that

$$y_{ab}^{R,Y,B} = [Z_{ab}^{R,Y,B}]^{-1} \tag{7.7}$$

and

$$j_{ab}^{R,Y,B} = - Y_{ab}^{R,Y,B} \cdot e_{ab}^{R,Y,B} \tag{7.8}$$

The above equation relates the source voltages to source currents of impedance form (Fig. 7.1) and the source currents of admittance form (Fig. 7.2).

7.1.1 STATIONARY NETWORK ELEMENT

The impedances or admittances of a three-phase stationary network element such as transformers are symmetric. Further, if the element is balanced then the diagonal elements are equal and also the off-diagonal elements are equal, i.e.,

$$z_{ab}^{RR} = z_{ab}^{YY} = z_{ab}^{BB} = z_{ab}^{R} \tag{7.9}$$

and

$$z_{ab}^{RY} = z_{ab}^{YB} = z_{ab}^{BR} = z_{ab}^{YR} = z_{ab}^{BY} = z_{ab}^{RB} = z_{ab}^{m} \tag{7.10}$$

corresponding relations hold true for the admittance values.

7.1.2 ROTATING NETWORK ELEMENT

For balanced three-phase rotating elements, such as synchronous generator, synchronous motor, and three-phase induction motor, the impedance and admittance matrices are not symmetric. Since the elements are constantly rotating the mutual impedance of phase R with respect to Y is not the same as impedance of Y with respect to R.

However

$$z_{ab}^{RY} = z_{ab}^{YB} = z_{ab}^{BR} = z_{ab}^{m_1} \tag{7.11}$$

and

$$z_{ab}^{YR} = z_{ab}^{BY} = z_{ab}^{RB} = z_{ab}^{m_2} \tag{7.12}$$

7.1.3 PERFORMANCE RELATIONS FOR PRIMITIVE THREE-PHASE NETWORK ELEMENT

For the three-phase network element of the primitive network, the following relations are true:

$$v^{R,Y,B} + e^{R,Y,B} = z^{R,Y,B} \cdot i^{R,Y,B} \tag{7.13}$$

and

$$i^{R,Y,B} + j^{R,Y,B} = y^{R,Y,B} \cdot v^{R,Y,B} \tag{7.14}$$

7.2 THREE-PHASE BALANCED NETWORK ELEMENTS

7.2.1 BALANCED EXCITATION

If the source voltages and source currents of all phases are equal in magnitude and displaced mutually by 120°, then the excitation system is said to be balanced. Mathematically balanced excitation is represented by

$$e_{ab}^{R,Y,B} = \begin{bmatrix} e_{ab}^{R} \\ e_{ab}^{Y} \\ e_{ab}^{B} \end{bmatrix} = \begin{bmatrix} 1 \\ a^2 \\ a \end{bmatrix} e_{ab}^{R} \tag{7.15}$$

and like wise,

$$i_{ab}^{R,Y,B} = \begin{bmatrix} j_{ab}^{R} \\ j_{ab}^{Y} \\ j_{ab}^{B} \end{bmatrix} = \begin{bmatrix} 1 \\ a^2 \\ a \end{bmatrix} j_{ab}^{R} \tag{7.16}$$

where the operator "a" is defined in Chapter 6.

For a stationary element, the performance equation is $e_{ab}^{RYB} + V_{ab}^{RYB} = Z_{ab}^{RYB} \cdot i_{ab}^{RYB}$

$$\begin{bmatrix} 1 \\ a^2 \\ a \end{bmatrix} v_{ab}^{R} + \begin{bmatrix} 1 \\ a^2 \\ a \end{bmatrix} e_{ab}^{R} = \begin{bmatrix} z_{ab}^{R} & z_{ab}^{m} & z_{ab}^{m} \\ z_{ab}^{m} & z_{ab}^{R} & z_{ab}^{m} \\ z_{ab}^{m} & z_{ab}^{m} & z_{ab}^{R} \end{bmatrix} \begin{bmatrix} 1 \\ a^2 \\ a \end{bmatrix} i_{ab}^{R} \tag{7.17}$$

Multiplying both sides of Eq. (7.17) by the conjugate transpose of

$$\begin{bmatrix} 1 \\ a^2 \\ a \end{bmatrix}$$

i.e., $\begin{bmatrix} 1 & a & a^2 \end{bmatrix}$

$$\begin{bmatrix} 1 & a & a^2 \end{bmatrix} \begin{bmatrix} 1 \\ a^2 \\ a \end{bmatrix} v_{ab}^{R} + \begin{bmatrix} 1 & a & a^2 \end{bmatrix} \begin{bmatrix} 1 \\ a^2 \\ a \end{bmatrix} e_{ab}^{R} = \begin{bmatrix} 1 & a & a^2 \end{bmatrix} \begin{bmatrix} z_{ab}^{R} & z_{ab}^{m} & z_{ab}^{m} \\ z_{ab}^{m} & z_{ab}^{R} & z_{ab}^{m} \\ z_{ab}^{m} & z_{ab}^{m} & z_{ab}^{R} \end{bmatrix} \begin{bmatrix} 1 \\ a^2 \\ a \end{bmatrix} i_{ab}^{R}$$

we obtain

$$3v_{ab}^{R} + 3e_{ab}^{R} = 3(z_{ab}^{R} - Z_{ab}^{m})i_{ab}^{R} \tag{7.18}$$

i.e.,

$$v_{ab}^{R} + e_{ab}^{R} = (z_{ab}^{R} - Z_{ab}^{m})i_{ab}^{R} \tag{7.19}$$

$(z_{ab}^{R} - Z_{ab}^{m})$ is called the positive sequence impedance designated by $Z_{ab}^{(1)}$ of the balanced stationery 3−ϕ element.

In this way, a balanced three-phase element with balanced excitation can be considered as a single-phase element in network analysis. Power in the element = 3 × (power per phase)

In a similar manner for a rotating element, we have the relation

$$\begin{bmatrix} 1 \\ a^2 \\ a \end{bmatrix} v_{ab}^R + \begin{bmatrix} 1 \\ a^2 \\ a \end{bmatrix} e_{ab}^R = \begin{bmatrix} Z_{ab}^R & Z_{ab}^{m_1} & Z_{ab}^{m_2} \\ Z_{ab}^{m_2} & Z_{cb}^R & Z_{ab}^{m_2} \\ Z_{cb}^{m_1} & Z_{cb}^{m_1} & Z_{ab}^R \end{bmatrix} \begin{bmatrix} 1 \\ a^2 \\ a \end{bmatrix} i_{ab}^R \tag{7.20}$$

Multiplying both sides of Eq. (7.20) by the conjugate transpose of

$$\begin{bmatrix} 1 \\ a^2 \\ a \end{bmatrix}$$

$$\begin{bmatrix} 1 & a & a^2 \end{bmatrix}\begin{bmatrix} 1 \\ a^2 \\ a \end{bmatrix} v_{ab}^R + \begin{bmatrix} 1 & a & a^2 \end{bmatrix}\begin{bmatrix} 1 \\ a^2 \\ a \end{bmatrix} e_{ab}^R = \begin{bmatrix} 1 & a & a^2 \end{bmatrix}\begin{bmatrix} Z_{ab}^R & Z_{ab}^{m_1} & Z_{ab}^{m_2} \\ Z_{ab}^{m_2} & Z_{cb}^R & Z_{ab}^{m_2} \\ Z_{cb}^{m_1} & Z_{cb}^{m_1} & Z_{ab}^R \end{bmatrix}\begin{bmatrix} 1 \\ a^2 \\ a \end{bmatrix} i_{ab}^R$$

i.e.,

$$v_{ab}^R + e_{ab}^R = (Z_{ab}^R + a^2 Z_{ab}^{m_1} + a\, Z_{ab}^{m_2}) I_{ab}^R \tag{7.21}$$

the quantity $(Z_{ab}^R + a^2 Z_{ab}^{m_1} + a\, Z_{ab}^{m_2})$ is called the positive sequence impedance of the three-phase rotating element.

The performance equation for stationary and rotating elements in admittance form are, respectively,

$$i_{ab}^R + j_{ab}^R = (Y_{ab}^R - Y_{ab}^m) v_{ab}^R \tag{7.22}$$

and

$$i_{ab}^R + j_{ab}^R = (Y_{ab}^R + a^2 Y_{ab}^m + a Y_{ab}^{m_2}) v_{ab}^R \tag{7.23}$$

Appropriate relations can be obtained for the admittances if needed.

7.2.2 TRANSFORMATION MATRICES

Consider the transformation matrix

$$T = \frac{1}{\sqrt{3}} \begin{bmatrix} 1 & 1 & 1 \\ 1 & a^2 & a \\ 1 & a & a^2 \end{bmatrix} \tag{7.24}$$

The matrix T is a unitary matrix [A unitary matrix T has the property that $[(T^*)^t T] = $ Unit matrix.].

Further, since T is a symmetric matrix $T^* = T^{-1}$ as discussed in Chapter 6.

The zero, positive, and negative sequence components of the three-phase network element can be defined by using the matrix T as follows:

$V_{ab}^{R,Y,B} = Z_{ab}^{R,Y,B} I_{ab}^{R,Y,B}$ replacing by sequence components on both sides

$T V_{ab}^{0,1,2} = Z_{ab}^{R,Y,B} T I_{ab}^{R,Y,B}$ Premultiplying by $(T^*)^t$

$$(T^*)^t\, T\, V_{ab}^{0,1,2} = (T^*)^t Z_{ab}^{R,Y,B} I_{ab}^{0,1,2}, \text{i.e., } V_{ab}^{0,1,2} = (T^*)^t\, Z_{ab}^{R,Y,B}\, T\, I_{ab}^{0,1,2}$$

since

$$(T^*)^t T = U$$

But in sequence components

$$V_{ab}^{0,1,2} = Z_{ab}^{0,1,2} \cdot I_{ab}^{0,1,2}$$

Hence

$$Z_{ab}^{(0,1,2)} = \frac{1}{\sqrt{3}} \begin{bmatrix} 1 & 1 & 1 \\ 1 & a & a^2 \\ 1 & a^2 & a \end{bmatrix} \begin{bmatrix} Z_{ab}^R & Z_{ab}^m & Z_{ab}^m \\ Z_{ab}^m & Z_{ab}^R & Z_{ab}^m \\ Z_{ab}^m & Z_{ab}^m & Z_{ab}^R \end{bmatrix} \frac{1}{\sqrt{3}} \begin{bmatrix} 1 & 1 & 1 \\ 1 & a^2 & a \\ 1 & a & a^2 \end{bmatrix}$$

$$= \begin{bmatrix} (Z_{ab}^R + 2Z_{ab}^m) & 0 & 0 \\ 0 & (Z_{ab}^R - Z_{ab}^m) & 0 \\ 0 & 0 & (Z_{ab}^R - Z_{ab}^m) \end{bmatrix} \tag{7.25}$$

From Eq. (7.25), the following are defined.
For a stationary three-phase network element

$$Z_{ab}^{(0)} = \text{Zero sequence impedance}$$
$$= Z_{ab}^R + 2Z_{ab}^m \tag{7.26}$$

$$Z_{ab}^{(1)} = \text{Positive sequence impedance}$$
$$= Z_{ab}^R - Z_{ab}^m \tag{7.27}$$

$$Z_{ab}^{(2)} = \text{Negative sequence impedance}$$
$$= Z_{ab}^R - Z_{ab}^m \tag{7.28}$$

$Z_{ab}^{(0,1,2)}$ matrix is diagonal for a balanced three-phase stationary element.
For a rotating three-phase element in a similar way, even though $Z_{ab}^{R,Y,B}$ is not symmetric we obtain

$$Z_{ab}^{(0,1,2)} = \frac{1}{\sqrt{3}} \begin{bmatrix} 1 & 1 & 1 \\ 1 & a & a^2 \\ 1 & a^2 & a \end{bmatrix} \begin{bmatrix} Z_{ab}^R & Z_{ab}^{m_1} & Z_{ab}^{m_2} \\ Z_{ab}^{m_2} & Z_{ab}^R & Z_{ab}^{m_1} \\ Z_{ab}^{m_1} & Z_{ab}^{m_2} & Z_{ab}^R \end{bmatrix} \frac{1}{\sqrt{3}} \begin{bmatrix} 1 & 1 & 1 \\ 1 & a^2 & a \\ 1 & a & a^2 \end{bmatrix} \tag{7.29}$$

$$= \begin{bmatrix} Z_{ab}^R + Z_{ab}^{m_1} + Z_{ab}^{m_2} & 0 & 0 \\ 0 & Z_{ab}^R + a^2 Z_{ab}^{m_1} + a Z_{ab}^{m_2} & 0 \\ 0 & 0 & Z_{ab}^R + a Z_{ab}^{m_1} + a^2 Z_{ab}^{m_2} \end{bmatrix} \tag{7.30}$$

The sequence components are defined by

$$Z_{ab}^{(0)} = Z_{ab}^R + Z_{ab}^{m_1} + Z_{ab}^{m_2} = \text{Zero sequence impedance} \tag{7.31}$$

$$Z_{ab}^{(1)} = Z_{ab}^R + a^2 Z_{ab}^{m_1} + a Z_{ab}^{m_2} = \text{Positive sequence impedance} \tag{7.32}$$

$$Z_{ab}^{(2)} = Z_{ab}^R + a Z_{ab}^{m_1} + a^2 Z_{ab}^{m_2} = \text{Negative sequence impedance} \tag{7.33}$$

In this way, the sequence impedances can be obtained for stationary and rotating three-phase balanced elements.

7.3 THREE-PHASE IMPEDANCE NETWORKS

For three-phase unbalanced network elements, it is not possible to determine a single transformation matrix T that will diagonalize the impedance matrix into three uncoupled impedances. It is, therefore, required to maintain the original three-phase quantities as they are and solve the network problems.

7.3.1 INCIDENCE AND NETWORK MATRICES FOR THREE-PHASE NETWORKS

The derivation of incidence and network matrices for three-phase network elements follows identical procedure as is adapted for single-phase network element in Chapter 2. However, the difference lies in the fact that in place of entries of 0, 1, and −1, we will have entries of null matrix and 3×3 unit matrices U and $-U$, respectively. The rows and columns of these matrices refer to phases R, Y, and B.

7.3.2 ALGORITHM FOR THREE-PHASE BUS IMPEDANCE MATRIX

7.3.2.1 Performance Equation of a Partial Three-Phase Network

The performance equation for a three-phase network in the impedance form is given by

$$V_{\mathrm{BUS}}^{\mathrm{R,Y,B}} = Z_{\mathrm{BUS}}^{\mathrm{R,Y,B}} \cdot I_{\mathrm{BUS}}^{\mathrm{R,Y,B}} \tag{7.34}$$

where

$V_{\mathrm{BUS}}^{\mathrm{R,Y,B}}$ is the three-phase bus voltage vector measured with respect to reference bus

$I_{\mathrm{BUS}}^{\mathrm{R,Y,B}}$ is the impressed three-phase current vector

$Z_{\mathrm{BUS}}^{\mathrm{R,Y,B}}$ is the three-phase bus impedance matrix.

7.3.2.2 Addition of a Branch

The performance equation of the partial network with an added branch $a-b$ in terms of three-phase quantities is given as follows. The manner of writing the equation is same as of equation except for the superscripts.

$$
\begin{bmatrix} V_1^{\mathrm{R,Y,B}} \\ V_2^{\mathrm{R,Y,B}} \\ \vdots \\ V_a^{\mathrm{R,Y,B}} \\ \vdots \\ V_m^{\mathrm{R,Y,B}} \\ \hline V_b^{\mathrm{R,Y,B}} \end{bmatrix}
=
\left[
\begin{array}{cccccccc|c}
Z_{11}^{\mathrm{R,Y,B}} & Z_{12}^{\mathrm{R,Y,B}} & \cdots & Z_{1a}^{\mathrm{R,Y,B}} & \cdots & Z_{1m}^{\mathrm{R,Y,B}} & Z_{1b}^{\mathrm{R,Y,B}} \\
Z_{21}^{\mathrm{R,Y,B}} & Z_{22}^{\mathrm{R,Y,B}} & \cdots & Z_{2a}^{\mathrm{R,Y,B}} & \cdots & Z_{2m}^{\mathrm{R,Y,B}} & Z_{2b}^{\mathrm{R,Y,B}} \\
\vdots & \vdots & \vdots & \vdots & \vdots & \vdots & \vdots \\
Z_{a1}^{\mathrm{R,Y,B}} & Z_{a2}^{\mathrm{R,Y,B}} & \cdots & Z_{aa}^{\mathrm{R,Y,B}} & \cdots & Z_{am}^{\mathrm{R,Y,B}} & Z_{ab}^{\mathrm{R,Y,B}} \\
\vdots & \vdots & \vdots & \vdots & \vdots & \vdots & \vdots \\
Z_{m1}^{\mathrm{R,Y,B}} & Z_{m2}^{\mathrm{R,Y,B}} & \cdots & Z_{ma}^{\mathrm{R,Y,B}} & \cdots & Z_{mm}^{\mathrm{R,Y,B}} & Z_{mb}^{\mathrm{R,Y,B}} \\
\hline
Z_{b1}^{\mathrm{R,Y,B}} & Z_{b2}^{\mathrm{R,Y,B}} & \cdots & Z_{bm}^{\mathrm{R,Y,B}} & \cdots & Z_{bm}^{\mathrm{R,Y,B}} & Z_{bb}^{\mathrm{R,Y,B}}
\end{array}
\right]
\begin{bmatrix} I_1^{\mathrm{R,Y,B}} \\ I_2^{\mathrm{R,Y,B}} \\ \vdots \\ I_a^{\mathrm{R,Y,B}} \\ \vdots \\ I_m^{\mathrm{R,Y,B}} \\ \hline I_b^{\mathrm{R,Y,B}} \end{bmatrix}
\tag{7.35}
$$

The elements of $Z_{bi}^{R,Y,B}$ $(i = 1, 2, \ldots, m)$ and $Z_{bb}^{R,Y,B}$ are determined by injecting a three-phase current at the ith bus just as in the case of single circuit element. This is shown in Fig. 7.3. The voltage at the bth bus with respect to reference node is measured.

In a similar manner, the elements of $Z_{ib}^{R,Y,B}$ are determined by injecting a three-phase current at the bth bus as in Fig. 7.4 and measuring the voltage at the ith bus with respect to the reference node.

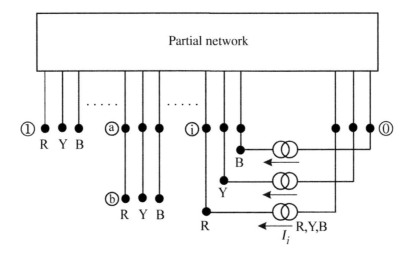

FIGURE 7.3

Injecting three-phase current at bus i.

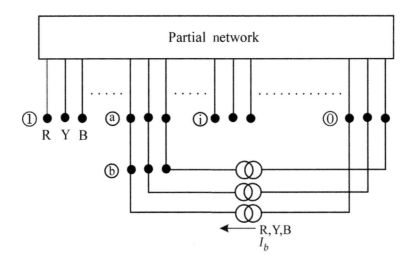

FIGURE 7.4

Injecting three-phase current at bus b.

If $I_i^{R,Y,B}$ is the current injected into bus I, then the voltage across the added element $a-b$ is

$$v_{ab}^{R,Y,B} = v_a^{R,Y,B} - v_b^{R,Y,B} \tag{7.36}$$

the voltage vector $v_{pq}^{R,Y,B}$ across the elements $p-q$ of the partial network

$$\bar{v}_{pq}^{R,Y,B} = \bar{v}_p^{R,Y,B} - \bar{v}_q^{R,Y,B} \tag{7.37}$$

The current in element $a-b$ in terms of the primitive admittances and the voltages across the elements can be expressed as

$$i_{ab}^{R,Y,B} = y_{ab-ab}^{R,Y,B} \, v_{ab}^{R,Y,B} + y_{ab-pq}^{R,Y,B} \, v_{pq}^{R,Y,B} \tag{7.38}$$

However, the current $i_{ab}^{R,Y,B} = 0$. Hence

$$v_{ab}^{R,Y,B} = \left(y_{ab-ab}^{R,Y,B} \right)^{-1} \left(y_{ab-pq}^{R,Y,B} \right) \bar{v}_{pq}^{R,Y,B} \tag{7.39}$$

Substituting into Eq. (7.36) we obtain using Eq. (7.37)

$$v_a^{R,Y,B} - v_b^{R,Y,B} = - \left(y_{ab-ab}^{R,Y,B} \right)^{-1} \left(y_{ab-pq}^{R,Y,B} \right) \left(\bar{v}_p^{R,Y,B} - \bar{v}_q^{R,Y,B} \right) \tag{7.40}$$

But

$$\bar{v}_p^{R,Y,B} = Z_{ai}^{R,Y,B} \, I_i^{R,Y,B} \tag{7.41}$$

and

$$V_b^{R,Y,B} = Z_{bi}^{R,Y,B} \, I_i^{R,Y,B} \tag{7.42}$$

Hence

$$v_a^{R,Y,B} - v_b^{R,Y,B} = Z_{ai}^{R,Y,B} \, I_{ai}^{R,Y,B} - Z_{bi}^{R,Y,B} \, I_{bi}^{R,Y,B}$$

At any bus k,

$$V_k^{R,Y,B} = Z_{ki}^{R,Y,B} \, I_i^{R,Y,B}$$

using the relations (7.41) and (7.42) then from the above equation:

$$Z_{bi}^{R,Y,B} I_i^{R,Y,B} = Z_{ai}^{R,Y,B} \, I_{ai}^{R,Y,B} + \left[\left(y_{ab-ab}^{R,Y,B} \right)^{-1} \left(y_{ab-pq}^{R,Y,B} \right) \left(Z_{pi}^{R,Y,B} - Z_{qi}^{R,Y,B} \right) \right] I_i^{R,Y,B} \tag{7.43}$$

For all $I_i^{R,Y,B}$ then

$$Z_{bi}^{R,Y,B} = Z_{ai}^{R,Y,B} + (y_{ab-ab}^{R,Y,B})^{-1} (y_{ab-pq}^{R,Y,B})(Z_{pi}^{R,Y,B} - Z_{qi}^{R,Y,B}) \tag{7.44}$$

To calculate $Z_{ib}^{R,Y,B}$, inject $I_b^{R,Y,B}$ at the bus b keeping currents at all other buses at zero value. If the element $a-b$ is not mutually coupled with elements of the partial network, the voltage at the ith bus will be the same whether $I_b^{R,Y,B}$ is injected at bus a or b.

Then

$$V_i^{R,Y,B} = Z_{ib}^{R,Y,B} \cdot I_i^{R,Y,B} = Z_{ia}^{R,Y,B} \cdot I_b^{R,Y,B} \qquad (7.45)$$

In case, the element $a-b$ is mutually coupled to one or more elements of the partial network, then

$$V_i^{R,Y,B} = Z_{ib}^{R,Y,B} \cdot I_b^{R,Y,B} = Z_{ia}^{R,Y,B} \cdot I_b^{R,Y,B} + \Delta V_i^{R,Y,B} \qquad (7.46)$$

where

$\Delta V_i^{R,Y,B}$ = change in voltage at bus i due to the effect of mutual coupling.

The induced voltage vector in element $p-q$ is

$$v_{pq}^{R,Y,B} = Z_{pq-ab}^{R,Y,B} \cdot I_b^{R,Y,B} \qquad (7.47)$$

Note that the series source voltages $V_{pq}^{R,Y,B}$ and parallel source currents $j_{pq}^{R,Y,B}$ are related by

$$j_{pq}^{R,Y,B} = - \left[z_{pq-pq}^{R,Y,B} \right]^{-1} v_{pq}^{R,Y,B} \qquad (7.48)$$

Substituting Eq. (7.47) into Eq. (7.48)

$$j_{pq}^{R,Y,B} = - \left[z_{pq-pq}^{R,Y,B} \right]^{-1} \left[z_{pq-ab}^{R,Y,B} \right] \cdot I_{pq}^{R,Y,B} \qquad (7.49)$$

$\therefore$

$$j_{pq}^{R,Y,B} = I_p^{R,Y,B} = -I_q^{R,Y,B} \qquad (7.50)$$

The change in voltage at the ith bus due to mutual coupling is

$$\Delta V_i^{R,Y,B} = Z_{ip}^{R,Y,B} \cdot I_p^{R,Y,B} + Z_{iq}^{R,Y,B} \cdot I_q^{R,Y,B} \qquad (7.51)$$

Substituting for $I_p^{R,Y,B}$ and from $I_q^{R,Y,B}$ (7.49) and (7.50)

$$\Delta V_i^{R,Y,B} = - \left(Z_{ip}^{R,Y,B} - Z_{iq}^{R,Y,B} \right) \left(z_{pq-pq}^{R,Y,B} \right)^{-1} (z_{pq-ab}^{R,Y,B}) I_b^{R,Y,B} \qquad (7.52)$$

substituting from Eq. (7.51) into Eq. (7.46)

$$Z_{ib}^{R,Y,B} = Z_{ia}^{R,Y,B} - \left(Z_{ip}^{R,Y,B} - Z_{iq}^{R,Y,B} \right) \left(z_{pq-pq}^{R,Y,B} \right)^{-1} \left(z_{pq-ab}^{R,Y,B} \right) \qquad (7.53)$$

we have

$$\begin{bmatrix} z_{ab-ab}^{R,Y,B} & z_{ab-pq}^{R,Y,B} \\ z_{pq-ab}^{R,Y,B} & z_{pq-pq}^{R,Y,B} \end{bmatrix} \begin{bmatrix} y_{ab-ab}^{R,Y,B} & y_{ab-pq}^{R,Y,B} \\ y_{pq-ab}^{R,Y,B} & y_{pq-pq}^{R,Y,B} \end{bmatrix} = \begin{bmatrix} U & \\ & U \end{bmatrix} \qquad (7.54)$$

Therefore

$$z^{R,Y,B}_{pq-ab} \cdot y^{R,Y,B}_{ab-ab} = - \left[z^{R,Y,B}_{pq-pq} \right] y^{R,Y,B}_{pq-ab} \qquad (7.55)$$

premultiplying by $[z^{R,Y,B}_{pq-pq}]^{-1}$ and postmultiplying by $[y^{R,Y,B}_{ab-ab}]^{-1}$, Eq. (7.55) becomes

$$\left[z^{R,Y,B}_{pq-pq} \right]^{-1} \left[z^{R,Y,B}_{pq-ab} \right] = - y^{R,Y,B}_{pq-ab} \left[y^{R,Y,B}_{ab-ab} \right]^{-1} \qquad (7.56)$$

substituting Eq. (7.56) into Eq. (7.53)

$$Z^{R,Y,B}_{ib} = Z^{R,Y,B}_{ia} + (Z^{R,Y,B}_{ip} - Z^{R,Y,B}_{iq}) y^{R,Y,B}_{pq-ab} (y^{R,Y,B}_{ab-ab})^{-1}$$

The element $Z^{R,Y,B}_{bb}$ can be determined by injecting a three-phase current at the bth bus and measuring the voltage at that bus with respect to the reference node.

Let the current at the bth bus be $I^{R,Y,B}_b$ and let all other currents be zero.

Since $i^{R,Y,B}_{ab} = - I^{R,Y,B}_b$. Using Eq. (7.38) and solving for $v^{R,Y,B}_{ab}$

$$V^{R,Y,B}_a - V^{R,Y,B}_b = - \left(y^{R,Y,B}_{ab-ab} \right)^{-1} \left[I^{R,Y,B}_b + y^{R,Y,B}_{ab-pq} \left(V^{R,Y,B}_p - V^{R,Y,B}_q \right) \right] \qquad (7.57)$$

Substituting from Eqs. (7.36) and (7.37) into Eq. (7.57)

$$V^{R,Y,B}_a - V^{R,Y,B}_b = - \left(y^{R,Y,B}_{ab-ab} \right)^{-1} \left[I^{R,Y,B}_b + y^{R,Y,B}_{ab-pq} \left(V^{R,Y,B}_p - V^{R,Y,B}_q \right) \right] \qquad (7.58)$$

From Eq. (7.35)

$$\left. \begin{array}{l} V^{R,Y,B}_a = Z^{R,Y,B}_{ab} \, I^{R,Y,B}_b \\[2mm] V^{R,Y,B}_b = Z^{R,Y,B}_{bb} \, I^{R,Y,B}_b \end{array} \right\} \qquad (7.59)$$

since, at any bus k

$$V^{R,Y,B}_k = Z^{R,Y,B}_{kb} \, I^{R,Y,B}_b$$

$$V^{R,Y,B}_{ab} = V^{R,Y,B}_a - V^{R,Y,B}_b$$

$$= - \left(y^{R,Y,B}_{ab-ab} \right)^{-1} \left[I^{R,Y,B}_b + y^{R,Y,B}_{ab-pq} \left(V^{R,Y,B}_p - V^{R,Y,B}_q \right) \right]$$

$$= - \left(y^{R,Y,B}_{ab-ab} \right)^{-1} \left[I^{R,Y,B}_b + y^{R,Y,B}_{ab-pq} \left(Z^{R,Y,B}_{pb} - Z_{qb} \right) I^{R,Y,B}_b \right]$$

Solving for Z^{RYB}_{bb} and canceling out I^{RYB}_b an either side by postmultiplication of $[I^{RYB}_b]^{-1}$

$$Z^{R,Y,B}_{bb} = Z^{R,Y,B}_{ab} + (y^{R,Y,B}_{ab-ab})^{-1} [U + y^{R,Y,B}_{ab-pq} (Z^{R,Y,B}_{pq} - Z^{R,Y,B}_{qb})] \qquad (7.60)$$

If there is no mutual coupling

$$\left. \begin{array}{l} Z^{R,Y,B}_{bi} = Z^{R,Y,B}_{ai} \\[2mm] Z^{R,Y,B}_{ib} = Z^{R,Y,B}_{ia} \\[2mm] Z^{R,Y,B}_{bb} = Z^{R,Y,B}_{ab} + Z^{R,Y,B}_{ab-ab} \end{array} \right\} \qquad (7.61)$$

in addition, of p is a reference node then

$$Z_{bb}^{R,Y,B} = z_{ab-ab}^{R,Y,B} \tag{7.62}$$

7.3.2.3 Addition of a Link

When the new element added is a link, it is connected in series with a voltage, source as shown in Fig. 7.5.

The three-phase voltage source $e_l^{R,Y,B}$ is selected such that the current through the added link is zero. Then the element $a-l$, where l is a fictitious node, can be treated as branch. Following the procedure developed for the single-phase case, and the addition of a branch for the three-phase case, we can derive the following formulae.

The elements $Z_{ai}^{R,Y,B}$ can be determined by injecting a three-phase current at the ith bus and measuring the voltage at the fictitious node l with respect to bus b.

$$Z_{li}^{R,Y,B} = Z_{ai}^{R,Y,B} - Z_{bi}^{R,Y,B} + \left(Z_{ab-ab}^{R,Y,B}\right) + y_{a-b,p-q}^{R,Y,B}\left(Z_{pi}^{R,Y,B} - Z_{qi}^{R,Y,B}\right) \tag{7.63}$$

The element $Z_{li}^{R,Y,B}$ can be determined by injecting a three-phase current between b and l measuring the voltage at bus i.

$$Z_{il}^{R,Y,B} = Z_{ia}^{R,Y,B} - Z_{ib}^{R,Y,B} + \left(Z_{ip}^{R,Y,B} - Z_{iq}^{R,Y,B}\right)\bar{y}_{1a-ab}^{R,Y,B}\left(y_{ab-ab}^{R,Y,B}\right)^{-1} \tag{7.64}$$

The element $Z_{ll}^{R,Y,B}$ can be determined by injecting a three-phase current between b and l and measuring the voltage at node l with respect to bus b.

$$Z_{ll}^{R,Y,B} = \left(Z_{ai}^{R,Y,B} - Z_{bi}^{R,Y,B}\right) + \left(y_{ab-ab}^{R,Y,B}\right)^{-1}\{U + y_{ab-pq}^{R,Y,B}(Z_{pl}^{R,Y,B} - Z_{ql}^{R,Y,B})\} \tag{7.65}$$

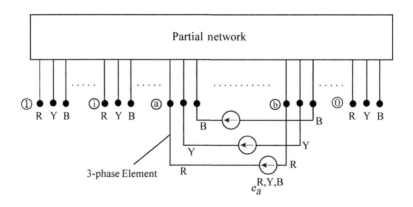

FIGURE 7.5

Addition of a link with voltage source in series.

If there is no mutual coupling

$$
\left.
\begin{aligned}
Z_{li}^{R,Y,B} &= Z_{ai}^{R,Y,B} - Z_{bi}^{R,Y,B} \\
Z_{il}^{R,Y,B} &= Z_{ia}^{R,Y,B} - Z_{ib}^{R,Y,B} \\
Z_{ll}^{R,Y,B} &= Z_{al}^{R,Y,B} - Z_{bl}^{R,Y,B} + Z_{ab-ab}^{R,Y,B}
\end{aligned}
\right\}
\tag{7.66}
$$

If a is also the reference node

$$
\left.
\begin{aligned}
Z_{li}^{R,Y,B} &= -Z_{ai}^{R,Y,B} \\
Z_{il}^{R,Y,B} &= -Z_{ib}^{R,Y,B} \\
Z_{li}^{R,Y,B} &= -Z_{bi}^{R,Y,B} + z_{ab-ab}^{R,Y,B}
\end{aligned}
\right\}
\tag{7.67}
$$

For balanced elements

$$
Z_{li}^{R,Y,B} = Z_{il}^{R,Y,B}
\tag{7.68}
$$

The fictitious node l is eliminated by short circuiting the link voltage source $e_l^{R,Y,B}$

$$
\begin{aligned}
Z_{ij}^{R,Y,B}(\text{modified}) &= Z_{ij}^{R,Y,B}(\text{before elimination of } l) \\
&= -Z_{il}^{R,Y,B} \cdot (Z_{ll}^{R,Y,B})^{-1} Z_{lj}^{R,Y,B}
\end{aligned}
\tag{7.69}
$$

SUMMARY OF THE FORMULAE

For the addition of a branch $a-b$

1. $Z_{bi}^{R,Y,B} = Z_{ai}^{R,Y,B} + \left(y_{ab-ab}^{R,Y,B}\right)^{-1}\left(y_{ab-pq}^{R,Y,B}\right)\left(Z_{pi}^{R,Y,B} - Z_{qi}^{R,Y,B}\right)$

2. $\left[z_{pq-pq}^{R,Y,B}\right]^{-1}\left[z_{pq-ab}^{R,Y,B}\right] = -y_{pq-ab}^{R,Y,B}\left[y_{ab-ab}^{R,Y,B}\right]^{-1}$

3. $Z_{bb}^{R,Y,B} = Z_{ab}^{R,Y,B} + \left(y_{ab-ab}^{R,Y,B}\right)^{-1}\left[U + y_{ab-pq}^{R,Y,B}\left(Z_{pq}^{R,Y,B} - Z_{qb}^{R,Y,B}\right)\right]$

4. $\begin{aligned} Z_{bi}^{R,Y,B} &= Z_{ai}^{R,Y,B} \\ Z_{ib}^{R,Y,B} &= Z_{ia}^{R,Y,B} \text{ and } Z_{bb}^{R,Y,B} = Z_{ab}^{R,Y,B} + Z_{ab-ab}^{R,Y,B} \end{aligned}$

For the addition of a link $a-b$

1. $Z_{li}^{R,Y,B} = Z_{ai}^{R,Y,B} - Z_{bi}^{R,Y,B} + \left(Z_{ab-ab}^{R,Y,B}\right) + y_{a-b,p-q}^{R,Y,B}\left(Z_{pi}^{R,Y,B} - Z_{qi}^{R,Y,B}\right)$

2. $Z_{il}^{R,Y,B} = Z_{ia}^{R,Y,B} - Z_{ib}^{R,Y,B} + \left(Z_{ip}^{R,Y,B} - Z_{iq}^{R,Y,B}\right)\bar{y}_{1a-ab}^{R,Y,B}\left(y_{ab-ab}^{R,Y,B}\right)^{-1}$

3. $Z_{ll}^{R,Y,B} = \left(Z_{ai}^{R,Y,B} - Z_{bi}^{R,Y,B}\right) + \left(y_{ab-ab}^{R,Y,B}\right)^{-1}\left\{U + y_{ab-pq}^{R,Y,B}\left(Z_{pl}^{R,Y,B} - Z_{ql}^{R,Y,B}\right)\right\}$

$$Z_{li}^{R,Y,B} = Z_{ai}^{R,Y,B} - Z_{bi}^{R,Y,B}$$

4. $Z_{il}^{R,Y,B} = Z_{ia}^{R,Y,B} - Z_{ib}^{R,Y,B}$

$$Z_{ll}^{R,Y,B} = Z_{al}^{R,Y,B} - Z_{bl}^{R,Y,B} + z_{ab-ab}^{R,Y,B}$$

$$Z_{li}^{R,Y,B} = - Z_{ai}^{R,Y,B}$$

5. $Z_{il}^{R,Y,B} = - Z_{ib}^{R,Y,B}$

$$Z_{li}^{R,Y,B} = - Z_{bi}^{R,Y,B} + Z_{ab-ab}^{R,Y,B}$$

6. $Z_{li}^{R,Y,B} = Z_{il}^{R,Y,B}$

7. $Z_{ij}^{R,Y,B}$ (modified) $= Z_{ij}^{R,Y,B}$ (before elimination of l) $- Z_{il}^{R,Y,B} \cdot \left(Z_{ll}^{R,Y,B} \right)^{-1}$

WORKED EXAMPLES

E.7.1. **Consider the three-phase network. Two generators are supplying power to a common load. Obtain the bus incidence matrix. Take ground as reference.**

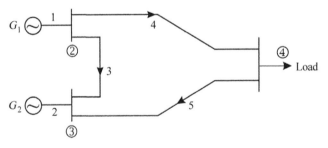

FIGURE E.7.1

Network for E.7.1.

Solution: The oriented connected graph is shown.

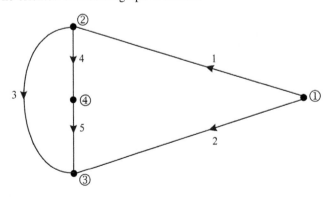

FIGURE E.7.2

Oriented graph.

The element-node incidence matrix is given by

$$\tilde{A} =$$

	(1)	(2)	(3)	(4)
1	U	$-U$		
2	U		$-U$	
3		U	$-U$	
4		U		$-U$
5			$-U$	U

where

$$U = \begin{bmatrix} 1 & 0 & 0 \\ 0 & 1 & 0 \\ 0 & 0 & 1 \end{bmatrix}$$

Identifying a tree, the branches and links are separated; 1−2−4 are branches 3−5 links

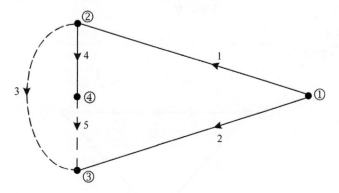

FIGURE E.7.3

Trees, branches, and links.

Deleting the reference node (1) and sequencing the branches and links in order, the bus incidence matrix is obtained as

$$A =$$

	(2)	(3)	(4)
1	$-U$		
2		$-U$	
4	U		$-U$
3	U	$-U$	
5		$-U$	U

E.7.2. A three bus system is shown in the figure. The self- and mutual impedances of the three transmission lines are given in tables. There are no mutual impedances between the transmission lines. The oriented connected graph is shown in the figure Taking ground as reference, obtain Y_{BUS} by transformation. There from obtain Z_{BUS}.

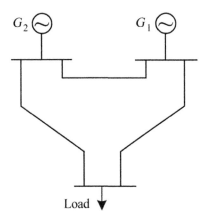

FIGURE E.7.4

Network for E.7.2.

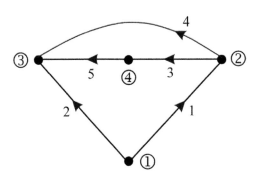

FIGURE E.7.5

Graph for network in E.7.2.

Bus Code	Line	$Z_{ab-ab}^{R,Y,B}$		
(1)–(2)	1	0.5	−0.05	−0.05
		−0.05	0.5	−0.05
		−0.05	−0.05	0.05
(1)–(3)	2	0.5	−0.05	−0.05
		−0.05	0.5	−0.05
		−0.05	−0.05	0.05

Continued		
Bus Code	**Line**	$Z_{ab-ab}^{R,Y,B}$
(2)–(3)	3	<table><tr><td>1.0</td><td>−0.1</td><td>−0.1</td></tr><tr><td>−0.1</td><td>1.0</td><td>−0.1</td></tr><tr><td>−0.1</td><td>−0.1</td><td>1.0</td></tr></table>
(2)–(4)	4	<table><tr><td>0.4</td><td>−0.04</td><td>−0.04</td></tr><tr><td>−0.04</td><td>0.4</td><td>−0.04</td></tr><tr><td>−0.04</td><td>−0.04</td><td>0.4</td></tr></table>
(4)–(3)	5	<table><tr><td>0.6</td><td>−0.06</td><td>−0.06</td></tr><tr><td>−0.06</td><td>0.6</td><td>−0.06</td></tr><tr><td>−0.06</td><td>−0.06</td><td>0.6</td></tr></table>

Solution: The bus incidence matrix A is obtained as

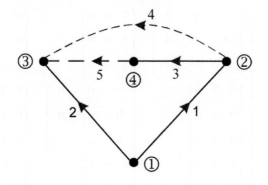

FIGURE E.7.6

Trees, branches, and links.

$$A = \begin{array}{c|ccc} & (2) & (3) & (4) \\ \hline 1 & -U & & \\ 2 & & -U & \\ 3 & U & & -U \\ 4 & U & -U & \\ 5 & & -U & U \end{array}$$

where

$$U = \begin{bmatrix} 1 & 0 & 0 \\ 0 & 1 & 0 \\ 0 & 0 & 1 \end{bmatrix}$$

The transpose of $[A]$ is $[A]^t =$

	1	2	4	3	5
(2)	$-U$		U	U	
(3)		$-U$		$-U$	$-U$
(4)			$-U$		U

$$Y_{BUS} = A^t y^{R,Y,B} A$$

where
$y^{R,Y,B}$ is the primitive admittance matrix of the network
$z^{R,Y,B}$ is the primitive impedance matrix $y^{R,Y,B} = [z^{R,Y,B}]^{-1}$.
    The bus incidence matrix is given by

$$A^{R,Y,B} = \begin{bmatrix}
\begin{bmatrix} -1 & 0 & 0 \\ 0 & -1 & 0 \\ 0 & 0 & -1 \end{bmatrix} & \begin{bmatrix} 0 & 0 & 0 \\ 0 & 0 & 0 \\ 0 & 0 & 0 \end{bmatrix} & \begin{bmatrix} 0 & 0 & 0 \\ 0 & 0 & 0 \\ 0 & 0 & 0 \end{bmatrix} \\
\begin{bmatrix} 0 & 0 & 0 \\ 0 & 0 & 0 \\ 0 & 0 & 0 \end{bmatrix} & \begin{bmatrix} -1 & 0 & 0 \\ 0 & -1 & 0 \\ 0 & 0 & -1 \end{bmatrix} & \begin{bmatrix} 0 & 0 & 0 \\ 0 & 0 & 0 \\ 0 & 0 & 0 \end{bmatrix} \\
\begin{bmatrix} 1 & 0 & 0 \\ 0 & 1 & 0 \\ 0 & 0 & 1 \end{bmatrix} & \begin{bmatrix} 0 & 0 & 0 \\ 0 & 0 & 0 \\ 0 & 0 & 0 \end{bmatrix} & \begin{bmatrix} -1 & 0 & 0 \\ 0 & -1 & 0 \\ 0 & 0 & -1 \end{bmatrix} \\
\begin{bmatrix} 1 & 0 & 0 \\ 0 & 1 & 0 \\ 0 & 0 & 1 \end{bmatrix} & \begin{bmatrix} -1 & 0 & 0 \\ 0 & -1 & 0 \\ 0 & 0 & -1 \end{bmatrix} & \begin{bmatrix} 0 & 0 & 0 \\ 0 & 0 & 0 \\ 0 & 0 & 0 \end{bmatrix} \\
\begin{bmatrix} 0 & 0 & 0 \\ 0 & 0 & 0 \\ 0 & 0 & 0 \end{bmatrix} & \begin{bmatrix} -1 & 0 & 0 \\ 0 & -1 & 0 \\ 0 & 0 & -1 \end{bmatrix} & \begin{bmatrix} 1 & 0 & 0 \\ 0 & 1 & 0 \\ 0 & 0 & 1 \end{bmatrix}
\end{bmatrix}$$

The primitive impedance matrix $Z^{R,Y,B}$ is given by

$$
Z^{R,Y,B} =
\begin{bmatrix}
0.500 & -0.0500 & -0.0500 & 0 & 0 & 0 & 0 & 0 & 0 & 0 & 0 & 0 & 0 & 0 & 0 \\
-0.050 & 0.5000 & -0.5000 & 0 & 0 & 0 & 0 & 0 & 0 & 0 & 0 & 0 & 0 & 0 & 0 \\
-0.050 & -0.500 & 0.5000 & 0 & 0 & 0 & 0 & 0 & 0 & 0 & 0 & 0 & 0 & 0 & 0 \\
0 & 0 & 0 & 0.5000 & -0.0500 & -0.0500 & 0 & 0 & 0 & 0 & 0 & 0 & 0 & 0 & 0 \\
0 & 0 & 0 & -0.0500 & 0.5000 & -0.0500 & 0 & 0 & 0 & 0 & 0 & 0 & 0 & 0 & 0 \\
0 & 0 & 0 & -0.0500 & -0.0500 & 0.5000 & 0 & 0 & 0 & 0 & 0 & 0 & 0 & 0 & 0 \\
0 & 0 & 0 & 0 & 0 & 0 & 1.0000 & -0.1000 & -0.1000 & 0 & 0 & 0 & 0 & 0 & 0 \\
0 & 0 & 0 & 0 & 0 & 0 & -0.1000 & 1.0000 & -0.1000 & 0 & 0 & 0 & 0 & 0 & 0 \\
0 & 0 & 0 & 0 & 0 & 0 & -0.1000 & -0.1000 & 1.0000 & 0 & 0 & 0 & 0 & 0 & 0 \\
0 & 0 & 0 & 0 & 0 & 0 & 0 & 0 & 0 & 0.4000 & -0.0400 & -0.0400 & 0 & 0 & 0 \\
0 & 0 & 0 & 0 & 0 & 0 & 0 & 0 & 0 & -0.0400 & 0.4000 & -0.0400 & 0 & 0 & 0 \\
0 & 0 & 0 & 0 & 0 & 0 & 0 & 0 & 0 & -0.0400 & -0.0400 & 0.4000 & 0 & 0 & 0 \\
0 & 0 & 0 & 0 & 0 & 0 & 0 & 0 & 0 & 0 & 0 & 0 & 0.6000 & -0.0600 & -0.0600 \\
0 & 0 & 0 & 0 & 0 & 0 & 0 & 0 & 0 & 0 & 0 & 0 & -0.0600 & 0.6000 & -0.0600 \\
0 & 0 & 0 & 0 & 0 & 0 & 0 & 0 & 0 & 0 & 0 & 0 & -0.0600 & -0.0600 & 0.6000
\end{bmatrix}
$$

The inverse of $Z^{R,Y,B}$, i.e., $y^{R,Y,B}$ the primitive admittance matrix is obtained as $[z^{R,Y,B}]^{-1} = y^{R,Y,B} =$

$$
\begin{bmatrix}
\begin{bmatrix} 2.0455 & 0.2273 & 0.2273 \\ 0.2273 & 2.0455 & 0.2273 \\ 0.2273 & 0.2273 & 2.0455 \end{bmatrix} & \begin{bmatrix} 0 & 0 & 0 \\ 0 & 0 & 0 \\ 0 & 0 & 0 \end{bmatrix} & \begin{bmatrix} 0 & 0 & 0 \\ 0 & 0 & 0 \\ 0 & 0 & 0 \end{bmatrix} & \begin{bmatrix} 0 & 0 & 0 \\ 0 & 0 & 0 \\ 0 & 0 & 0 \end{bmatrix} & \begin{bmatrix} 0 & 0 & 0 \\ 0 & 0 & 0 \\ 0 & 0 & 0 \end{bmatrix} \\
\begin{bmatrix} 0 & 0 & 0 \\ 0 & 0 & 0 \\ 0 & 0 & 0 \end{bmatrix} & \begin{bmatrix} 2.0455 & 0.2273 & 0.2273 \\ 0.2273 & 2.0455 & 0.2273 \\ 0.2273 & 0.2273 & 2.0455 \end{bmatrix} & \begin{bmatrix} 0 & 0 & 0 \\ 0 & 0 & 0 \\ 0 & 0 & 0 \end{bmatrix} & \begin{bmatrix} 0 & 0 & 0 \\ 0 & 0 & 0 \\ 0 & 0 & 0 \end{bmatrix} & \begin{bmatrix} 0 & 0 & 0 \\ 0 & 0 & 0 \\ 0 & 0 & 0 \end{bmatrix} \\
\begin{bmatrix} 0 & 0 & 0 \\ 0 & 0 & 0 \\ 0 & 0 & 0 \end{bmatrix} & \begin{bmatrix} 0 & 0 & 0 \\ 0 & 0 & 0 \\ 0 & 0 & 0 \end{bmatrix} & \begin{bmatrix} 1.0227 & 0.1136 & 0.1136 \\ 0.1136 & 1.0227 & 0.1136 \\ 0.1136 & 0.1136 & 1.0227 \end{bmatrix} & \begin{bmatrix} 0 & 0 & 0 \\ 0 & 0 & 0 \\ 0 & 0 & 0 \end{bmatrix} & \begin{bmatrix} 0 & 0 & 0 \\ 0 & 0 & 0 \\ 0 & 0 & 0 \end{bmatrix} \\
\begin{bmatrix} 0 & 0 & 0 \\ 0 & 0 & 0 \\ 0 & 0 & 0 \end{bmatrix} & \begin{bmatrix} 0 & 0 & 0 \\ 0 & 0 & 0 \\ 0 & 0 & 0 \end{bmatrix} & \begin{bmatrix} 0 & 0 & 0 \\ 0 & 0 & 0 \\ 0 & 0 & 0 \end{bmatrix} & \begin{bmatrix} 2.5568 & 0.2841 & 0.2841 \\ 0.2841 & 2.5568 & 0.2841 \\ 0.2471 & 0.2841 & 2.5568 \end{bmatrix} & \begin{bmatrix} 0 & 0 & 0 \\ 0 & 0 & 0 \\ 0 & 0 & 0 \end{bmatrix} \\
\begin{bmatrix} 0 & 0 & 0 \\ 0 & 0 & 0 \\ 0 & 0 & 0 \end{bmatrix} & \begin{bmatrix} 0 & 0 & 0 \\ 0 & 0 & 0 \\ 0 & 0 & 0 \end{bmatrix} & \begin{bmatrix} 0 & 0 & 0 \\ 0 & 0 & 0 \\ 0 & 0 & 0 \end{bmatrix} & \begin{bmatrix} 0 & 0 & 0 \\ 0 & 0 & 0 \\ 0 & 0 & 0 \end{bmatrix} & \begin{bmatrix} 1.7045 & 0.1894 & 0.1894 \\ 0.1894 & 1.7045 & 0.1894 \\ 0.1894 & 0.1894 & 1.7045 \end{bmatrix}
\end{bmatrix}
$$

The y_{BUS} computed from $[A^{R,Y,B}]^t \, y^{R,Y,B} \, [A^{R,Y,B}] =$

$$
\begin{bmatrix}
5.6250 & 0.6250 & 0.6250 & -2.5568 & -0.2841 & -0.2841 & -1.0227 & -0.1136 & -0.1136 \\
0.6250 & 5.6250 & 0.6250 & -0.2841 & -2.5568 & -0.2841 & -0.1136 & -1.0227 & -0.1136 \\
0.6250 & 0.6250 & 5.6250 & -0.2841 & -0.2841 & -2.5568 & -0.1136 & -0.1136 & -1.0227 \\
-2.5568 & -0.2841 & -0.2841 & 6.3068 & 0.7008 & 0.7008 & -1.7045 & -0.1894 & -0.1894 \\
-0.2841 & -2.5568 & -0.2841 & 0.7008 & 6.3068 & 0.7008 & -0.1894 & -1.7045 & -0.1894 \\
-0.2841 & -0.2841 & -2.5568 & 0.7008 & 0.7008 & 6.3068 & -0.1894 & -0.1894 & -1.7045 \\
-1.0227 & -0.1136 & -0.1136 & -1.7045 & -0.1894 & -0.1894 & 2.7273 & 0.3030 & 0.3030 \\
-0.1136 & -1.2007 & -0.1136 & -0.1894 & -1.7045 & -0.1894 & 0.3030 & 2.7273 & 0.3030 \\
-0.1136 & -0.1136 & -1.0227 & -0.1894 & -0.1894 & -1.7045 & 0.3030 & 0.3030 & 2.7273
\end{bmatrix}
$$

The inverse of $y_{BUS}^{R,Y,B}$ so obtained will be the $Z_{BUS}^{R,Y,B}$

$$
Z_{BUS}^{R,Y,B}
\begin{bmatrix}
0.3106 & -0.0325 & -0.031 & 0.1893 & -0.0198 & -0.0189 & 0.234 & 0.024 & -0.023 \\
-0.0325 & 0.3240 & -0.032 & -0.0198 & 0.1975 & -0.0198 & -0.024 & 0.244 & -0.024 \\
-0.0310 & -0.0325 & 0.310 & -0.0189 & -0.0198 & 0.1893 & -0.023 & -0.024 & 0.234 \\
0.1894 & -0.0206 & -0.018 & 0.3106 & -0.0321 & -0.0311 & 0.265 & -0.027 & -0.0265 \\
-0.0206 & 0.2045 & -0.0206 & -0.0321 & 0.3198 & -0.0321 & -0.027 & 0.276 & -0.027 \\
-0.0189 & -0.0206 & 0.1894 & -0.0311 & -0.0321 & 0.3106 & -0.026 & -0.027 & 0.264 \\
0.2350 & -0.0273 & -0.0233 & 0.2652 & -0.0289 & -0.0264 & 0.628 & -0.065 & -0.062 \\
-0.0273 & 0.2709 & -0.0273 & -0.0289 & 0.2871 & -0.0289 & -0.065 & 0.656 & -0.0658 \\
-0.0233 & -0.0273 & 0.2350 & -0.0264 & -0.0289 & 0.2652 & -0.062 & -0.065 & 0.6288
\end{bmatrix}
$$

E.7.3. **Form the bus impedance matrix $Z_{BUS}^{R,Y,B}$ using the algorithm for the system shown.**

Element	Self-impedance
1	$\begin{bmatrix} 0.5 & -0.05 & -0.05 \\ -0.05 & 0.05 & -0.05 \\ -0.05 & -0.05 & 0.05 \end{bmatrix}$
2	$\begin{bmatrix} 0.5 & -0.05 & -0.05 \\ -0.05 & 0.05 & -0.05 \\ -0.05 & -0.05 & 0.05 \end{bmatrix}$
3	$\begin{bmatrix} 1.0 & -0.1 & -0.1 \\ -0.1 & 1.0 & -0.1 \\ -0.1 & -0.1 & 1.0 \end{bmatrix}$
4	$\begin{bmatrix} 0.4 & -0.04 & -0.04 \\ -0.04 & 0.4 & -0.04 \\ -0.04 & -0.04 & 0.4 \end{bmatrix}$
5	$\begin{bmatrix} 0.6 & -0.06 & -0.06 \\ -0.06 & 0.6 & -0.06 \\ -0.06 & -0.06 & 0.6 \end{bmatrix}$

Solution:

The graph is shown in the figure.

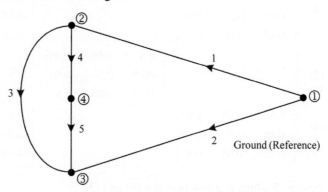

FIGURE E.7.7

Graph for E.7.3.

Step 1: $a = 1$; $b = 2$; a is the reference bus

$$Z_{BUS}^{R,Y,B} = \begin{matrix} & (2) \\ (2) & \begin{bmatrix} 0.5 & -0.05 & -0.05 \\ -0.05 & 0.5 & -0.05 \\ -0.05 & -0.05 & 0.5 \end{bmatrix} \end{matrix}$$

Step 2: Add element 2 which is a branch from $a = 1$ to $b = 3$. a is the reference bus A new bus is created.

$$
Z_{BUS}^{R,Y,B} = \begin{array}{c} (2) \\ (3) \end{array} \left[\begin{array}{ccc}
\overset{(2)}{\begin{bmatrix} 0.5 & -0.05 & -0.05 \\ -0.05 & 0.5 & -0.05 \\ -0.05 & -0.05 & 0.5 \end{bmatrix}} & & \\
& & \overset{(3)}{\begin{bmatrix} 0.5 & -0.05 & -0.05 \\ -0.05 & 0.5 & -0.05 \\ -0.05 & -0.05 & 0.5 \end{bmatrix}}
\end{array} \right]
$$

Step 3: Add element 4. A new bus

$$Z_{bi}^{RYB} = Z_{ai}^{RYB}; \quad i = 2, 3$$
$$Z_{ib}^{RYB} = Z_{ia}^{RYB}; \quad i = 2, 3$$
$$Z_{bb}^{RYB} = Z_{ab}^{RYB} + Z_{ab-ab}^{RYB}$$

$$
Z_{BUS}^{R,Y,B} = \begin{array}{c} (2) \\ \\ (3) \\ \\ (4) \end{array} \left[\begin{array}{ccc}
\overset{(2)}{\begin{bmatrix} 0.5 & -0.05 & -0.05 \\ -0.05 & 0.5 & -0.05 \\ -0.05 & -0.05 & 0.5 \end{bmatrix}} & & \overset{(4)}{\begin{bmatrix} 0.5 & -0.05 & -0.05 \\ -0.05 & 0.5 & -0.05 \\ -0.05 & -0.05 & 0.5 \end{bmatrix}} \\
& \overset{(3)}{\begin{bmatrix} 0.5 & -0.05 & -0.05 \\ -0.05 & 0.5 & -0.05 \\ -0.05 & -0.05 & 0.5 \end{bmatrix}} & \\
\begin{bmatrix} 0.5 & -0.05 & -0.05 \\ -0.05 & 0.5 & -0.05 \\ -0.05 & -0.05 & 0.5 \end{bmatrix} & & \begin{bmatrix} 0.9 & -0.09 & -0.09 \\ -0.09 & 0.9 & -0.09 \\ -0.09 & -0.09 & 0.9 \end{bmatrix}
\end{array} \right]
$$

Step 4: Add element 5 which is a link between (4) and (3)

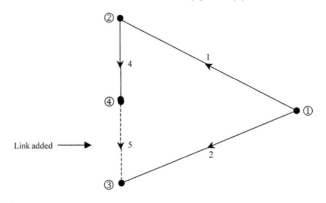

FIGURE E.7.8

Application of building algorithm to partial network.

$$a = 4; b = 3$$

$$Z_{li} = Z_{ai} - Z_{bi}$$

$$Z_{il} = Z_{ia} - Z_{ib}$$

$$Z_{ll} = Z_{al} - Z_{bl} + z_{ab-ab}$$

Substituting sequentially $i = 2$, 3, and 4

$$Z_{12} = Z_{42} - Z_{32}$$

$$Z_{13} = Z_{43} - Z_{33}$$

$$Z_{14} = Z_{44} - Z_{34}$$

and

$$Z_{ll} = Z_{44} - Z_{34} + Z_{43-43}$$

$$Z_{l2} = \begin{bmatrix} 0.5 & -0.05 & -0.05 \\ -0.05 & 0.5 & -0.05 \\ -0.05 & -0.05 & 0.5 \end{bmatrix} = Z_{2l}$$

$$Z_{l3} = \begin{bmatrix} -0.5 & 0.05 & 0.05 \\ 0.05 & -0.5 & 0.05 \\ 0.05 & 0.05 & -0.5 \end{bmatrix} = Z_{3l}$$

$$Z_{l4} = \begin{bmatrix} 0.9 & -0.09 & -0.09 \\ -0.09 & 0.9 & -0.09 \\ -0.09 & -0.09 & 0.9 \end{bmatrix} = Z_{4l}$$

$$Z_{ll} = \begin{bmatrix} 0.95 & -0.05 & -0.14 \\ -0.14 & 0.95 & -0.14 \\ -0.14 & -0.14 & 0.95 \end{bmatrix}$$

Then Z_{BUS} is obtained with the fictitious node l.

	(2)	(3)	(4)	(l)
(2)	$\begin{bmatrix} 0.5 & -0.05 & -0.05 \\ -0.05 & 0.5 & -0.05 \\ -0.05 & -0.05 & 0.5 \end{bmatrix}$	$- - -$	$\begin{bmatrix} 0.5 & -0.05 & -0.05 \\ -0.05 & 0.5 & -0.05 \\ -0.05 & -0.05 & 0.5 \end{bmatrix}$	$\begin{bmatrix} 0.5 & -0.05 & -0.05 \\ -0.05 & 0.5 & -0.05 \\ -0.05 & -0.05 & 0.5 \end{bmatrix}$
(3)	$- - -$	$\begin{bmatrix} 0.5 & -0.05 & -0.05 \\ -0.05 & 0.5 & -0.05 \\ -0.05 & -0.05 & 0.5 \end{bmatrix}$	$- - -$	$\begin{bmatrix} 0.5 & -0.05 & -0.05 \\ -0.05 & 0.5 & -0.05 \\ -0.05 & -0.05 & 0.5 \end{bmatrix}$
(4)	$\begin{bmatrix} 0.5 & -0.05 & -0.05 \\ -0.05 & 0.5 & -0.05 \\ -0.05 & -0.05 & 0.5 \end{bmatrix}$	$- - -$	$\begin{bmatrix} 0.9 & -0.09 & -0.09 \\ -0.09 & 0.9 & -0.06 \\ -0.09 & -0.09 & 0.6 \end{bmatrix}$	$\begin{bmatrix} 0.9 & -0.09 & -0.09 \\ -0.09 & 0.9 & -0.06 \\ -0.09 & -0.09 & 0.6 \end{bmatrix}$
(l)	$\begin{bmatrix} 0.5 & -0.05 & -0.05 \\ -0.05 & 0.5 & -0.05 \\ -0.05 & -0.05 & 0.5 \end{bmatrix}$	$\begin{bmatrix} -0.5 & 0.05 & 0.05 \\ 0.05 & -0.5 & 0.05 \\ 0.05 & 0.05 & -0.5 \end{bmatrix}$	$\begin{bmatrix} 0.9 & -0.09 & -0.09 \\ -0.09 & 0.9 & -0.06 \\ -0.09 & -0.09 & 0.6 \end{bmatrix}$	$\begin{bmatrix} 2.0 & -0.2 & -0.2 \\ -0.2 & 2.0 & -0.2 \\ -0.2 & -0.2 & 2.0 \end{bmatrix}$

Step 5: The fictitious node "*l*" is eliminated

$$Z_{ij}^{R,Y,B} = Z_{ij}^{R,Y,B}(\text{before modification}) - [Z_{il}^{R,Y,B}][Z_{ll}^{R,Y,B}]^{-1}[Z_{lj}^{R,Y,B}]$$

Hence

$$Z_{22}^l = [Z_{22}] - [Z_{2l}][Z_{ll}]^{-1}[Z_{l2}]$$

$$Z_{23}^l = [Z_{23}] - [Z_{2l}][Z_{ll}]^{-1}[Z_{l3}]$$

$$Z_{24}^l = [Z_{24}] - [Z_{2l}][Z_{ll}]^{-1}[Z_{l4}]$$

$$Z_{33}^l = [Z_{33}] - [Z_{3l}][Z_{ll}]^{-1}[Z_{l3}]$$

$$Z_{33}^l = [Z_{33}] - [Z_{3l}][Z_{ll}]^{-1}[Z_{l3}]$$

$$Z_{32}^l = [Z_{32}] - [Z_{3l}][Z_{ll}]^{-1}[Z_{l2}]$$

$$Z_{34}^l = [Z_{34}] - [Z_{3l}][Z_{ll}]^{-1}[Z_{l4}]$$

$$Z_{43}^l = [Z_{43}] - [Z_{4l}][Z_{ll}]^{-1}[Z_{l3}]$$

$$Z_{44}^l = [Z_{44}] - [Z_{4l}][Z_{ll}]^{-1}[Z_{l4}]$$

so that
$Z_{BUS}^{R,Y,B}$ after *l* is eliminated is given by

$$Z_{BUS}^{R,Y,B} = \begin{bmatrix} Z_{22}^l & Z_{23}^l & Z_{24}^l \\ Z_{32}^l & Z_{33}^l & Z_{34}^l \\ Z_{42}^l & Z_{43}^l & Z_{44}^l \end{bmatrix}$$

where Z_{ab}^l is $Z_{ab}^{R,Y,B}$ after *l* is eliminated and Z_{ab} is $Z_{ab}^{R,Y,B}$ before *l* is eliminated.

$$Z_{BUS}^{R,Y,B} = \begin{bmatrix} Z_{22}^l & Z_{23}^l & Z_{24}^l \\ Z_{42}^l & Z_{33}^l & Z_{34}^l \\ Z_{42}^l & Z_{43}^l & Z_{44}^l \end{bmatrix} =$$

	(2)			(3)			(4)		
(2)	0.3750	−0.0375	−0.0375	0.125	−0.0125	−0.0125	0.2750	−0.0275	−0.0275
	−0.0375	0.3750	−0.0375	−0.0125	0.125	−0.0125	−0.0275	0.2750	−0.0275
	−0.0375	−0.0375	0.3750	−0.0125	−0.0125	0.125	−0.0275	−0.0275	0.2750
(3)	−0.125	−0.0125	−0.0125	0.3750	−0.0375	−0.0375	0.2250	−0.0225	−0.0225
	−0.0125	−0.0125	0.125	−0.0375	0.3750	−0.0375	−0.0225	0.2250	−0.0225
	−0.0125	−0.125	−0.0125	−0.0375	−0.0375	0.3750	−0.0225	−0.0225	0.2250
(4)	0.2750	−0.0125	−0.0275	0.2250	−0.0225	−0.0225	0.4950	−0.0495	−0.0495
	−0.0275	0.2750	−0.0275	−0.0225	0.2250	−0.0225	−0.0495	0.4950	−0.0495
	−0.0275	−0.0275	0.2750	−0.0225	−0.0225	0.2250	−0.0495	−0.0495	0.4950

Step 6: Finally element 3 is added across the nodes (2) and (3) forming a link as shown
$a = (2)$ and $b = (3)$

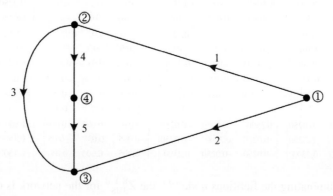

FIGURE E.7.9

Application of building algorithm when link is added.

$$Z_{li} = Z_{ai} - Z_{bi}$$

$$Z_{il} = Z_{ia} - Z_{ib}$$

$$Z_{ll} = Z_{al} - Z_{bl} + z_{ab-ab}$$

Substituting $i = 2, 3, 4$ once again we obtain

$$Z_{l2} = Z_{22} - Z_{32} = Z_{21}$$

$$Z_{l3} = Z_{23} - Z_{33} = Z_{31}$$

$$Z_{l4} = Z_{24} - Z_{34}$$

$$Z_{l1} = Z_{21} - Z_{31} + 1.0$$

The $Z_{BUS}^{R,Y,B}$ is obtained as follows:
$$\begin{bmatrix} Z_{22}^1 & Z_{23}^1 & Z_{24}^1 & Z_{2l}^1 \\ Z_{32}^1 & Z_{33}^1 & Z_{34}^1 & Z_{3l}^1 \\ Z_{42}^1 & Z_{43}^1 & Z_{44}^1 & Z_{4l}^1 \\ Z_{l2}^1 & Z_{l3}^1 & Z_{l4}^1 & Z_{ll}^1 \end{bmatrix}$$

	(2)	(3)	(4)	(l)
(2)	$\begin{bmatrix} 0.33 & 0.033 & 0.033 \\ -0.033 & 0.22 & -0.033 \\ -0.03 & -0.033 & 0.33 \end{bmatrix}$	$\begin{bmatrix} 0.167 & 0.0167 & 0.0167 \\ 0.0167 & -0.1667 & 0.0167 \\ 0.0167 & -0.012 & -0.1667 \end{bmatrix}$	$\begin{bmatrix} 0.2 & 0.02 & 0.02 \\ -0.02 & 0.2 & -0.02 \\ -0.02 & -0.02 & 0.2 \end{bmatrix}$	$\begin{bmatrix} 0.2500 & -0.0250 & -0.0250 \\ -0.0250 & 0.2500 & -0.0250 \\ -0.0250 & -0.0250 & 0.2500 \end{bmatrix}$
(3)	$\begin{bmatrix} -0.167 & 0.0167 & 0.0167 \\ 0.0167 & -0.1667 & 0.0167 \\ 0.0167 & -0.012 & -0.1667 \end{bmatrix}$	$\begin{bmatrix} 0.33 & -0.033 & -0.033 \\ -0.033 & 0.33 & -0.033 \\ -0.03 & -0.033 & 0.33 \end{bmatrix}$	$\begin{bmatrix} -0.3 & 0.03 & 0.03 \\ 0.03 & -0.3 & 0.03 \\ 0.03 & 0.03 & -0.3 \end{bmatrix}$	$\begin{bmatrix} 0.2500 & -0.0250 & -0.0250 \\ -0.0250 & -0.250 & 0.0250 \\ -0.0250 & -0.0250 & 0.2500 \end{bmatrix}$
(4)	$\begin{bmatrix} 0.2 & -0.02 & -0.02 \\ -0.02 & 0.2 & -0.02 \\ -0.02 & -0.02 & 0.2 \end{bmatrix}$	$\begin{bmatrix} -0.3 & 0.03 & 0.03 \\ 0.03 & -0.3 & 0.03 \\ 0.03 & 0.03 & -0.3 \end{bmatrix}$	$\begin{bmatrix} 0.36 & -0.036 & -0.036 \\ -0.036 & 0.36 & -0.036 \\ -0.036 & -0.036 & 0.36 \end{bmatrix}$	$\begin{bmatrix} 0.05 & -0.005 & -0.005 \\ -0.005 & 0.05 & -0.05 \\ -0.005 & -0.005 & 0.05 \end{bmatrix}$
(l)	$\begin{bmatrix} 0.2500 & -0.0250 & -0.0250 \\ -0.0250 & 0.2500 & -0.0250 \\ -0.0250 & -0.0250 & 0.2500 \end{bmatrix}$	$\begin{bmatrix} 0.2500 & -0.0250 & -0.0250 \\ -0.0250 & 0.2500 & -0.0250 \\ -0.0250 & -0.0250 & 0.2500 \end{bmatrix}$	$\begin{bmatrix} 0.05 & -0.005 & -0.005 \\ -0.005 & 0.05 & -0.05 \\ -0.005 & -0.005 & 0.05 \end{bmatrix}$	$\begin{bmatrix} 1.500 & -0.1500 & -0.1500 \\ -0.1500 & 1.500 & -0.1500 \\ -0.1500 & -0.1500 & 1.500 \end{bmatrix}$

As before eliminating the fictitious node "l," the $Z_{BUS}^{R,Y,B}$ for the network is obtained.

$$= \begin{bmatrix} Z_{22}^l & Z_{23}^l & Z_{24}^l \\ Z_{32}^l & Z_{33}^l & Z_{34}^l \\ Z_{42}^l & Z_{43}^l & Z_{44}^l \end{bmatrix}$$

	(2)	(3)	(4)
(2)	$\begin{bmatrix} 0.3333 & -0.0333 & -0.0333 \\ -0.0333 & 0.3333 & -0.0333 \\ -0.0333 & -0.0333 & 0.3333 \end{bmatrix}$	$\begin{bmatrix} 0.1667 & -0.0167 & -0.0167 \\ -0.0167 & 0.1667 & -0.0167 \\ -0.0167 & -0.0167 & 0.1667 \end{bmatrix}$	$\begin{bmatrix} 0.2667 & -0.0267 & -0.0267 \\ -0.0267 & 0.2667 & -0.0267 \\ -0.0267 & -0.0267 & 0.2667 \end{bmatrix}$
(3)	$\begin{bmatrix} 0.1667 & -0.0167 & -0.0167 \\ -0.0167 & 0.1667 & -0.0167 \\ -0.0167 & -0.0167 & 0.1667 \end{bmatrix}$	$\begin{bmatrix} 0.3333 & -0.0333 & -0.0333 \\ -0.0333 & 0.3333 & -0.0333 \\ -0.0333 & -0.0333 & 0.3333 \end{bmatrix}$	$\begin{bmatrix} 0.2333 & -0.0233 & -0.0233 \\ -0.0233 & -0.02333 & -0.0233 \\ -0.0233 & -0.0233 & 0.2333 \end{bmatrix}$
(4)	$\begin{bmatrix} 0.2667 & -0.0267 & -0.0267 \\ 0.0267 & 0.2667 & -0.0267 \\ -0.0267 & -0.0267 & 0.2667 \end{bmatrix}$	$\begin{bmatrix} -0.175 & 0.0175 & 0.0175 \\ 0.0175 & -0.175 & 0.0175 \\ 0.0175 & 0.0175 & -0.175 \end{bmatrix}$	$\begin{bmatrix} 0.4933 & -0.0493 & -0.0493 \\ -0.0493 & 0.4933 & -0.0493 \\ -0.0493 & -0.0493 & 0.4933 \end{bmatrix}$

QUESTIONS

7.1. Derive the transformation matrices with symmetrical components.

7.2. Give the algorithm for the formation of three-phase Z_{BUS}.

7.3. Explain how the impedance matrices of stationary and rotating elements are represented in a three-phase system.

7.4. Show that the sequence impedance matrices obtained for stationary and rotating elements are diagonalized. Derive the formula used.

7.5. Develop the expressions for forming of three-phase Z_{BUS} for the element which is added between two existing buses in partial network.

7.6. Show that although phase components primitive impedance matrix is not symmetric, the symmetrical component transformation matrix decouples the sequence impedances for stationary and rotating elements.

7.7. Derive the necessary expression for building up algorithm of a three-phase bus impedance matrix when the added element is a branch.

   Explain the primitive network three-phase representation of a component in admittance form?

PROBLEMS

P.7.1. Find Z_{pq}^{012} given $Z_{pq}^{abc} = \begin{bmatrix} j0.4 & j.1 & j0.1 \\ j0.2 & j0.6 & j0.2 \\ j0.3 & j0.3 & j0.8 \end{bmatrix}$

P.7.2. Given the element I with $Z_{BUS}^{R,Y,B} = \begin{matrix} & \begin{matrix} R & Y & B \end{matrix} \\ \begin{matrix} R \\ Y \\ B \end{matrix} & \begin{bmatrix} 0.08 & -0.02 & -0.02 \\ -0.02 & 0.08 & -0.02 \\ -0.02 & -0.02 & 0.08 \end{bmatrix} \end{matrix}$ if another element z with

self-impedance 0.06 and mutual impedance -0.01 is added as a branch to the given element, obtain the $Z_{BUS}^{R,Y,B}$ for the combination.

P.7.3. Given a three-phase circuit element with self-impedance 0.05 and mutual impedance -0.01 and if this element is connected across the two series connected elements of Problem 7.2, as link obtain the $Z_{BUS}^{R,Y,B}$ for three element combination.

SYNCHRONOUS MACHINE

The synchronous machine is the most important component in a power system. In fact the very existence of the power system depends on it.

A synchronous machine consists of two major components viz. the stator and the rotor which are in relative motion.

While the field carries the direct current, the generated e.m.f. is alternating. The detailed analysis of a synchronous machine is very complicated due to the presence of harmonics on account of variation in air gap permeance. The important quantities are the various voltages generated, the corresponding flux linkages, and the currents. The various reactances invariably play a vital role in the performance of the machine. It is an established fact that the machine analysis is better achieved through the two-axis theory. At first a brief derivation of the two-axis voltages will be given based on the transformer and speed voltages produced. Later, a detailed analysis is given.

Much of the work carried out by early pioneers of synchronous machine analysis like Doherty and Nickle, Park, Blondel, and others will be too difficult at this stage to understand. Hence, a much simplified analysis is presented that will be sufficient to obtain steady-state and transient-state phasor diagrams.

8.1 THE TWO-AXIS MODEL OF SYNCHRONOUS MACHINE

In the following, the two-axis model of a three-phase armature winding of a synchronous machine will be obtained. We can replace the three-phase winding by a two-phase winding with the same exciting effect at any point around the machine air gap between the stator and the rotor. Consider the time variation of the m.m.f. of the three-phase windings R, Y, and B.

$$\overline{M}_R = M_m \sin \omega t \cdot \cos \theta \tag{8.1}$$

$$\overline{M}_y = M_m \sin\left(\omega t - \frac{2\pi}{3}\right) \cdot \cos\left(\theta - \frac{2\pi}{3}\right) \tag{8.2}$$

$$\overline{M}_B = M_m \sin\left(\omega t - \frac{4\pi}{3}\right) \cdot \cos\left(\theta - \frac{4\pi}{3}\right) \tag{8.3}$$

where the time variation of m.m.f. resulting from the currents is represented by sine functions. The time zero is chosen such that the R-phase current is zero and increasing. The space distributions of the m.m.f.s are represented by cosine functions; q is a space angle the space origin being

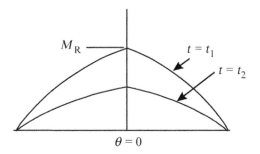

FIGURE 8.1

Rotating m.m.f. waves.

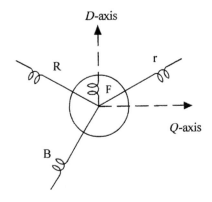

FIGURE 8.2

Symmetrical three-phase windings.

the axis of R-winding. The stator windings are symmetrical and carry balanced three-phase currents (Figs. 8.1 and 8.2).

$$\overline{M}_R + \overline{M}_y + \overline{M}_B = \frac{3}{2}M_m \sin(\omega t - \theta) \tag{8.4}$$

Consider an equivalent two-axis windings as shown in Fig. 8.3 displaced mutually in space by 90°. The m.m.f.s are

$$M_d = \frac{3}{2}M_m \sin \omega t \cdot \cos \theta \tag{8.5}$$

$$M_q = \frac{3}{2}M_m \sin\left(\omega t - \frac{\pi}{2}\right) \cdot \cos\left(\theta - \frac{\pi}{2}\right) \tag{8.6}$$

$$\overline{M}_d + \overline{M}_q = \frac{3}{2}\overline{M}_m \sin(\omega t - \theta)$$

$$\overline{M}_R + \overline{M}_y + \overline{M}_B \tag{8.7}$$

Thus the two windings d and q will give the same resulting m.m.f. at any point around the air gap as the three-phase windings.

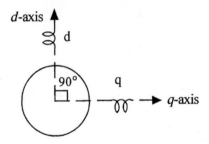

FIGURE 8.3

Two-axis representation.

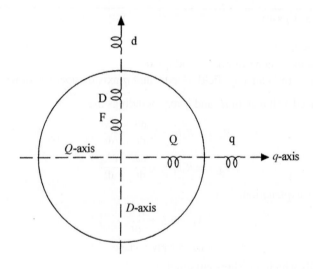

FIGURE 8.4

Two-axis model with one damper coil in each axis.

8.2 DERIVATION OF PARK'S TWO-AXIS MODEL

Consider the schematic of the two-axis model of the synchronous machine shown in Fig. 8.4. For simplicity only one damper coil is considered on each axis; d and q represent the fictitious armature windings along d and q axes.

D and Q represent the short-circuited damper coils along the d and q axes. F is the field winding on the main polar axis or d axis.

Let

V = voltage in volts

I = current in amperes

λ = flux linkages in webers with suffix f, d and q for field, d and q axes

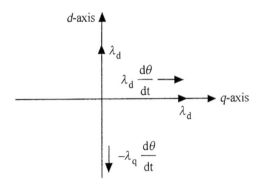

FIGURE 8.5

Induced voltages along d and q axes.

ω = speed in rad/s = $d\theta/dt$

R_a = armature resistance, same in both d and q axes

R_f, R_d, and R_q refer to resistance of field, d axis, and q axis damper windings, respectively.

The generator-induced voltages in d- and q-axis windings are

$$V_d = -R_a I_d - \lambda_q \frac{d\theta}{dt} + \frac{d\lambda_d}{dt} \tag{8.8}$$

$$V_q = -R_a I_q + \lambda_d \frac{d\theta}{dt} + \frac{d\lambda_q}{dt} \tag{8.9}$$

Look at Fig. 8.5 for explanation.

$$V_f = R_f I_f + \frac{d\lambda_f}{dt} \tag{8.10}$$

$$= \text{d.c. supply voltage}$$

For the damper coils which are short circuited

$$0 = R_d I_d + \frac{d\lambda_d}{dt} \tag{8.11}$$

$$0 = R_q I_q + \frac{d\lambda_q}{dt} \tag{8.12}$$

In the following section, a detailed analysis of the machine performance is given.

8.3 SYNCHRONOUS MACHINE ANALYSIS

In the analysis, it is always assumed that the rotor magnetic paths and all of its electrical circuits are symmetrical with respect to pole and interpole axes. The field winding has its axis in line with the pole axis. In reality, all the damper bars are connected together to form a closed mesh. For analysis, the current paths in these bars may be assumed to be symmetrical with respect to both the pole and interpole axes. Fig. 8.6 shows the arrangement.

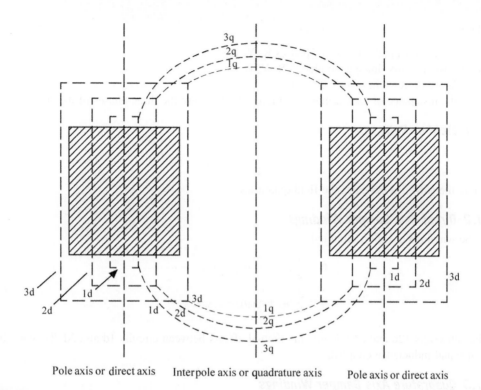

Pole axis or direct axis   Interpole axis or quadrature axis   Pole axis or direct axis

FIGURE 8.6

Arrangement of rotor circuits.

All mutual inductances between stator and rotor circuits are periodic functions of rotor angular positions. Also, the mutual inductances between any two stator phases are also periodic functions of rotor angular position. For ease of analysis, saturation is neglected.

It is assumed that the stator windings are sinusoidally distributed. Further stator slots do not cause any change in rotor inductances with rotor angle.

8.3.1 VOLTAGE RELATIONS—STATOR OR ARMATURE

$$
\left.
\begin{aligned}
e_R &= \frac{d}{dt}\Psi_R - ri_R \\[2ex]
e_y &= \frac{d}{dt}\Psi_y - ri_y \\[2ex]
e_B &= \frac{d}{dt}\Psi_B - ri_B
\end{aligned}
\right\}
\tag{8.13}
$$

where

e_R, e_y, e_B are terminal voltages of phases R, Y, B
Ψ_R, Ψ_y, Ψ_B are flux linkages in phases R, Y, B
i_R, i_y, i_B are currents in phases R, Y, B
"r" is the resistance of each armature winding. R, Y, B are the three phases of the stator.

8.3.1.1 Field or Rotor

$$e_{fd} = \frac{d}{dt}\psi_{fd} + r_{fd}i_{fd} \tag{8.14}$$

where the suffix fd indicates the field quantities.

8.3.1.2 Direct Axis Damper Windings

Since the windings are short circuited

$$0 = \frac{d}{dt}\psi_{1d} + r_{11d}i_{1d} + r_{12d}i_{2d} + \cdots$$

$$0 = \frac{d}{dt}\psi_{2d} + r_{21d}i_{1d} + r_{22d}i_{2d} + \cdots \tag{8.15}$$

The subscripts 12d and 21d denote the mutual effects between circuits 1d and 2d. They are both resistance and inductance coupled.

8.3.1.3 Quadrature Axis Damper Windings

Here also, as the windings are short circuited.

$$0 = \frac{d}{dt}\psi_{1q} + r_{11q}\,i_{1q} + r_{12q}\,i_{2q} + \cdots$$

$$0 = \frac{d}{dt}\psi_{2q} + r_{21q}\,i_{1q} + r_{22q}\,i_{2q} + \cdots \tag{8.16}$$

8.3.2 FLUX LINKAGE RELATIONS

8.3.2.1 Armature

$$\psi_R = -x_{RR}i_R - x_{Ry}i_y - x_{RB}i_B + x_{Rfd}i_{fd} + x_{R1d}i_{1d} + x_{R2d}i_{2d} + \cdots + x_{R1q}i_{1q} + x_{R2q}i_{2q} + \cdots$$

$$\psi_y = -x_{yR}i_R - x_{yy}i_y - x_{yB}i_B + x_{yfd}i_{fd} + x_{y1d}i_{1d} + x_{y2d}i_{2d} + \cdots + x_{y1q}i_{1q} + x_{y2q} + \cdots$$

$$\psi_B = -x_{BR}i_R - x_{By}i_y - x_{BB}i_B + x_{Bfd}i_{fd} + x_{B1d}i_{1d} + x_{B2d}i_{2d} + \cdots + x_{B1q}i_{1q} + x_{B2q}i_{2q} + \cdots \tag{8.17}$$

8.3.2.2 Field

$$\psi_{fd} = -x_{fRd}i_R - x_{fyd}i_y - x_{fBd}i_B + x_{ffd}i_{fd} + x_{f1d}i_{1d} + x_{f2d}i_{2d} + \cdots + x_{f1q}i_{1q} + x_{f2q}i_{2q} + \cdots \tag{8.18}$$

8.3.2.3 Direct Axis Damper Winding

$$\psi_{1d} = -x_{1Rd}i_R - x_{1yd}i_y - x_{1Bd}i_B + x_{1fd}i_{fd} + x_{11d}i_{1d} + x_{12d}i_{2d} + \cdots + x_{id1q}i_{1q} + x_{1d2q}i_{2q} + \cdots \qquad (8.19)$$

8.3.2.4 Quadrature Axis Damper Winding

$$\psi_{1q} = -x_{1Rq}i_R - x_{1yq}i_y - x_{1Bq}i_B + x_{1q\,fd}i_{fd} + x_{1q1d}i_{1d} + x_{1q2d}i_{2d} + \cdots + x_{11q}i_{1q} + x_{12q}i_{2q} + \cdots \qquad (8.20)$$

8.3.3 INDUCTANCE RELATIONS

8.3.3.1 Self-Inductance of the Armature Windings

The self-inductance of any armature winding varies periodically from a maximum value when the pole axis is in line with the phase axis to a minimum value when the quadrature pole axis is in line with the phase axis (Fig. 8.7). Because of the symmetry of the rotor, sinusoidal distribution of winding with inductance having a period of 180° (elec).

$$
\begin{aligned}
x_{RR} &= x_{RR0} = x_{RR2} \cos 2q_R \\
x_{yy} &= x_{yy0} = x_{yy2} \cos 2q_y \\
x_{BB} &= x_{BB0} = x_{BB1} \cos 2q_B
\end{aligned}
\qquad (8.21)
$$

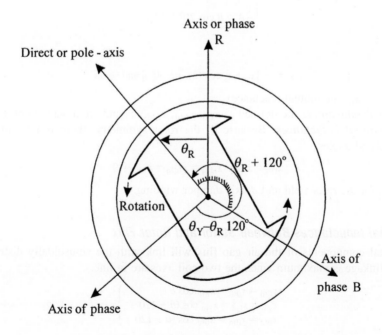

FIGURE 8.7

The rotor and stator axes.

8.3.3.2 Mutual Inductances of the Armature Windings

These inductances can be written as

$$X_{Ry} = X_{yR} = -[x_m + x_s \cos 2(\theta + 30°)]$$
$$X_{yB} = X_{By} = -[x_m + x_s \cos 2(\theta + 90°)] \qquad (8.22)$$
$$X_{BR} = X_{RB} = -[x_m + x_s \cos(\theta + 150°)]$$

The above expressions are obtained from the following considerations:

A component of mutual flux of armature phases does not link the rotor and is therefore independent of angle

$$f_d = k_d \cos \theta_R$$
$$f_q = -k_q \sin \theta_R$$

The linkage with phase "y" due to these components is proportional to

$$f_d \cos \theta_y - f_d \sin \theta_y = k_d \cos \theta_R \cos \theta_y + k_q \cos \theta_R \cos \theta_y$$
$$= k_d \cos \theta_R \cos(\theta - 120) + k_d \sin \theta_R \sin (\theta - 120°)$$
$$= \frac{k_d + k_q}{4} + \frac{k_d + k_q}{2} \cos 2(\theta - 60°)$$
$$= \frac{1}{2}A + B \cos 2(\theta - 60°)$$
$$= \left[\frac{1}{2}A + B \cos 2(\theta + 30°)\right]$$

Hence

$$x_{Ry} = -[x_y + x_s \cos (2)(q + 30°)] \text{ and so on}$$

(Note that $x_y = x_m$, the mutual reactance).

The rotor self-inductances are defined as x_{ffd}, x_{11d}, x_{22d}, ... and are assumed constants.

The rotor mutual inductances: Because of the rotor symmetry, there is no mutual coupling between rotor d and q axes.

$$x_{f1q} = x_{f2q} = x_{id1q} = x_{1d2q} = x_{1qfd} = 0$$

$x_{f1d} = x_{1fd}$, etc., for the rotor field and d-axis damper winding.

8.3.3.3 Mutual Inductances Between Stator and Rotor Flux

The fundamental components of the air gap flux will link with the sinusoidally distributed stator flux. The flux linkage is maximum when the two coil axes are in line.

$$\left.\begin{array}{l} x_{Rfd} = x_{fRd} = x_{Rfd} \cos \theta \\ x_{yfd} = x_{fyd} = x_{yfd} \cos (\theta - 120°) \\ x_{Bfd} = x_{fBd} = x_{Bfd} \cos (\theta + 120°) \end{array}\right\} \qquad (8.23)$$

$$\left.\begin{array}{l} x_{R1d} = x_{1Rd} = x_{R1d} \cos \theta \\ x_{y1d} = x_{1yd} = x_{y1d} \cos (\theta - 120°) \\ x_{B1d} = x_{1Bd} = x_{B1d} \cos (\theta + 120°) \end{array}\right\} \qquad (8.24)$$

$$\left.\begin{array}{l} x_{R1q} = x_{1Rq} = -x_{R1q}\sin\theta \\ x_{y1q} = x_{1yq} = -x_{R1q}\sin(\theta - 120°) \\ x_{B1q} = x_{1Bq} = -x_{B1q}\sin(\theta + 120°) \end{array}\right\} \tag{8.25}$$

8.3.4 FLUX LINKAGE EQUATIONS

Using the mutual inductance relation (8.23), the flux linkage equations are rewritten as follows.

8.3.4.1 Field

$$[i_R\cos\theta + i_y\cos(\theta - 120°) + i_B\cos(\theta + 120°) + x_{ffd}i_{fd} + x_{f1d}i_{fd} + x_{f2d}i_{fd} + \cdots] \tag{8.26}$$

8.3.4.2 Direct Axis Damper Winding

$$[i_R\cos\theta + i_y\cos(\theta - 120°) + i_B\cos(\theta + 120°) + x_{1fd}i_{fd} + x_{11d}i_{fd} + x_{12d}i_{fd} + \cdots] \tag{8.27}$$

8.3.4.3 Quadrature Axis Damper Winding

$$[i_R\sin\theta + i_y\sin(\theta - 120°) + i_B\sin(\theta + 120°) + x_{11q}i_{1q} + x_{12q}i_{2q} + \cdots] \tag{8.28}$$

8.4 THE TRANSFORMATIONS

Now, it is possible to consider a transformation of phase quantities into "*d-*" and "*q*"-axis components. However, since the transformed variables for three-phase system must also be three, we introduce the zero-sequence components for consistent transformation of variables. Thus it is defined that

$$\left.\begin{array}{l} i_d = \dfrac{2}{3}\left[i_R\cos\theta + i_y\cos(\theta - 120°) + i_B\cos(\theta + 120°)\right] \\[2ex] i_q = -\dfrac{2}{3}\left[i_R\sin\theta + i_y\sin(\theta - 120°) + i_B\sin(\theta + 120°)\right] \\[2ex] i_0 = \dfrac{1}{3}(i_R + i_y + i_B) \end{array}\right\} \tag{8.29}$$

substituting Eq. (8.7) into Eqs. (8.14), (8.15), and (8.16)

$$\left.\begin{array}{l} \psi_{fd} = \dfrac{3}{2}x_{Rfd}i_d + x_{ffd}i_{fd} + x_{f1d}i_d + \cdots \\[2ex] \psi_{1d} = \dfrac{3}{2}x_{R1d}i_d + x_{1fd}i_{fd} + x_{11d}i_d + x_{12d}i_d + \cdots \\[2ex] \psi_{1q} = \dfrac{3}{2}x_{R1q}i_d + x_{11q}i_{1q} + x_{12q}i_{2q} + \cdots \end{array}\right\} \tag{8.30}$$

substituting Eqs. (8.21), (8.22), (8.23), (8.24), and (8.25) into Eq. (8.17)

$$\psi_R = -x_{RR0}i_R + x_{Ry0}(i_y + i_B) - x_{RR2}i_R \cos 2\theta + x_{RR2}i_R \cos 2(\theta + 30°) + x_{RR2}i_B \cos 2(\theta + 150°)$$
$$+ (x_{Rfd}i_{fd} + x_{R1d}i_{1d} + x_{R2d}i_{2d} + \cdots)\cos \theta - (x_{R1q}i_{1q} + x_{R2q}i_{2q} + \cdots)\sin \theta \tag{8.31}$$

$$\psi_y = -x_{RR0}i_y + x_{Ry0}(i_B + i_R) - x_{RR2}i_R \cos 2(\theta + 30°) - x_{RR2}i_y \cos 2(\theta + 120°) + x_{RR2}i_B \cos 2(\theta + 90°)$$
$$+ (x_{Rfd}i_{fd} + x_{R1d}i_{1d} + x_{R2d}i_{2d} + \cdots) \cos (\theta - 120°) - (x_{R1q}i_{1q} + x_{R2q}i_{2q} + \cdots) \sin (\theta - 120°)$$
$$\tag{8.32}$$

$$\psi_B = -x_{RR0}i_B + x_{Ry0}(i_R + i_y) + x_{RR2}i_R \cos 2(\theta + 150°) - x_{RR2}i_y \cos 2(\theta - 90°) - x_{RR2}i_B \cos 2(\theta + 120°)$$
$$+ (x_{Rfd}i_{fd} + x_{R1d}i_{1d} + x_{R2d}i_{2d} + \cdots)\cos (\theta + 120°) - (x_{R1q}i_{1q} + x_{R2q}i_{2q} + \cdots) \sin (\theta + 120°)$$
$$\tag{8.33}$$

Eliminating the armature phase currents i_R, i_y, and i_B by i_d, i_q, and i_0, we can obtain Ψ_R, Ψ_y, and Ψ_B. Further, defining Ψ_d, Ψ_q, and Ψ_0 by a similar transformation that is used for currents.

$$\psi_d = \frac{2}{3}\left[\psi_R \cos \theta + \psi_y \cos (\theta - 120°) + \psi_B \cos (\theta + 120°)\right]$$

$$\psi_q = -\frac{2}{3}\left[\psi_R \sin \theta + \psi_y \sin (\theta - 120°) + \psi_B \sin (\theta + 120°)\right] \tag{8.34}$$

$$\psi_0 = \frac{1}{3}\left[\psi_R + \psi_y + \psi_B\right]$$

Substituting Eqs. (8.19), (8.20), (8.21) into Eq. (8.34)

$$\psi_d = -\left(x_{RR0} + x_{Ry0} + \frac{3}{2}x_{RR2}\right)i_d + x_{Rfd}i_{fd} + x_{R1d}i_{1d} + x_{R2d}i_{2d} + \cdots \tag{8.35}$$

$$\psi_q = -\left(x_{RR0} + x_{Ry0} + \frac{3}{2}x_{RR2}\right)i_q + x_{R1q}i_{1q} + x_{R2q}i_{2q} + \cdots \tag{8.36}$$

$$\psi_0 = -(x_{RR0} - 2x_{Ry0})i_0 \tag{8.37}$$

Thus, for direct and quadrature axes we can define

$$x_d = \left(x_{RR0} + x_{Ry0} + \frac{3}{2}x_{RR2}\right) \tag{8.38}$$

$$x_q = \left(x_{RR0} + x_{Ry0} + \frac{3}{2}x_{RR2}\right) \tag{8.39}$$

Further

$$x_0 = x_{RR0} + 2x_{Ry0} \tag{8.40}$$

8.5 STATOR VOLTAGE EQUATIONS

Now, it is desired to eliminate i_R, i_y, and i_B from Eq. (8.13).

Just as in the case of currents and flux linkages, voltage transformation will also be now defined as

$$e_d = \frac{2}{3} \left[e_R \cos \theta + e_y \cos (\theta - 120°) + e_B \cos (\theta + 120°) \right]$$

$$e_q = -\frac{2}{3} \left[e_R \sin \theta + e_y \sin (\theta - 120°) + e_B \sin (\theta + 120°) \right] \qquad (8.41)$$

$$e_0 = \frac{1}{3} \left[e_R + e_y + e_B \right]$$

substituting Eq. (8.13) into Eq. (8.40) and using Eq. (8.29)

$$e_d = -\frac{2}{3} \left[\cos \theta \frac{d\psi_R}{dt} + \cos (\theta - 120°) \frac{d\psi_y}{dt} + \cos (\theta + 120°) \frac{d\psi_B}{dt} \right] - r i_d$$

$$e_q = -\frac{2}{3} \left[\sin \theta \frac{d\psi_R}{dt} + \sin (\theta - 120°) \frac{d\psi_y}{dt} + \sin (\theta + 120°) \frac{d\psi_B}{dt} \right] - r i_q \qquad (8.42)$$

$$e_0 = \frac{d}{dt} \psi_0 - r i_0$$

differentiating Ψ_d from Eq. (8.34)

$$\frac{d\psi_d}{dt} = \frac{2}{3} \left[\cos \theta \frac{d\psi_R}{dt} + \cos (\theta - 120°) \frac{d\psi_y}{dt} + \cos (\theta + 120°) \frac{d\psi_B}{dt} \right]$$

$$- \frac{2}{3} \left[\psi_R \sin \theta \frac{d\vartheta}{dt} + \psi_y \sin (\theta - 120°) \frac{d\vartheta}{dt} + \psi_B \sin (\theta + 120°) \frac{d\vartheta}{dt} \right] \qquad (8.43)$$

Substituting Ψ_d from Eq. (8.34) further

$$\frac{d\psi_d}{dt} = \frac{2}{3} \left[\cos \theta \frac{d\psi_R}{dt} + \cos (\theta - 120°) \frac{d\psi_y}{dt} + \cos (\theta + 120°) \frac{d\psi_B}{dt} + \psi_q \frac{d\vartheta}{dt} \right] \qquad (8.44)$$

In a similar way

$$\frac{d}{dt} \psi_q = -\frac{2}{3} \left[\sin \theta \frac{d\psi_R}{dt} + \sin (\theta - 120°) \frac{d\psi_y}{dt} + \sin (\theta + 120°) \frac{d\psi_B}{dt} \right] \qquad (8.45)$$

Eq. (8.42) reduces to

$$e_d = \dot{\psi}_d = \psi_q \dot{\theta} - r i_d$$

$$e_q = \dot{\psi}_q = \psi_d \dot{\theta} - r i_q \qquad (8.46)$$

$$e_0 = \dot{\psi}_0 - r i_0$$

Eq. (8.32) is the same as Eqs. (8.8) and (8.9).

8.6 STEADY-STATE EQUATION

In the steady state, the flux linkages are

$$\left.\begin{array}{l} \psi_d = -x_d i_d + x_{rfd} I_{fd} \\ \psi_{fd} = -x_{rfd} i_d + x_{ffd} I_{fd} \\ \psi_q = -x_q i_q \end{array}\right\} \tag{8.47}$$

the voltage relation for the field is then

$$e_{fd} = \dot{\psi}_{fd} + R_{fd} I_{fd}$$

In the steady state from Eq. (8.46) at $w = 1.0$ p.u.

$$e_d = -\dot{\psi}_d - ri_d (\dot{\psi}_d = 0) \tag{8.48}$$

$$e_q = -\dot{\psi}_d - ri_q (\dot{\psi}_d = 0) \tag{8.49}$$

$$\psi_d = x_{Rfd} I_{fd} - x_d i_d = \frac{x_{Rfd} e_{fd}}{R_{fd}} - x_d i_d$$

$$= E - x_d i_d \tag{8.50}$$

where E is the field excitation measured in terms of terminal voltage that it produces on open circuit at normal speed.

$$e_d = x_q i_q - ri_d \tag{8.51}$$

$$e_q = E - x_d i_d - ri_q \tag{8.52}$$

so that, neglecting resistances

$$i_d = \frac{E - e \cos \delta}{x_d} \tag{8.53}$$

and

$$i_q = \frac{e \sin \delta}{x_q} \tag{8.54}$$

where the open-circuit voltage E is ahead of the corresponding bus voltages by an angle δ.

8.7 STEADY-STATE VECTOR DIAGRAM

$$e_d = x_q i_q - ri_d \tag{8.55}$$

$$e_q = E - x_d i_d - ri_q \tag{8.56}$$

An voltage E_q is defined as

$$E_q = E - (x_d - x_q) i_d \tag{8.57}$$

Then Eq. (8.56) can be put

$$e_q = E_q - x_d i_q - r i_d \tag{8.58}$$

Defining

$$\left. \begin{aligned} \bar{e} &= e_d + j l_q \\ \bar{i} &= i_d + j i_q \end{aligned} \right\} \tag{8.59}$$

Then

$$\bar{e} = j E_q - (r + j x_q)\bar{i} \tag{8.60}$$

It may be noted that $\bar{e}$ is also the terminal voltage of the machine v_t.

E is the voltage behind excitation (Fig. 8.8).

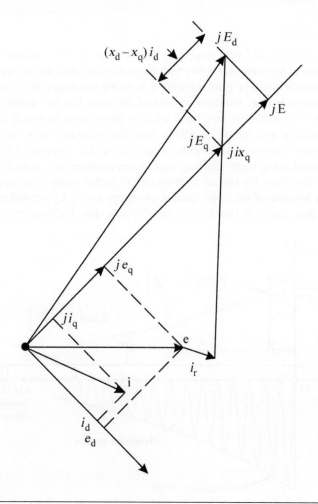

FIGURE 8.8

Synchronous generator in steady-state phasor diagram.

8.8 REACTANCES

Whenever a three-phase short circuit occurs at the terminals of an alternator, the current in the armature circuit increases suddenly to a large value and since the resistance of the circuit then is small compared to its reactance, the current is highly lagging and the p.f. is approximately zero. Due to this sudden switching, there are two components of currents.

1. a.c. component
2. d.c. component (decaying).

The oscillogram of current variation as a function of time is shown in Fig. 8.9 for a three-phase fault at the terminals of an alternator.

O_a—subtransient current
O_b—transient current
O_c—steady-state current.

The rotor rotates at zero speed with respect to the field due to a.c. component of current in the stator whereas it rotates at synchronous speed with respect to the field due to the d.c. component of current in the stator conductors. The rotor winding acts as the secondary of a transformer for which the primary is the stator winding. Similarly in case of the rotor that has damper winding fixed on its poles, the whole system will work as a three winding transformer in which stator is the primary and the rotor field winding and damper windings form the secondary of a transformer. It is to be noted that the transformer action is there with respect to the d.c. component of current only. The a.c. component of current being highly lagging tries to demagnetize, i.e., reduce the flux in air gap.

This reduction of flux from the instant of short circuit to the steady-state operation cannot take place instantaneously because of the large amount of energy stored by the inductance of the corresponding system. So this change in flux is slow and depends upon the time constant of the system.

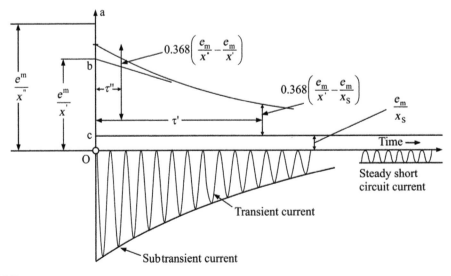

FIGURE 8.9

Analysis of symmetrical short circuit current.

In order to balance the suddenly increased demagnetizing m.m.f. of the armature current, the exciting current, i.e., the field winding current must increase in the same direction of flow as before the fault.

This happens due to the transformer action. At the same time, the current in the damper and the eddy currents in the adjacent metal parts increase in obedience to the Lenz's law, this assisting the rotor field winding to sustain the flux in the air gap. At the instant of the short circuit, there is mutual coupling between the stator winding, the rotor winding, and the damper winding, and the equivalent circuit is represented in Fig. 8.10A–C.

Since the equivalent resistance of the damper winding when referred to the stator is more as compared to the rotor winding, the time constant of damper winding t'' is smaller than the rotor field winding. Therefore, the effect of damper winding and the eddy current in the pole faces disappears after the first few cycles. Accordingly, the equivalent circuit after first few cycles reduces to the one shown in Fig. 8.10B. After a few more cycles depending upon the time constant of the field

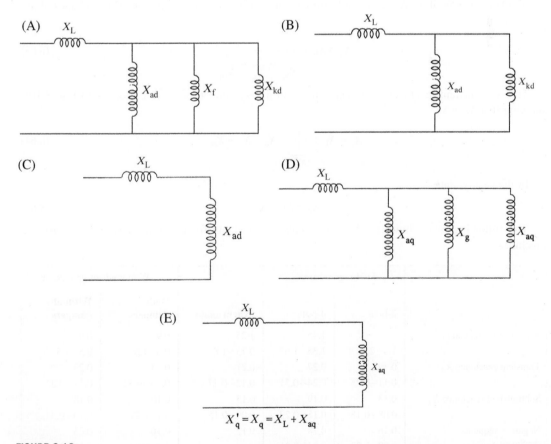

FIGURE 8.10

(A) Equivalent circuit for subtransient state (*d*- axis), (B) equivalent circuit for transient state (*d* axis), (C) equivalent circuit under steady state (*d* axis), (D) equivalent circuit subtransient state (*q* axis), and (E) equivalent circuit for transient state (*q* axis).

winding t' the effect of the d.c. component will die down and steady-state condition will prevail for which the equivalent circuit is shown in Fig. 8.10C. The inductance increases from the initial state to the final steady state.

i.e., synchronous reactance > transient reactance > subtransient reactance.

In the subtransient state

$$X_d'' = X_L + \cfrac{1}{\cfrac{1}{X_{ad}} + \cfrac{1}{X_f} + \cfrac{1}{X_{kd}}} \tag{8.61}$$

In the transient state

$$X_d' = X_L + \cfrac{1}{\cfrac{1}{X_{ad}} + \cfrac{1}{X_f}} \tag{8.62}$$

In the steady state, $X_d = X_L + X_{ad}$

In a similar way, if X_g represents a fictitious coil g to represent the effect of flux linkages along q-axis, then

$$X_q'' = X_L + \cfrac{1}{\cfrac{1}{X_{aq}} + \cfrac{1}{X_g} + \cfrac{1}{X_{kq}}} \tag{8.63}$$

This subtransient reactance X_q'' is very small. In a very short time circuit changes to Fig. 8.10E showing that $X_q = X_q'$

$$X_q' = X_L + \cfrac{1}{\cfrac{1}{X_{aq}}} = X_q = X_L + X_{aq} \tag{8.64}$$

In the steady state

$$X_q = X_L + X_{aq} \tag{8.65}$$

The following table shows the typical values of constants for different types of synchronous machines

	Turbo generator			Water wheel generator	
	2-pole	4-pole	Synchronous	With dampers	Without dampers
Synchronous reactance x_{sd}	2.0	1.45	1.25	0.9	0.9
	1.4–2.5	1.35–1.65	0.75–1.8	0.5–1.5	0.5–1.5
Transient reactance X_d'	0.19	0.27	0.21	0.23	0.23
	0.11–0.25	0.24–0.31	0.12–0.27	0.14–0.32	0.14–0.32
Subtransient reactance X_d''	0.13	0.19	0.13	0.16	0.18
	0.08–0.18	0.15–0.23	0.09–0.15	0.1–0.27	0.16–0.3
Negative-sequence reactance x_2	0.16	0.28	0.12	0.16	0.23
	0.09–0.23	0.24–0.31		0.1–0.27	0.12–0.37
Zero-sequence reactance x_0	0.08	0.28	0.03	0.08	0.08
	0.02–0.15	0.22–0.31		0.06–0.1	0.06–0.1

Continued					
	Turbo generator			**Water wheel generator**	
	2-pole	**4-pole**	**Synchronous**	**With dampers**	**Without dampers**
Inertia constant *H*	4.7	4.8	1.2	3.4	3.4
	2.6–6	4.3–5.7	0.7–1.8	2–5	2–5
Open-circuit time constant t_{ds}	9.5	5.5			
	3.5–16				

Note: *Average values are given in each block.*

8.9 EQUIVALENT CIRCUITS AND PHASOR DIAGRAMS

8.9.1 MODEL FOR TRANSIENT STABILITY

For study periods of the order of 1 s or less, the synchronous machine can be represented by a voltage behind a transient reactance that is constant in magnitude but changes its angular position. This is shown in Fig. 8.11.

The voltage relation is expressed by

$$E' = V_t + r_a I_t + jX' I_t \tag{8.66}$$

I_t = terminal current
v_t = machine terminal voltage. The phasor diagram is shown in Fig. 8.12.

This model can be extended to include the effect of saliency and the effect of field flux linkages considering the *d*- and *q*-axis quantities.

The direct axis is chosen as the center line passing through the machine pole and the quadrature axis 90° (clc) behind this axis. The position of the *q*-axis can be determined by calculating a fictitious voltage located on the *q*-axis called E_q (voltage behind *q*-axis reactance). The equivalent circuit and the phasor diagram are shown in Figs. 8.13 and 8.14.

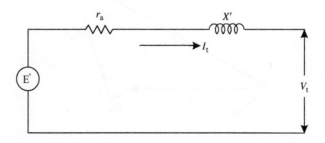

FIGURE 8.11

Equivalent circuit for transient stability study.

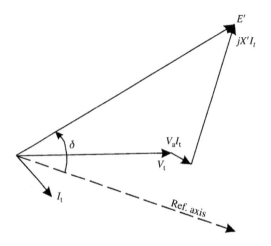

FIGURE 8.12

Phasor diagram for transient stability.

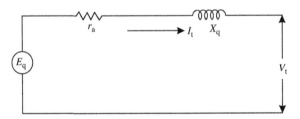

FIGURE 8.13

The q-axis equivalent circuit.

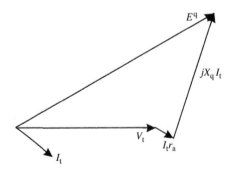

FIGURE 8.14

Phasor diagram with induced voltage along q axis.

It may be noted that the sinusoidal flux produced by the field current acts along the main or d axis. The voltage induced by the field current lags behind this flux by 90° and is therefore on the q axis. This voltage can be obtained by adding the voltage drop across r_a to v_t and the voltage drops representing the demagnetizing effects along the d and q axes.

8.10 TRANSIENT STATE PHASOR DIAGRAM

$$\left.\begin{array}{l} \psi_d = -x_d i_d + x_{Rfd}\, i_{fd} \\ \psi_{fd} = -x_{Rfd} i_d + x_{ffd}\, i_{fd} \end{array}\right\} \tag{8.67}$$

Eliminating the field current i_{fd}

$$i_{fd} = \frac{\psi_d + x_d i_d}{x_{Rfd}} \tag{8.68}$$

Again from Eq. (8.47)

$$\psi_{fd} = \frac{x_{ffd}}{x_{Rfd}}\left[\psi_d + \left(x_d - \frac{x_{Rfd}^2}{x_{ffd}}\right)i_d\right] \tag{8.69}$$

the quantity $\left(x_d - (x_{Rfd}^2/x_{ffd})\right)$ is a short circuit reactance of the armature direct axis circuit and is defined.

Direct axis transient reactance

$$x_d^| = x_d - \frac{x_{Rfd}^2}{x_{ffd}} \tag{8.70}$$

The quantity

$$\left(\frac{X_{Rfd}}{x_{ffd}}\right)\psi_d = E_d^| = \text{is the voltage behind transient reactance } x_d^|. \tag{8.71}$$

Hence, Eq. (8.71) becomes

$$E_q^| = \psi_d + x_d^|\, i_d \tag{8.72}$$

Since $e_q = Y_d - r i_q$

$$e_q = E_q^| - x_d^|\, i_d - r i_q \tag{8.73}$$

Again from Eq. (8.72)

$$E_q^| = \frac{x_{Rfd}}{x_{ffd}}\psi_{fd} = -\frac{x_{Rfd}^2}{x_{ffd}}i_d + E \tag{8.74}$$

$$= (x_d - x_d^|)i_d \tag{8.75}$$

The transient state phasor diagram is given in Fig. 8.15.
Also

$$E_q^| = E_q - \bar{I}(x_q - x_d^|) \tag{8.76}$$

which can be verified from the phasor diagram.

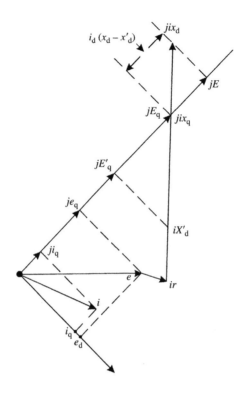

FIGURE 8.15

Transient state phasor diagram of synchronous machine.

8.11 POWER RELATIONS

Now, expressions for the real and reactive powers developed by the synchronous generator will be derived.

The phasor diagram for a salient pole synchronous machine neglecting the resistance is given in Fig. 8.16. The complex power generated is given by

$$S = P + jQ \tag{8.77}$$

$$= |e| \, |i| \cos \phi + j \, |e| \, |i| \sin \phi \tag{8.78}$$

From the figure

$$e \cos \delta = jE - i_d x_d \tag{8.79}$$

$$e \sin \delta = i_q x_q \tag{8.80}$$

$$i_d = i \sin \beta \tag{8.81}$$

$$i_q = i \cos \beta \tag{8.82}$$

where

$$\beta - \delta = \phi$$

$$\cos \phi = \cos(\beta - \delta) = \cos \beta \cos \delta + \sin \beta \sin \delta$$

$$\sin \phi = \sin(\beta - \delta) = \sin \beta \cos \delta + \cos \beta \sin \delta$$

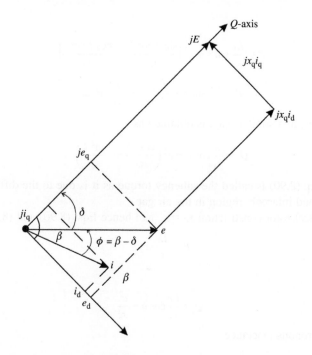

FIGURE 8.16

Salient pole synchronous machine resistance neglected.

Multiplying the above equation by i

$$i \cos \phi = i \cos \beta \cos \delta + i \sin \beta \sin \delta \tag{8.83}$$

and also

$$i \sin \phi = i \sin \beta \cos \delta - i \cos \beta + \sin \delta \tag{8.84}$$

utilizing Eqs. (8.80) and (8.81)

$$i \cos \phi = i_q \cos \delta = i_d \sin \delta \tag{8.85}$$

$$i \sin \phi = i_d \cos \delta = i_q \sin \delta \tag{8.86}$$

From Eq. (8.77)

$$S = e(i_q \cos \delta + i_d \sin \delta) + j\, e(i_d \cos \delta - i_q \sin \delta) \tag{8.87}$$

From Eqs. (8.78) and (8.79)

$$i_d = \frac{E - e \cos \delta}{x_d} \tag{8.88}$$

and

$$i_q = \frac{e \sin \delta}{x_d} \tag{8.89}$$

$$S = e\left[\frac{e \sin \delta}{x_q}\cos \delta + \frac{E - e \cos \delta}{x_d}\sin \delta\right] + je\left[\frac{E - e \cos \delta}{x_d}\cos \delta - \frac{e \sin \delta}{x_q}\sin \delta\right] \tag{8.90}$$

The real power

$$P = \frac{Ee}{x_d}\sin\delta + e^2\left[\frac{\sin\delta\cos\delta}{x_q} - \frac{\cos\delta\sin\delta}{x_d}\right]$$

$$P = \frac{Ee}{x_d}\sin\delta + \frac{e^2}{2}\left[\frac{1}{x_q} - \frac{1}{x_d}\right]\sin 2\delta \tag{8.91}$$

The reactive power, in a similar way, is obtained as

$$Q = \frac{Ee}{x_d}\cos\delta - e^2\left(\frac{\cos^2\delta}{x_d} + \frac{\sin^2\delta}{x_q}\right) \tag{8.92}$$

the second term in Eq. (8.90) is called the saliency torque as it is due to the difference in the reluctance of pole region and interpole region in the air gap.

In case of cylindrical rotor construction $x_d = x_q$ and hence Eqs. (8.90) and (8.91) will reduce to

$$P = \frac{Ee}{x_s}\sin\delta \tag{8.93}$$

and

$$Q = \frac{Ee}{x_s}\cos\delta - \frac{e^2}{x_s} \tag{8.94}$$

where x_s is the synchronous reactance.

8.12 SYNCHRONOUS MACHINE CONNECTED THROUGH AN EXTERNAL REACTANCE

Consider a synchronous generator connected to a bus through an external reactance x_1 (Fig. 8.17). The machine may have a synchronous reactance of x_s or d- and q-axis reactances x_d and x_q. The voltage phasor diagram of Fig. 8.16 will simply change into Fig. 8.18.

Thus it is possible to include the effect of an external reactance by considering the values in Fig. 8.19 in the following way.

$$\delta \to \delta_1$$
$$x_d \to x_d + x_1$$
$$x_q \to x_q + x_1$$

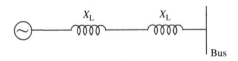

FIGURE 8.17

Synchronous machine with external reactance.

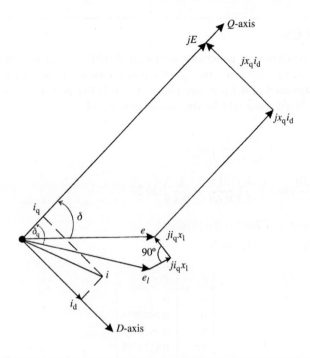

FIGURE 8.18

Effect of external reactance x_l.

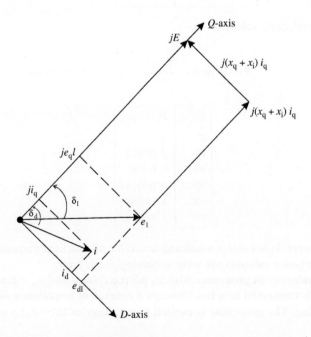

FIGURE 8.19

Equivalent phasor diagram with external reactance x_l.

WORKED EXAMPLES

E.8.1. A salient pole synchronous generator is operated with $E = 1.1$ p.u. and $e = 1.0$ p.u. (line). Given that $x_d = 0.8$ p.u. and $x_q = 0.6$ p.u. for each phase. The synchronous reactance is estimated at 0.7 p.u. per phase. What is the power developed. If the saliency is neglected what will be the power generated?

Ans.

$$P = \frac{Ev}{x_d}\sin\delta + \frac{e^2}{2}\left(\frac{1}{x_q} - \frac{1}{x_d}\right)\sin 2\delta$$

$$= \frac{(1.1)(1.0)}{0.8}\sin\delta + \frac{1}{2}\left(\frac{1}{0.65} - \frac{1}{0.8}\right)\sin 2\delta$$

$$= 1.375\sin\delta + (0.7692 - 0.625)\sin 2\delta$$

$$= 1.375\sin\delta + 0.1442\sin 2\delta$$

δ	P
0	0
10°	0.2880879
20°	0.5629663
30°	0.8123772
40°	1.0258305
50°	1.195259

Taking x_s, neglecting saliency

$$P = \frac{E_e}{x_s}\sin\delta = \frac{(1.1)(1.0)}{(0.7)}\sin\delta = 1.5714286\sin\delta$$

δ	P
0	0
10°	0.272875
20°	0.53746
30°	0.7857143
40°	1.0100829
50°	1.2037143

The two powers do not differ much and hence for all practical purposes saliency can be neglected in power calculations without causing much error.

E.8.2. Consider a synchronous generator with x_d 0.8 p.u./phase and $x_q = 0.6$ p.u./phase. The machine is connected to a bus through a reactor of impedance $j0.1$ p.u. on generator rating. The generator is excited to a voltage of $|E| = 1.2$ p.u. and delivers a

load of 10 MW to the bus at 1 p.u. voltage. The generator is rated at 15 MW with a line voltage of 13.6 kV.

Determine δ, δ_l, i, and e

Solution:

$$P = \frac{E\,e}{x_d + x_l}\sin\delta_l + \frac{e^2}{2}\left(\frac{1}{x_q + x_l} - \frac{1}{x_d + x_l}\right)\sin 2\delta_l$$

In per unit system $P = \frac{10\ MW}{15\ MW} = 0.0667$ p.u.

$$0.667 = \frac{1.2 \times 1.0}{0.8 + 0.1}\sin\delta_l + \frac{1.0^2}{2}\left(\frac{1}{0.6 + 0.1} - \frac{1}{0.8 + 0.1}\right)\sin 2\delta_l$$

$$= \frac{1.2}{0.9}\sin\delta_l + \frac{1}{2}(1.4285714 - 1.111111)\sin 2\delta_l$$

$$\delta_l = 24.4°$$

$$I_q = \frac{1.0(\sin 24.4°)}{0.6 + 0.1} = \frac{0.4131}{0.7} = 0.590$$

$$I_d = \frac{1.2 - 1.0\cos 24.4°}{0.8 + 0.1} = \frac{1.2 - 0.9107}{0.9} = \frac{0.2893}{0.9} = 0.3214$$

$$I = \sqrt{(0.3214)^2 + (0.59)^2}\tan^{-1}\left(\frac{0.3214}{0.59}\right)$$

$$= 0.67186\angle\tan^{-1}0.5447457$$

$$= 0.67186\angle - 28.58°$$

The terminal voltage $v_t = v + jx_1^1$

$$= 1.0\angle - 24.4° + j0.1 \times 0.67186\angle - 28.58$$

$$= (0.9106836 - j0.4131) + 0.03214 + j0.0589993$$

$$= 0.9428236 - j0.3541$$

$$= 1.0071264\angle - 20.58°$$

QUESTIONS

8.1. Explain the steady-state modeling of a synchronous machine

8.2. State and derive the swing equation

8.3. Explain the steady state and transient modes of generator modeling

8.4. Clearly explain how a synchronous generator is modeled for steady-state analysis. Draw the phasor diagram and obtain the power angle equation for a nonsalient pole synchronous generator connected to an infinite bus. Sketch the power angle curve

8.5. What is the advantage of transformation of phase quantities into d–q–0 components?

8.6. What is the significance of inertia constant?

8.7. In swing equation, what is the reactance used for computation of electrical power?

8.8. Sketch the oscillogram of synchronous machine currents on short circuit

8.9. From the oscillogram of Q. no. (8.8), explain the following quantities:

    1. subtransient reactance

    2. transient reactance

    3. steady-state reactance

8.10. Represent the reactances in Q. no. (8.9) by equivalent circuits for both direct and quadrate axes.

PROBLEMS

P.8.1. A synchronous generator is rated at 50 Hz, 22 kV, and 100 MVA. Based upon its own ratings, the reactances of the machine are $x_d = 1.00$ p.u. and $x_q = 0.65$ p.u.

    The machine is delivering 50 MW at 1.0 p.u. terminal voltage with an excitation of 1.4 p.u. Find δ and the reactive power generated.

P.8.2. A synchronous generator has $x_d = 0.95$ p.u. and $x'_d = 0.3$ p.u. on its own ratings. The excitation is maintained at 1.45 p.u. The generator delivers 0.7 p.u. MW to a bus at 1.0 p.u. voltage. If the generator voltage falls by 50% suddenly determine an expression for the power generated subsequently.

LINES AND LOADS

9

9.1 LINES

Transmission lines are important components in the formation of a power system. The generating station and the load centers are interconnected by transmission lines, operating at different levels of voltages constituting primary, secondary, and subtransmission lines. The distance and voltage level of the line determines the model. In the modeling of the lines, the line charging capacitance plays an important role and is dependent on the line voltage. In case of d.c. lines, the reactance is, however, not present.

For ease of modeling, the lines can be classified as short, medium, and long lines depending upon the length and the voltage level.

9.1.1 SHORT LINES

Lines up to 50—60 km and voltage not exceeding about 20 kV may be classified as short lines. In such lines, the line charging capacitance can be neglected. The resistance and reactance of the lines are considered as lumped parameters. The single phase equivalent circuit is shown in Fig. 9.1.

9.1.2 MEDIUM LINES

These are lines having a length between 50 and 150 km and voltages not exceeding 110 kV. In such lines, the line charging capacitance can be lumped either at both ends of the line or at the center of the line to give reasonably good analysis of the line. These models are called π and T type by virtue of the shape of the model (Fig. 9.2A and B).

9.1.3 LONG LINES

Lines operating at 110 kV or above and in length exceeding 150 km come under this category. In the analysis of such lines, treating the line charging capacitance as lumped value at one or two locations may give erroneous results.

The distributed nature of all the line parameters viz., resistance, inductance, and charging capacitance are considered through exact equations, see Fig. 9.3.

The voltage and current relations can be modeled through a general network model with A, B, C, D parameters. Consider Fig. 9.4.

Power Systems Analysis. DOI: http://dx.doi.org/10.1016/B978-0-08-101111-9.00009-4

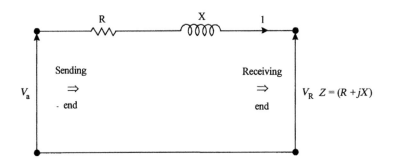

FIGURE 9.1

Short transmission line.

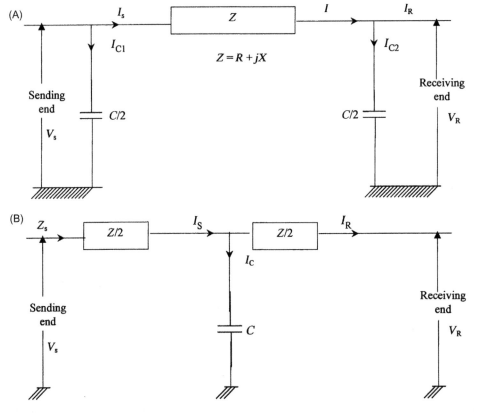

FIGURE 9.2

(A) Medium line—π equivalent. (B) Medium line—T equivalent.

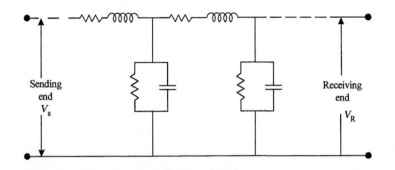

FIGURE 9.3

Long line with distributed parameters.

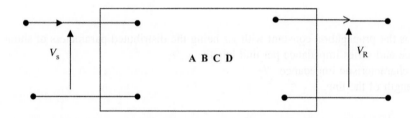

FIGURE 9.4

Two-part network model.

For the two-part network model, the following equations can be written:

$$V_s = AV_R + BI_R \qquad (9.1)$$

$$I_S = CV_R + DI_R \qquad (9.2)$$

For a short line:

$$A = 1; \quad B = Z; \quad C = 0; \quad D = 1$$

For a medium line:
Nominal T model

$$A = D = 1 + \frac{1}{2}YZ$$

$$B = Z + \frac{YZ^2}{4}$$

$$C = Y$$

Nominal π-model

$$A = 1 + \frac{YZ}{2} = D$$

$$B = Z$$

$$C = Y\left(1 + \frac{1}{4}YZ\right)$$

where Y is the admittance of the line changing capacitance branch.

Long lines

$$A = D = \cosh \gamma l$$

$$B = Z_c \sinh \gamma l$$

$$C = \frac{1}{Z_c} \sinh \gamma l$$

where

$\gamma = \sqrt{yz}$ is the propagation constant with y,z being the distributed parameters of shunt admittance and series impedance per unit length.
Z_c is the characteristic impedance $\sqrt{\frac{z}{y}}$
l is the length of the line.

9.2 TRANSFORMERS

Transformers operate with taps on lines. The tap setting will alter the line flows and the voltage. Modeling of different types of transformers with off nominal turns ratio is required in load flow solution.

Further, a phase shifting transformer may be present on the lines for control purpose. The exact modeling of these devices will be presented now.

9.2.1 TRANSFORMER WITH NOMINAL TURNS RATIO

Consider a transformer with turns ratio $a{:}1$. This can be represented as an ideal autotransformer in series with an admittance. Let $p-q$ represents the input and output buses of the transformer. The ideal autotransformer is shown between p and t buses, while the series admittance is shown between t and q (see Fig. 9.5).

$$I_{tq} = \text{current flowing from } t \text{ to } q$$

$$I_{tq} = (v_t - v_q)y_{pq} \tag{9.3}$$

The terminal current at p,

$$I_p = (V_t - V_q)\frac{Y_{pq}}{a} \tag{9.4}$$

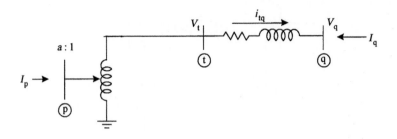

FIGURE 9.5

Transformer with taps.

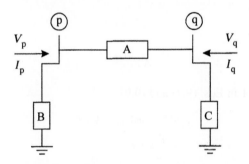

FIGURE 9.6

Equivalent π network for variable turn ratio transformer.

But

$$V_t = \frac{V_p}{a} \tag{9.5}$$

The terminal current at q is similarly

$$I_q = (V_q - V_t)y_{pq} \tag{9.6}$$

Substituting for

$$v_t \ I_q = \left(V_q - \frac{V_p}{a}\right) \ Y_{pq} = (av_q - v_p) \cdot \frac{y_{pq}}{a} \tag{9.7}$$

Now, let us consider an equivalent π-network mode "l" for the transformer as shown in Fig. 9.6. For the π-network

$$I_p = (V_p - V_q)A + V_p B \tag{9.8}$$

$$I_q = (V_q - V_p)A + V_q C \tag{9.9}$$

Let $V_p = 0$ and $V_q = 1$ in Eqs. (9.4) and (9.8).
Then

$$I_p = \frac{-Y_{pq}}{a} \quad \text{and} \quad I_p = -A$$

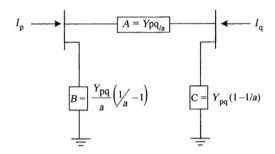

FIGURE 9.7

Equivalent π-network model.

So that,

$$A = \frac{Y_{pq}}{a} \tag{9.10}$$

Letting $E_p = 0$ and $E_q = 1$ in Eqs. (9.7) and (9.9)

$$I_q = Y_{pq} \quad \text{and} \quad I_q = A + C$$
$$= \frac{Y_{pq}}{a} + C$$

hence

$$C = \left(1 - \frac{1}{a}\right) Y_{pq} \tag{9.11}$$

Equating the currents in Eqs. (9.3) and (9.8) and substituting for A from equation

$$B = \frac{1}{a}\left(\frac{1}{a} - 1\right) Y_{pq} \tag{9.12}$$

Thus we obtain the equivalent π-model in terms of admittance and off-nominal turns ratio as shown in Fig. 9.7.

9.2.2 PHASE SHIFTING TRANSFORMERS

A phase shifting transformer can be represented by its impedance or admittance in series with an ideal autotransformer having a complex turns ratio as shown in Fig. 9.8.

$$\frac{V_p}{V_s} = a + jb \tag{9.13}$$

Since there is no power loss in an ideal autotransformer

$$V_p^* i_{pr} = V_s^* i_{sq} \tag{9.14}$$

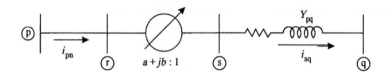

FIGURE 9.8

Phase shifting transformer.

i.e.,

$$\frac{i_{pr}}{i_{sq}} = \frac{V_s^*}{V_p^*} = \frac{1}{a - jb} \tag{9.15}$$

Also, $i_{sq} = (V_s - V_q)\,Y_{pq}$ and hence

$$i_{pr} = (V_s - V_q)\frac{Y_{pq}}{a - jb} \tag{9.16}$$

Substituting for V_s from Eq. (9.13)

$$i_{pr} = \left(\frac{V_p}{a + jb} - V_q\right)\frac{Y_{pq}}{a - jb} \tag{9.17}$$

$$\left[V_p - V_q(a + jb)\right]\frac{Y_{pq}}{a^2 + b^2} \tag{9.18}$$

similarly, we can prove that

$$i_{qs} = (V_q - V_s)Y_{pq} \quad \text{and substituting} \tag{9.19}$$

for V_s again from Eq. (9.13)

$$i_{qs} = \left[(a + jb)V_q - V_p\right] \cdot \frac{Y_{pq}}{a + jb} \tag{9.20}$$

To evaluate the constants, we shall substitute known boundary conditions into relevant equations.

Let $V_p = 0$; let all other buses be short circuited. The phase shifting transformer lies between buses p and q. The total bus admittance

$$Y_{PP} = i_{P1} + i_{P2} + \cdots + i_{Pr} + \cdots + i_{Pn} \tag{9.21}$$

where n is the number of buses connected to bus p.

Note: $I_p = \sum_{k=1}^{n} i_{pk} = V_p Y_{pp}$ and $V_p = 1.0$ p.u.

Hence

$$\left.\begin{array}{l} i_{p1} = Y_{p1} \\ i_{p2} = Y_{p2} \\ \overline{i_{pn} = Y_{pn}} \end{array}\right\} \tag{9.22}$$

and

$$i_{pr} = \frac{Y_{pq}}{a^2 + b^2} \tag{9.23}$$

from Eq. (9.17) with $V_p = 1.0$ and since all other buses are short circuited $V_q = 0$.

The current flowing out of bus p is $-i_{sq}$, the mutual admittance

$$y_{qp} = -i_{sq} \tag{9.24}$$

Then

$$y_{qp} = -i_{sq} = -(V_s - V_q)Y_{pq} \tag{9.25}$$

Since

$$V_q = 0$$

we obtain

$$y_{qp} = -Y_{pq}V_s \tag{9.26}$$

Similarly letting $V_q = 1.0$ p.u. and short circuiting all other buses, the self-admittance at bus q is

$$Y_{qq} = i_{q1} + i_{q2} + \cdots + i_{qs} + \cdots + i_{qn} \tag{9.27}$$

i.e.,

$$Y_{qq} = i_{q1} + i_{q2} + \cdots + i_{qp} + \cdots + i_{qn} \tag{9.28}$$

The current flowing out of bus p to bus q is given by

$$i_{pq} = +i_{pr} \tag{9.29}$$

Therefore, the mutual admittance

$$Y_{pq} = V_q \, i_{pr} = i_{pr} \tag{9.30}$$

Then

$$Y_{pq} = i_{pr} = (V_s - V_q)\frac{Y_{pq}}{a - jb} = \frac{-Y_{pq}}{a - jb} \tag{9.31}$$

The complex terms ratio $a + jb$ can be completed for a specified angular displacements and tap setting from

$$A + jb = A(\cos\theta + j\sin\theta) \tag{9.32}$$

where

$$|V_p| = A|V_s| \tag{9.33}$$

Thus all the required parameters are determined.

9.3 LOAD MODELING

The term load refers to a group of devices that consume energy from the supply lines. The loads may be of several types:

1. Motor loads, ranging in capacity from factional h.p. to hundreds of h.p. for a variety of drive applications
2. Heating, welding, and electrolysis applications
3. A large variety of electronic devices
4. Lighting for domestic, street lighting, and industrial purposes.

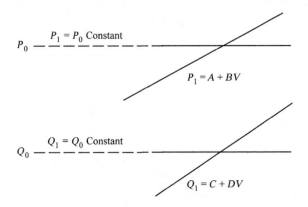

FIGURE 9.9

Voltage-dependent real and reactive powers.

Traction can be identified as a separate type of load, even though it is only the motors in traction that consume power.

The characteristics of major loads again are dependent upon the nature of the load and its susceptibility to various factors such as frequency and voltage and its variation with time. Further, the occurrence of load demand could be a regular feature or of random nature.

System loads viz. heating, welding, traction, industrial motors of a.c. and d.c. types, etc., are in general very different in nature. The real and reactive power characteristics of these loads differ from one another.

Further, the characteristics will be different under dynamic and static conditions of operation. The system loads change continuously and never remain static.

Consider Fig. 9.9. It is assumed that the active and reactive load demand changes linearly with the applied voltage or remains constant as the voltage changes.

On the other hand, load demand can be represented as function of voltage squared (Fig. 9.10).

This is equivalent to the assumption that the load is represented by a combination of fixed resistance and reactance.

The resistance and reactance connected in parallel are given by

$$R_l = \frac{V^2}{P_l} \text{ and } x_l = \frac{V^2}{Q_l} \tag{9.34}$$

For the series connection

$$Z_l = R_l + jx_l = \frac{V^2}{S} = \frac{V^2}{|S|}(\cos \varphi \pm j \sin \varphi)$$

Linear models can be used if the voltage and frequency fluctuation are small. It is usual to describe the steady-state condition of the system as quasi-static state. In general, it is possible to describe the static characteristics for composite loads by

$$\Delta P = \frac{\partial P}{\partial V} \Delta V + \frac{\partial P}{\partial f} \Delta f \tag{9.35}$$

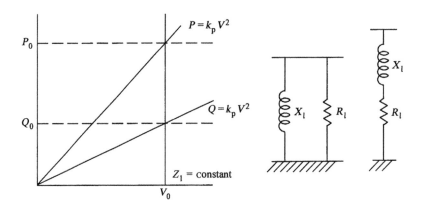

FIGURE 9.10

Constant impedance load models.

and

$$\Delta Q = \frac{\partial Q}{\partial V}\Delta V + \frac{\partial Q}{\partial f}\Delta f \tag{9.36}$$

where ΔP and ΔQ are the variations in real and reactive power demands. The system voltage and system frequency are the variables and

$$\frac{\partial P}{\partial V}; \quad \frac{\partial P}{\partial f}; \quad \frac{\partial Q}{\partial V} \quad \text{and} \quad \frac{\partial Q}{\partial f}$$

are called regulation coefficients.

For composite loads, the following are typical values for the range of regulation coefficients.

$$\frac{\partial P}{\partial V} = 1.5 \text{ to } 3.5$$

$$\frac{\partial P}{\partial f} = 1.5 \text{ to } 3$$

$$\frac{\partial Q}{\partial V} = 0.3 \text{ to } 0.75$$

$$\frac{\partial Q}{\partial f} = 1 \text{ to } 6$$

Another way of describing the system loads, where frequency does not influence much, is given by exponential representation.

For 80%−120% variation of voltage around its nominal value, the voltage dependence of composite loads can be represented by the relationships

$$P = \left(\frac{V}{V_{\mathrm{n}}}\right)^{\alpha} P_{\mathrm{N}} \tag{9.37}$$

and

$$Q = \left(\frac{V}{V_n}\right)^\beta Q_N \qquad (9.38)$$

where P and Q are the real and reactive powers V and V_n are the actual and nominal voltage magnitudes, and P_N and Q_N are the real and reactive nominal powers.

From measurements taken on networks from several countries such as United States, Sweden, Germany, and Poland, it is found that the exponents α and β lie in the range.

$$\alpha = 0.6 \text{ to } 1.4$$

and

$$\beta = 1.5 \text{ to } 3.2$$

With such a representation, which is adequate for most of the practical cases, where frequency does not influence much (e.g., load flow studies), an average value of $\alpha = 1$ and $\beta = 2$ works out satisfactorily.

This assumption means that the active component of current remains constant and reactance of load remains constant.

Another way of dealing with loads for specific situations involves in dividing the loads into

1. Constant current model
2. Constant impedance model
3. Constant power model.

9.3.1 CONSTANT CURRENT MODEL

Here, only voltage is assumed to vary while frequency stays fixed. The variation of system voltage does not affect the current drawn by the load so that

$$I_1 = I_2 \qquad (9.39)$$

The power at voltage $V_1 = P_1 = V_1 I_1 \cos \varphi_1$
and the power at voltage $V_2 = P_2 = V_2 I_2 \cos \varphi_2$
Substituting Eq. (9.39): $I_1 = \frac{P_1}{V_1 \cos \varphi} = I_2 = \frac{P_2}{V_2 \cos \varphi_2}$

$$P_2 = P_1 \frac{V_2 \cos \phi_2}{V_1 \cos \phi_1}$$

load power factor remaining the same

$$P_2 = \frac{P_1 V_2}{V_1}$$

which means that $P \propto V$
In a similar fashion

$$Q_1 = V_1 I_1 \sin \varphi_1$$

and

$$Q_2 = V_2 I_2 \sin \varphi_2$$

$$I_1 = I_2 \text{ yields } Q_2 = Q_1 \frac{V_2 \sin \phi_2}{V_1 \sin \phi_1}$$

With constant power factor

$$Q \propto V$$

For implementation of this type of load in system studies, the initial value for the current is obtained from

$$I_{P0} = \frac{P_{LP} - jQ_{LP}}{V_P^*} \tag{9.40}$$

where P_{LP} and Q_{LP} are scheduled bus real and reactive load powers at bus P. V_P is the voltage at bus P and I_{PO} is the current that flows from bus P to ground, i.e., to bus "0." The magnitude and power factor angle of I_{PO} remains constant.

9.3.2 CONSTANT IMPEDANCE MODEL

In this case, the impedance of the load remains constant

$$z = \frac{V}{I} = \text{constant}$$

$$P_1 = V_1 I_1 \cos \varphi_1 \tag{9.41}$$

i.e.,

$$P_1 = Z I_1^2 \cos \varphi_1 \text{ and similarly } P_2 = Z I_2^2 \cos \varphi_2$$

or

$$P_1 = \frac{V_1^2}{Z} \cos \phi_1 \text{ and } P_2 = \frac{V_2^2}{Z} \cos \varphi_2 \tag{9.42}$$

$$P_2 = P_1 \left(\frac{I_2}{I_1}\right)^2 \cdot \frac{\cos \phi_2}{\cos \phi_1} \tag{9.43}$$

$$= P_1 \left(\frac{V_2}{V_1}\right)^2 \cdot \frac{\cos \phi_2}{\cos \phi_1} \tag{9.44}$$

for constant power factor

$$P \propto V^2 \tag{9.45}$$

In a similar manner

$$Q \propto V^2 \tag{9.46}$$

If the load is represented at any bus p by a static admittance then we have

$$(V_p - V_0)Y_{p0} = I_{p0} \tag{9.47}$$

where V_{p0} is the ground or reference voltage (say, equal to zero), V_p is the bus voltage, and Y_{p0} is the constant admittance

$$y_{p0} = \frac{I_{p0}}{V_p} \tag{9.48}$$

Multiplying the numerator and denominator by V_p^*

$$Y_{p0} = \frac{I_{p0}V_p^*}{V_p V_p^*} = \frac{P_{LP} - jQ_{LP}}{\left(v_p^| + jv_p^{||}\right)\left(v_p^| - jv_p^{||}\right)} \tag{9.49}$$

Where $v_p^| + jv_p^{||} = v_p$ and P_{LP} and Q_{LP} are the load powers at bus p.

$$Y_{p0} = \frac{P_{LP}}{v_p^{|\,2} + v_p^{||\,2}} - \frac{jQ_{LP}}{v_p^{|\,2} + v_p^{||\,2}} = G_{p0} - jB_{p0} \tag{9.50}$$

where

$$G_{p0} = \frac{P_{LP}}{v_p^{|\,2} + v_p^{||\,2}} \tag{9.51}$$

and

$$B_{p0} = \frac{Q_{LP}}{v_p^{|\,2} + v_p^{||\,2}} \tag{9.52}$$

Note that the constant impedance load at bus P is given by $Z_P = R_p + jX_p$ and the admittance is then $Y_p = G_p - jB_p$.

9.3.3 CONSTANT POWER MODEL

For power invariance

$$P_1 = P_2$$

and

$$Q_1 = Q_2$$

i.e.,

$$V_1 I_1 \cos \varphi_1 = V_2 I_2 \cos \varphi_2$$

and

$$V_1 I_1 \sin \varphi_1 = V_2 I_2 \sin \varphi_2$$

Constant power load is either equal to the scheduled real and reactive power (base powers) or it may be a percentage of specified values in case of combined representation.

9.4 COMPOSITE LOAD

A typical composite load may have the following composition:

Small induction motors	35%
Large induction motors	15%
Lighting load	25%
Rectifiers, inverters, heaters	8%
Synchronous motors and other load	10%
Network losses	7%

Example: Lighting Load: Incandescent lamp power consumption is independent of frequency and varies with voltage as $V^{1.6}$. Such a load consumes no reactive power. For fluorescent lamps, the active consumption changes very little with voltage but reduces 0.5%–0.8% with a change of 1% frequency.

9.4.1 DYNAMIC CHARACTERISTICS

These are similar to static characteristics but are valid for fast variations, in the operating conditions. Hence, the rate of change of the variables becomes important then

$$P = \varphi\left[v, f, \frac{dv}{dt}, \frac{df}{dt}, \frac{d^2v}{dt^2}, \frac{d^2f}{dt^2}, \ldots\right]$$
(9.53)

They are to be evaluated for composite loads from field studies only.

9.5 INDUCTION MACHINE MODELING

Induction Motor as Simple Impedance

In most of the power system studies, the system loads are represented by shunt impedances. We have already considered the changes in real and reactive power of these shunt impedances with changes in system voltage and frequency. An approximate representation to analyze its steady-state behavior is shown in Fig. 9.11.

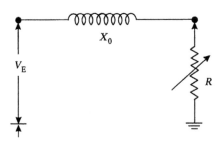

FIGURE 9.11

Simple equivalent circuit for induction motor.

Before any disturbance takes place, X_0 and R are set to motor terminal conditions. During the disturbance, X_0 is kept fixed and R is varied to maintain constant power.

The exact equivalent circuit of the induction motor shown in Fig. 9.12 does not include the core loss.

Letting

$$N_s = \text{synchronous speed of the motor}$$

$$= \frac{120f}{p}$$

and

$$N = \text{actual rotor speed}$$

the slip

$$S = \frac{N_s - N}{N_s}$$

The two resistances on the rotor side may be combined as in Fig. 9.13.

In the above equivalent circuit diagram:

R_S = Stator resistance per phase
X_s = Stator reactance per phase

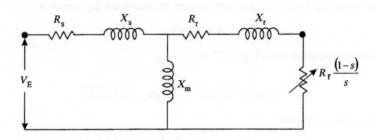

FIGURE 9.12

Exact equivalent circuit for induction motor.

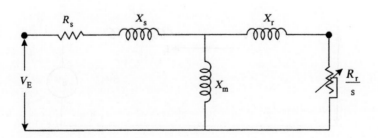

FIGURE 9.13

Modified equivalent circuit for induction motor.

R_r = Rotor resistance per phase
X_r = Rotor reactance per phase
V_T = Terminal voltage
I_T = Terminal or motor input current
X_m = Magnetizing reactance.

9.6 MODEL WITH MECHANICAL TRANSIENTS

If the system disturbance is such that the induction motor is partially or completely interrupted from its power source due to a short circuit, the time constant that matters is rotor time constant. The value of effective rotor time constant of an induction motor is dependent on the external system parameters in addition to motor constants. If this is small, then both stator and rotor transients can be neglected. In all cases where this is not negligible we have to use the transient equations. It is usual even here to neglect the stator transients and consider only the rotor transients. The equivalent circuit is shown in Fig. 9.14 for such an analysis.

X' = transient reactance
E' = voltage behind transient reactance.

R_r, the rotor resistance is small compared to the reactances and hence is neglected. From the equivalent circuit shown in Fig. 9.13, the open circuit reactance of the motor is

$$X = X_s + X_m$$

The blocked rotor reactance from Fig. 9.15 is

$$X_s + \frac{1}{(1/X_m) + (1/X_r)} = X_s + \frac{1}{(X_m + X_r)/X_m X_r}$$

which is also the transient reactance X'.

$$X' = X_s + \frac{X_m X_r}{X_m + X_r}$$

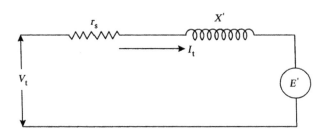

FIGURE 9.14

Equivalent circuit for mechanical transients.

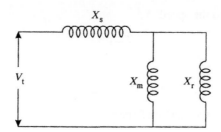

FIGURE 9.15

Circuit for rotor blocked induction motor.

Consider the transient condition with changing speed. The rotor induced e.m.f. SE′ changes at the rate $\frac{\partial}{\partial t} sE'$. It is well known that the partial derivative of a field quantity can be replaced by j.ω. in frequency analysis.

Hence the transient-induced voltage in the rotor is j.ω.s. $E^{l} = j\, 2\pi f\, sE'$.

Again, if the reactance changes from X to X' the rate of change of voltage $= j\,(X - X')I_{t}/T^{0}$ where I_{t} is the current in the circuit and T^{0} the rotor open circuit time constant. This is changing voltage under the transient condition with voltage behind transient reactance E'. The net rate of change of voltage on the rotor side is $(1/T^{0})[E' - j(X - X')I_{t}]$.

Thus the differential equation describing the rate of change of voltage is

$$\frac{dE'}{dt} = -j2\pi fsE' - \frac{1}{T_{0}}[E' - j(X - X')I_{r}] \tag{9.54}$$

It may be noted that

$$I_{t} = \frac{V_{t} - E'}{r_{s} + jx'} \tag{9.55}$$

as given by circuit diagram in Fig. 9.14. Thus the complete equivalent circuit which represents the induction motor under transient conditions can be identified.

9.6.1 POWER TORQUE AND SLIP

The active power consumed by an induction motor and hence the electromagnetic torque developed by the motor are determined by the mechanical power required for the driven mechanism and by its characteristics. The torque characteristic can be represented by

$$T_{mech} = f(\omega)$$

The mechanical system can have three basic types of torque characteristics:

1. The torque is constant or independent of speed

$$T_{mech} = \text{const}$$

   therefore

$$P_{mech} \propto \omega$$

2. The torque is proportional to the speed

$$T_{mech} \propto \omega$$

therefore

$$P_{mech} \propto \omega^2$$

3. The torque is proportional to the square of speed

$$T_{mech} \propto \omega^2$$

Hence

$$P_{mech} \propto \omega^3$$

Hoisting crane motors, nonreversible rolling mill motors, piston compressor motors, crushers have constant torque characteristics (Fig. 9.16). Motors used in textile industry for calendar mills have linear characteristics. Motors driving centrifugal pumps have power law, i.e., $T \propto \omega^2$.

$$T_{mech} = T = \frac{P}{\omega_s} = \frac{I^2 R_2}{\omega_s s} = constant$$

If

$$\omega_s = 1 \text{ p.u.}$$

then

$$T = \frac{I^2 R_2}{s} \quad \text{or} \quad s = \frac{I^2 R_2}{P}$$

which means that $s \propto I^2$.

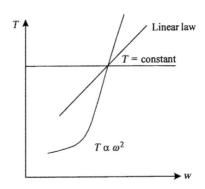

FIGURE 9.16

Variation of torque with frequency.

9.6.2 REACTIVE POWER AND SLIP

From the equivalent circuit in Fig. 9.12, the reactive power consumed by the motor has two components:

1. Q_m, the reactive power in the magnetizing branch
2. Q_s, the component that depends upon the stator and rotor leakage.

$$Q = Q_m + Q_s$$

$$Q_m = \frac{V^2}{X_m} = I_m V$$

where I_m is the no-load current while $Q_s = I^2 x_s$ and V is the applied voltage.

In fact x_m decreases with saturation as shown in Fig. 9.17. Hence, the relation between Q_m and V may not be square law.

The active or real power developed by the motor.

$$P = I^2 \frac{R_2}{s} = \frac{V^2 R_2}{\left[(R_2/s)^2 + x_s^2\right]s}$$

$$= \frac{V^2 R_2 s}{R_2^2 + (x_s s)^2} \tag{9.56}$$

The plots of the above equation(9.56) showing the relation between P and s on one hand s and V on the other hand are shown in Fig. 9.18. The latter curve is derived from the $P = f(s)$ curve correlating V and s. It may be seen that each motor has a critical condition corresponding to which there are critical voltage V_{critical} and critical slip S_{Cr}. During this critical condition, the maximum power developed by the motor is equal to the mechanical power required by the drive. For a voltage of value less than V_{critical}, operation is not feasible as the electrical torque developed will be less than the mechanical torque required.

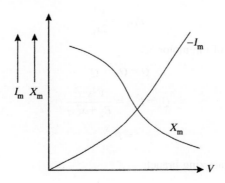

FIGURE 9.17

Magnetizing current with voltage variation.

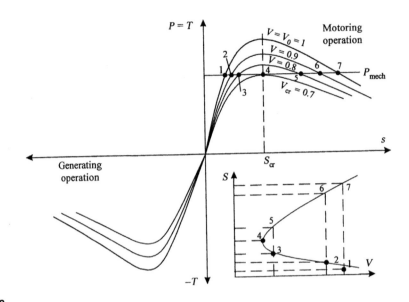

FIGURE 9.18

Variation of power and slip with voltage of operation.

Differentiating Eq. (9.56) with respect to s and equating the result to zero

$$\frac{dp}{ds} = V^2 R_2 \frac{R_2^2 - x_s^2 s^2}{\left(R_2^2 + x_s^2 s^2\right)^2} = 0 \tag{9.57}$$

given

$$s = s_{cr} = \frac{R_2}{X_s} \quad \text{(see Fig. 9.14)} \tag{9.58}$$

the maximum power corresponding to this slip is given by

$$P_{max} = \frac{V^2}{2x_s} \tag{9.59}$$

Again, the reactive power of the motor

$$Q = Q_m + Q_s$$

$$Q_s = I^2 x_s = \frac{V^2 s^2 x_s^2}{R_2^2 + x_s^2 s_2} \tag{9.60}$$

$$= P \frac{s}{s_{cr}} \tag{9.61}$$

the reactive power in the magnetizing branch

$$Q_m = \frac{V^2}{x_m} \tag{9.62}$$

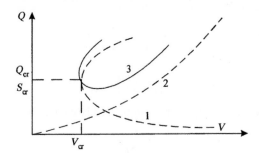

FIGURE 9.19

Variation of reactive power with voltage in induction motor.

The total reactive power as a function of voltage is shown in Fig. 9.19.

Curve (1) shows Q_s as a function of V. Curve (2) shows Q_m as a function of V^2, curve (3) is the resultant curve X.

At

$$V = V_{cr}$$

$$\frac{dQ}{dV} = -\infty$$

or $\frac{dV}{dQ} = 0$. This is the same point as (4) in Fig. 9.18, corresponding to $dP/ds = 0$. The motor operation till this point is stable.

Expression for critical slip and the maximum torque P_{max} can be obtained by differentiating

$$P = \frac{V^2 R^2 s}{R_2^2 + (Sx_3)^2}$$

with respect to s

$$\frac{dP}{ds} = 0$$

gives

$$s_{cr} = \frac{P_2}{x_s}$$

For this slip, the maximum power $P_{max} = \frac{V^2}{2X_s}$.

It is convenient to express the critical slip in terms of rated slip s_0 and the relative maximum torque $b_0 = P_m/P_0$

Then

$$s_{cr} = s_0\left(b_0 + \sqrt{b_0^2 - 1}\right)$$

The value of s_0 depends upon motor capacity. For low power motor of the order of 5 hp s_0 amounts to 5%–6%. For medium power motor (from 10 hp to 50 hp), it is about 2%–4% and for very large motors (100 hp) s_0 falls to 1%–1.5%.

Hence

$$s_{cr} = \frac{R_2}{x_S} \text{ and } P_m = \frac{V^2}{2x_s}$$

$$P = \frac{2P_m}{(s/s_{cr}) + (s_{cr}/s)}$$

At the synchronous speed, $\omega_0 = 1$ (power = torque in relative units)
i.e.,

$$T = \frac{2T_m}{(s/s_{cr}) + (s_{cr}/s)}$$

The mechanical power developed at the shaft

$$P_{max} = T(1 - s) = P(1 - s)$$

Taking into account the stator losses, the power consumed $= P_1 = P + I^2R$.

During transients, an induction motor may operate as a generator ($s < 0$) and as a brake ($s > 1$). The three possible modes of operation are shown in Fig. 9.20.

In electrical system, large fluctuations of frequency and voltage may occur. Frequency fluctuations are generally caused by the imbalance between the electrical power of the generators and the mechanical output of the generators and the mechanical output of their prime movers. These fluctuations change the reactive power consumed by the load as well as active and reactive power losses in the supply network and consequently cause changes in the supply voltage. A decrease in frequency generally results in voltage reduction.

Let us consider the case, where the torque and reactive power vary due to simultaneous change in voltage and frequency.

Let the motor shaft torque T_{shaft} remains constant. The power consumed by the motor varies in proportion to frequency according to the relation.

$$P = \omega T.$$

A decrease in frequency causes a decrease in slip that can be determined from

$$T = \frac{V^2 R_2 S}{\left[R_2^2 + \left[(x_{s0}/\omega_0)\omega s \right]^2 \right]}$$

where $x_{s0} = x_s$ at $\omega = \omega_0$ the synchronous frequency.
with T kept constant
$s = f(\omega)$ can be represented by a straight line
i.e.,

$$s \propto f$$

However, with decreasing frequency, the critical slip increases since

$$s_{cr} = \frac{R_2\omega_0}{\omega x_{s0}}$$

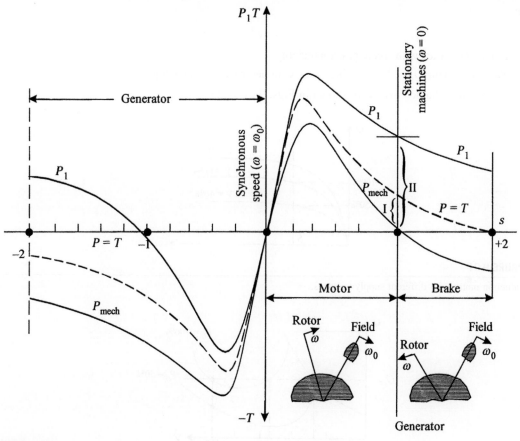

FIGURE 9.20

Generating and braking mode operation of induction motor.

The maximum value of T also increases giving better stability.

The reactive power of motor depends upon frequency

$$Q = j(\omega)$$

To study the effect of f on Q, the components Q_s and Q_μ are to be analyzed separately. Let

$$s = \frac{I^2 R_2}{\omega T}$$

so that $I^2 \propto \omega s$

$$Q_s = I^2 \frac{x_{s0}}{\omega_0} \omega$$

i.e.,

$$Q_s \propto \omega^2 s$$

Hence, Q_s varies with frequency appreciably

The components $Q_\mu = \frac{V^2 \omega_0}{x_{\mu 0} \omega} \propto \frac{1}{f}$ increases with decrease of frequency.

The various characteristics are shown in Figs. 9.21−9.23.

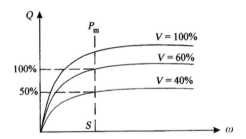

FIGURE 9.21

Induction motor with different supply voltages.

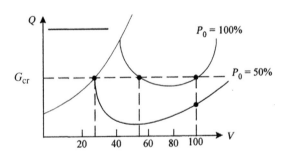

FIGURE 9.22

Induction motor $Q-V$ characteristics.

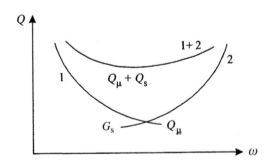

FIGURE 9.23

Variation of reactive power Q with change in frequency.

9.6.3 **SYNCHRONOUS MOTOR**

The static and dynamic characteristics of synchronous motors are determined from the expressions for torque and power.

When the applied voltage to the motor is changed, power and torque of the motor changes in proportion to the voltage.

$$P = \frac{VE_q}{x_d} \sin \delta \ \ and \ \ T = \frac{P}{\omega_0}$$

This is true if the stator losses are not taken into account.

The synchronous motor usually operates with a load angle of $\delta = 25-30°$. It has a large overload capacity of $2-2.5$.

Neglecting saturation, the magnitude of the torque is proportional to rotor current.

9.7 **RECTIFIERS AND INVERTER LOADS**

The active and reactive components of converter and inverter loads change with voltage and depend upon the control used. The output power depends on d.c. system parameters and firing angles. Furnace loads using converters supplying an electrolyzer have characteristics similar to induction motors. Different types of furnaces have different characteristics. The arc furnace consumes purely active power which is proportional to square of the applied voltage. The characteristics $P = f(v)$ is similar to that of a lighting load.

9.7.1 **STATIC LOAD MODELING FOR LOAD FLOW STUDIES**

Of the various models suggested for load representation under steady-state conditions, the exponential relationship appears to be the most suitable for computational purposes. In a region of 80%−120% of nominal voltage, the voltage dependence of composite system loads can be expressed by the relationship.

$$P = \left(\frac{V}{V_n}\right)^a P_n \ \ (\text{for real power}) \tag{9.63}$$

$$Q = \left(\frac{V}{V_n}\right)^b Q_n \ \ (\text{for reactive power}) \tag{9.64}$$

where the suffix n denotes nominal values.

9.7.2 **VOLTAGE DEPENDENCE OF EQUIVALENT LOADS**

System loads in large interconnected power networks are reduced to equivalent loads at few important nodes. This reduces the entire system to a simplified network with fewer nodes.

If this is not done, hundreds of radial lines from major buses will have to be treated as independent branches resulting in a large increase in the dimension of the problem (Fig. 9.24) refers to such a line.

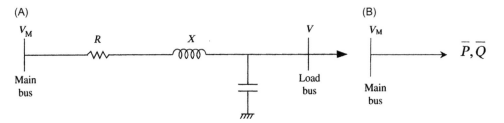

FIGURE 9.24

(A) Main bus and load bus. (B) Equivalent load representation.

The load at the end of the radial line $P + jQ$ is to be replaced by equivalent load at the bus by $\overline{P} + j\overline{Q}$. R and X are the resistance and reactance of the line from the main bus to the load bus; X_c is the capacitive reactance of the line including any capacitance used for reactive power compensation on the line.

Let

$$\overline{P} = A_0 + A_1 \left(\frac{V}{V_n}\right)^{k_1} + A_2 \left(\frac{V}{V_n}\right)^{k_2} \tag{9.65}$$

$$\overline{Q} = B_0 + B_1 \left(\frac{V}{V_n}\right)^{k_3} + B_2 \left(\frac{V}{V_n}\right)^{k_4} \tag{9.66}$$

where A_0, A_1, A_2 and B_0, B_1, B_2 are constants and V_n is the nominal voltage.

9.7.3 DERIVATION FOR EQUIVALENT LOAD POWERS

For the constant power model, the equivalent powers as indicated in Eqs. (9.65) and (9.66) can be derived as follows:

$$\overline{P} = P + I^2 R$$

$$= P + \left(\frac{P}{V}\right)^2 R + \left(\frac{Q}{V} - \frac{V}{X_c}\right)^2 R \tag{9.67}$$

$$= P - \frac{2Q}{X_c} R + \frac{P^2 + Q^2}{V_n^2} R \left(\frac{V}{V_n}\right)^{-2} + \frac{V_n^2}{X_c} \frac{R}{X_c} \left(\frac{V}{V_n}\right)^2$$

$$\overline{Q} = Q + \left(\frac{P}{V}\right)^2 X + \left(\frac{Q}{V} - \frac{V}{X_c}\right)^2 X - \left(\frac{V}{X_c}\right)^2 X_c$$

$$= Q - 2Q\frac{X}{X_c} + \frac{P^2 + Q^2}{V_n^2} X \left(\frac{V_n}{V}\right)^2 - \frac{V_n^2}{X_c}\left(1 - \frac{X}{X_c}\right)\left(\frac{V}{V_n}\right)^2 \tag{9.68}$$

$$= \left(Q - 2Q\frac{X}{X_c}\right) + \frac{P^2 + Q^2}{V_n^2} X \left(\frac{V}{V_n}\right)^{-2} - \frac{V_n^2}{X_c}\left(1 - \frac{X}{X_c}\right)\left(\frac{V}{V_n}\right)^2$$

If the load model is represented by the exponential relationships of Eqs. (9.63) and (9.64) with $a = 1$ and $b = 2$, then the equivalent powers can be derived as follows.

Let I_w and I_r be the watt component and reactive component of the live current I, respectively. Then

$$\bar{P} = P + I^2 R$$

$$= P + (I_w^2 + I_r^2)R$$

$$= I_w V + I_w^2 R + \left(\frac{V}{X_L} - \frac{V}{X_c}\right) R \tag{9.69}$$

$$= I_w^2 R + I_w V_n \left(\frac{V}{V_n}\right) + \left(\frac{V_n^2}{X_L} - \frac{V_n^2}{X_c}\right)\left(\frac{R}{X_L} - \frac{R}{X_c}\right)\left(\frac{V}{V_n}\right)^2$$

$$\bar{Q} = Q + I^2 X$$

$$= I_r V + \left(\frac{V}{X_L} - \frac{V}{X_c}\right)^2 X + I_w^2 X$$

$$= \left(\frac{V}{X_L} - \frac{V}{X_c}\right) V + \left(\frac{V_n}{X_L} - \frac{V_n}{X_c}\right)^2 \left(\frac{V}{V_n}\right)^2 X + I_w^2 \tag{9.70}$$

$$= \left(\frac{V_n^2}{X_L} - \frac{V_n^2}{X_c}\right)\left(\frac{V}{V_n}\right)^2 - \left(\frac{V_n}{X_L} - \frac{V_n}{X_c}\right)\left(\frac{V}{V_n}\right)^2 X + I_w^2 X$$

$$= I_w^2 R + I_w V_n \left(\frac{V_n^2}{X_L} - \frac{V_n^2}{X_c}\right)\left[1 + \left(\frac{X}{X_L} - \frac{X}{X_c}\right)\right]\left(\frac{V}{V_n}\right)^2$$

For a simple exponential relationship of the form indicated by Eqs. (9.63) and (9.64)

$$\bar{P} = (A_0 + A_1 + A_2)V_M^a \tag{9.71}$$

$$\bar{Q} = (B_0 + B_1 + B_2)V_M^b \tag{9.72}$$

The exponents can be determined from the approximate expressions

$$a = \frac{\ln \dfrac{A_0 + A_1 V_b^{k_1} + A_2 V_b^{k_2}}{A_0 + A_1 + A_2}}{\ln V_M} \tag{9.73}$$

$$b = \frac{\ln \dfrac{B_0 + B_1 V_b^{k_3} + B_2 V_b^{k_4}}{B_0 + B_1 + B_2}}{\ln V_M} \tag{9.74}$$

where

$$V_b = \frac{V}{V_n}$$

For a wide range of line impedance and shunt capacitance values that exist in practical networks, the exponents a and b are found to be very near to 1 and 2 for real and reactive powers, respectively.

The Coefficients in Eqs. (9.64) and (9.65)

Load characteristics with exponents	Load characteristics with exponents
$a = b = 0$	$a = 1; b = 2$
$A_0 = P - \dfrac{20}{X_c} R$	$A_0 = I_w^2 R$
$A_1 = \dfrac{P^2 + Q^2}{V^2} R; \ k_1 = -2$	$A_1 = I_w V_n; \ k_1 = 1$
$A_2 = \dfrac{V_n^2}{X_c^2} R; \ k_2 = 2$	$A_2 = \left(\dfrac{V_n^2}{X_L} - \dfrac{V_n^2}{X_c}\right)\left(\dfrac{R}{X_L} - \dfrac{R}{X_c}\right); \ k_2 = 2$
$B_0 = Q\left(1 - 2\dfrac{X}{X_c}\right)$	$B_0 = I_w^2 X$
$B_1 = \dfrac{P^2 + Q^2}{V^2} X; \ k_3 = -2$	$B_1 = 0; \ k_3 = 0$
$B_2 = \dfrac{-V_n^2}{X_c}\left(1 - \dfrac{X}{X_c}\right); \ k_4 = 2$	$B_2 = \left(\dfrac{V_n^2}{X_L} - \dfrac{V_n^2}{X_c}\right)\left[1 + \left(\dfrac{X}{X_L} - \dfrac{X}{X_c}\right)\right]; \ k_4 = 2$

WORKED EXAMPLES

E.9.1. **An inductive load is subjected to a voltage variation of 1%. By how much percent will the real load change?**

Solution:

Complex Power

$$S = P + jQ = V \cdot I^* = |V|^2 Y^*$$

$$= |V|^2 = \frac{1}{R - jx} = |V|^2 \frac{R + jX}{R^2 + X^2}$$

so that

$$P = \frac{V^2 R}{R^2 + X^2}$$

$$P \propto V^2$$

$$\frac{\Delta P}{\Delta |V|} = \frac{\partial P}{\partial V} = 2|V|\frac{R}{(R^2 + X^2)} = \frac{2}{|V|} P$$

$$\frac{\Delta P}{P} - 2\frac{\Delta V}{|V|}$$

Hence 1% voltage variation causes 2% variation in real load power.

E.9.2. **An inductive load is subjected to a frequency variation of 1%. By how much percent will the real load change? Given cos $\phi = 0.8$**

Solution:

$$\frac{\Delta P}{\Delta f} = \frac{\partial P}{\partial V} = \frac{\partial}{\partial F}\left[|V|^2 \frac{R}{R^2 + (2\pi L)^2}\right]$$

$$= \frac{-|V|^2 R(2\pi fL) \cdot 2.2\pi L}{(R^2 + X^2)^2}$$

$$= \frac{-V^2 R}{(R^2 + X^2)} * 2 * X * \frac{X}{f} * \frac{1}{(R^2 + X^2)} = -\frac{P2X^2}{f(R^2 + X^2)}$$

$$\frac{\Delta P}{P} = -\frac{\Delta f}{f}\frac{2X^2}{(R^2 + X^2)}$$

If cos $\phi = 0.8$ sin $\phi = \frac{X}{\sqrt{R^2 + X^2}}$ and

$$\frac{X^2}{(R^2 + X^2)} = \sin^2 \varphi = (0.6)^2 = 0.36$$

$$\frac{\Delta P}{P} = -0.36X2\frac{\Delta f}{f} = -0.72\frac{\Delta f}{f}$$

The percentage change in real load will be 0.72%. If the frequency falls by 1% load power will increase by 0.72%.

QUESTIONS

9.1. For a transformer with turns ratio 1: n, obtain a π-equivalent model?

9.2. What is a phase shifting transformer? Explain its modeling.

9.3. Write short notes on composite load modeling.

9.4. Explain modeling of the following type of loads
   1. lighting
   2. heating
   3. welding
   4. fan loads

9.5. Explain modeling of induction motor for transient studies

9.6. Explain the torque-frequency model for induction motor

9.7. Discuss the voltage-slip model for induction motor

9.8. Explain the behavior of induction motor with voltage

9.9. Explain the possible modes of operation of induction motor during transients

9.10. Discuss the $Q-V$ and $Q-f$ characteristics of induction motor.

PROBLEMS

P.9.1. Obtain a π-model for a transformer with turn ratio 1:5; the equivalent impedance of the transformer is $j0.01$ p.u.

P.9.2. The candle power of a filament lamp is proportional to I^5 where I is the current and proportional to V^4 where V is the voltage. The power consumed in watts is also proportional to $I^{2.5}$.

For an increase of current by 1% what will be the change in watts per candle power?

POWER FLOW STUDIES

10

Power flow studies are performed to determine voltages, active and reactive power, etc. at various points in the network for different operating conditions subject to the constraints on generator capacities and specified net interchange between operating systems and several other restraints. Power flow or load flow solution is essential for continuous evaluation of the performance of the power systems so that suitable control measures can be taken in case of necessity. In practice it will be required to carry out numerous power flow solutions under a variety of conditions.

10.1 NECESSITY FOR POWER FLOW STUDIES

Power flow studies are undertaken for various reasons, some of which are the following:

1. The line flows
2. The bus voltages and system voltage profile
3. The effect of change in configuration and incorporating new circuits on system loading
4. The effect of temporary loss of transmission capacity and/or generation on system loading and accompanied effects
5. The effect of in-phase and quadrative boost voltages on system loading
6. Economic system operation
7. System loss minimization
8. Transformer tap setting for economic operation
9. Possible improvements to an existing system by change of conductor sizes and system voltages.

For the purpose of power flow studies, a single phase representation of the power network is used, since the system is generally balanced. When systems had not grown to the present size, networks were simulated on network analyzers for load flow solutions. These analyzers are of analog type, scaled down miniature models of power systems with resistances, reactances, capacitances, autotransformers, transformers, loads, and generators. The generators are just supply sources operating at a much higher frequency than 50 Hz to limit the size of the components. The loads are represented by constant impedances. Meters are provided on the panel board for measuring voltages, currents, and powers. The power flow solution is obtained directly from measurements for any system simulated on the analyzer.

With the advent of the modern digital computers possessing large storage and high speed, the mode of power flow studies has changed from analog to digital simulation. A large number of algorithms are developed for digital power flow solutions. The methods basically distinguish

Power Systems Analysis. DOI: http://dx.doi.org/10.1016/B978-0-08-101111-9.00010-0

between themselves in the rate of convergence, storage requirement, and time of computation. The loads are generally represented by constant power.

Network equations can be solved in a variety of ways in a systematic manner. The most popular method is node voltage method. When nodal or bus admittances are used, complex linear algebraic simultaneous equations will be obtained in terms of nodal or bus currents. However, as in a power system since the nodal currents are not known, but powers are known at almost all the buses, the resulting mathematical equations become nonlinear and are required to be solved by interactive methods. Load flow studies are required as has been already explained for power system planning, operation, and control as well as for contingency analysis. The bus admittance matrix is invariably utilized in power flow solutions.

10.2 CONDITIONS FOR SUCCESSFUL OPERATION OF A POWER SYSTEM

The conditions for successful operation of a power system are as follows:

1. There should be adequate real power generation to supply the power demand at various load buses and also the losses
2. The bus voltage magnitudes are maintained at values very close to the rated values
3. Generators, transformers, and transmission lines are not overloaded at any point of time or the load curve.

10.3 THE POWER FLOW EQUATIONS

Consider an *n*-bus system, the bus voltages are given by

$$\underline{V} = \begin{bmatrix} V_1 \ \angle \delta_1 \\ \vdots \\ V_n \ \angle \delta_n \end{bmatrix} \tag{10.1}$$

The bus admittance matrix

$$[Y] = [G] + j[B] \tag{10.2}$$

where

$$\begin{aligned} Y_{ik} &= |\ y_{ik}\ |\ \theta_{i-k} \\ &= g_{ik} + jb_{ik} \end{aligned}$$

$$\underline{V_i} = |V_i| \angle \delta_i = |V_i|(\cos \delta_i + j \sin \delta_i) \tag{10.3}$$

$$V_k^* = |V_k| \angle \ \ \delta_k = |V_k|(\cos \delta_k - j \sin \delta_k) \tag{10.4}$$

The current injected into the network at bus "i"

$$I_i = Y_{i1}V_1 + Y_{i2}V_2 + \cdots + Y_{in}V_n$$

where n is the number of buses.

$$\therefore \quad I_i = \sum_{k=1}^{n} Y_{ik} \ V_k \tag{10.5}$$

The complex power into the system at bus i

$$S_i = P_i + j \ Q_i = V_i \ I_i^*$$
$$= V_i \sum_{k=1}^{n} Y_{ik}^* \ V_k^* \tag{10.6}$$
$$= \sum_{k=1}^{n} |V_i \ V_k \ Y_{ik}| \exp(\delta_i - \delta_k - \theta_{ik})$$

Equating the real and imaginary parts

$$P_i = \sum_{k=1}^{n} |V_i \ V_k \ Y_{ik}| \cos(\delta_i - \delta_k - \theta_{ik}) \tag{10.7}$$

and

$$Q_i = \sum_{k=1}^{n} |V_i \ V_k \ Y_{ik}| \ \sin(\delta_i - \delta_k - \theta_{ik}) \tag{10.8}$$

where

$$i = 1, 2, \ldots, n$$

Excluding the slack bus, the above power flow equations are $2(n-1)$ and the variables are P_i, Q_i, $|V_i|$, and $\angle \delta_i$.

Simultaneous solution to the $2(n-1)$ equations:

$$P_{Gi} - P_{Di} - \sum_{k=1}^{n} |V_i \ V_k \ Y_{ik}| \cos(\delta_i - \delta_k - \theta_{ik}) = 0 \tag{10.9}$$
$$k \neq \text{slack bus}$$

$$Q_{Gi} - Q_{Di} - \sum_{k=1}^{n} |V_i \ V_k \ Y_{ik}| \sin(\delta_i - \delta_k - \theta_{ik}) = 0 \tag{10.10}$$
$$k \neq \text{slack bus}$$

constitutes the power flow or load flow equations.

The voltage magnitudes and the phase angles at all load buses are the quantities to be determined. They are called state variables or dependent variables. The specified or scheduled values at all buses are the independent variables.

Y-matrix interactive methods are based on solution to power flow equations using their current mismatch at a bus given by

$$\Delta I_i = I_i - \sum_{k=1}^{n} Y_{ik} V_k \tag{10.11}$$

or using the voltage form

$$\Delta V_i = \frac{\Delta I_i}{Y_{ii}} \tag{10.12}$$

At the end of the interactive solution to power flow equation, ΔI_i or more usually ΔV_i should become negligibly small so that they can be neglected.

10.4 CLASSIFICATION OF BUSES

1. *Load bus*: A bus where there is only load connected and no generation exists is called a load bus. At this bus, real and reactive load demands P_d and Q_d are drawn from the supply. The demand is generally estimated or predicted as in load forecast or metered and measured from instruments. Quite often, the reactive power is calculated from real power demand with an assumed power factor. A load bus is also called a P, Q bus. Since the load demands P_d and Q_d are known values at this bus. The other two unknown quantities at a load bus are voltage magnitude and its phase angle at the bus. In a power balance equation, P_d and Q_d are treated as negative quantities since generated powers P_g and Q_g are assumed positive.

2. *Voltage controlled bus or generator bus*: A voltage controlled bus is any bus in the system where the voltage magnitude can be controlled. The real power developed by a synchronous generator can be varied by changing the prime mover input. This in turn changes the machine rotor axis position with respect to a synchronously rotating or reference axis or a reference bus. In other words, the phase angle of the rotor δ is directly related to the real power generated by the machine. The voltage magnitude on the other hand is mainly influenced by the excitation current in the field winding. Thus at a generator bus the real power generation P_g and the voltage magnitude $|V_g|$ can be specified. It is also possible to produce vars by using capacitor or reactor banks too. They compensate the lagging or leading vars consumed and then contribute to voltage control. At a generator bus or voltage controlled bus, also called a PV bus, the reactive power Q_g and δ_g are the values that are not known and are to be computed.

3. *Slack bus*: In a network as power flows from the generators to loads through transmission lines, power loss occurs due to the losses in the line conductors. These losses when included, we get the power balance relations

$$P_g - P_d - P_L = 0 \tag{10.13}$$

$$Q_g - Q_d - Q_L = 0 \tag{10.14}$$

where P_g and Q_g are the total real and reactive generations, P_d and Q_d are the total real and reactive power demands, and P_L and Q_L are the power losses in the transmission network. The values of P_g, Q_g, P_d, and Q_d are either known or estimated. Since the flow of currents in the various lines in the transmission lines are not known in advance, P_L and Q_L remain unknown before the analysis of the network. But, these losses have to be supplied by the generators in the system. For this purpose, one of the generators or generating bus is specified as "slack bus" or "swing bus." At this bus, the generation P_g and Q_g are not specified. The voltage magnitude is specified at this bus. Further, the voltage phase angle δ is also fixed at this bus.

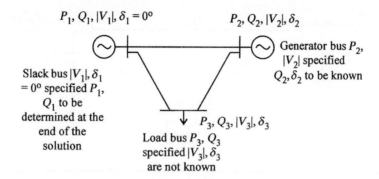

FIGURE 10.1

Three-bus power system.

Table 10.1 Bus Classification and Variables		
Bus	**Specified Variables**	**Computed Variables**
Slack bus	Voltage magnitude and its phase angle	Real and reactive powers
Generator bus (PV bus or voltage controlled bus)	Magnitudes of bus voltages and real powers (limit on reactive powers)	Voltage phase angle and reactive power
Load bus	Real and reactive powers	Magnitude and phase angle of bus voltages

Generally it is specified as $0°$ so that all voltage phase angles are measured with respect to voltage at this bus. For this reason, slack bus is also known as reference bus. All the system losses are supplied by the generation at this bus. Further the system voltage profile is also influenced by the voltage specified at this bus. The three types of buses are illustrated in Fig. 10.1.

Bus classification is summarized in Table 10.1.

10.5 BUS ADMITTANCE FORMATION

Consider the transmission system shown in Fig. 10.2.

The line impedances joining buses 1, 2, and 3 are denoted by z_{12}, z_{22}, and z_{31}, respectively. The corresponding line admittances are y_{12}, y_{22}, and y_{31}.

The total capacitive susceptances at the buses are represented by y_{10}, y_{20}, and y_{30}.

Applying Kirchoff's current law at each bus

$$\left.\begin{array}{l} I_1 = V_1 y_{10} + (V_1 - V_2)y_{12} + (V_1 - V_3)y_{13} \\ I_2 = V_2 y_{20} + (V_2 - V_1)y_{21} + (V_2 - V_3)y_{23} \\ I_3 = V_3 y_{30} + (V_3 - V_1)y_{31} + (V_3 - V_2)y_{32} \end{array}\right\} \qquad (10.15)$$

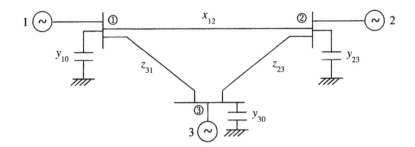

FIGURE 10.2

A three bus network.

In matrix form

$$
\begin{bmatrix} I_1 \\ I_2 \\ I_3 \end{bmatrix} = \begin{bmatrix} y_{10} + y_{12} + y_{13} & -y_{12} & -y_{13} \\ -y_{12} & y_{20} + y_{12} + y_{23} & -y_{23} \\ -y_{13} & -y_{23} & y_{30} + y_{13} + y_{23} \end{bmatrix}
$$

(10.16)

$$
\times \begin{bmatrix} V_1 \\ V_2 \\ V_3 \end{bmatrix} = \begin{bmatrix} y_{11} & y_{12} & y_{13} \\ y_{21} & y_{22} & y_{23} \\ y_{31} & y_{32} & y_{33} \end{bmatrix} \cdot \begin{bmatrix} V_1 \\ V_2 \\ V_3 \end{bmatrix}
$$

where

$$
\left.\begin{array}{l} Y_{11} = y_{10} + y_{12} + y_{13} \\ Y_{22} = y_{20} + y_{12} + y_{23} \\ Y_{33} = y_{30} + y_{13} + y_{23} \end{array}\right\}
$$

(10.17)

are the self-admittances forming the diagonal terms and

$$
\left.\begin{array}{l} Y_{12} = Y_{21} = -y_{12} \\ Y_{13} = Y_{31} = -y_{13} \\ Y_{23} = Y_{32} = -y_{23} \end{array}\right\}
$$

(10.18)

are the mutual admittances forming the off-diagonal elements of the bus admittance matrix. For an n-bus system, the elements of the bus admittance matrix can be written down merely by inspection of the network as

diagonal terms

$$
Y_{ii} = y_{i0} + \sum_{\substack{k=1 \\ k \neq i}}^{n} y_{ik}
$$

(10.19)

off and diagonal terms

$$
Y_{ik} = -y_{ik}
$$

If the network elements have mutual admittance (impedance), the above formulae will not apply. For a systematic formation of the y-bus, linear graph theory with singular transformations may be used.

10.6 SYSTEM MODEL FOR LOAD FLOW STUDIES

The variable and parameters associated with bus i and a neighboring bus k are represented in the usual notation as follows. The voltage at a typical bus i of the system is given by

$$V_i = |V_i| \exp j\, \delta_i = V_i(\cos \delta_i + j \sin \delta_i) \tag{10.20}$$

The bus admittance

$$Y_{ik} = |Y_{ik}| \exp j\, \theta_{ik} = |Y_{ik}|(\cos q_{ik} + j \sin \theta_{ik})$$
$$= G_{ik} + j\, B_{ik} \tag{10.21}$$

Complex power

$$S_i = P_i + j\, Q_i = V_i\, I_i^* \tag{10.22}$$

Using the indices G and L for generation and load

$$P_i = P_{Gi} - P_{Li} = \text{Re}[V_i\, I_i^*] \tag{10.23}$$

$$Q_i = Q_{Gi} - Q_{Li} = \text{Im}[V_i\, I_i^*] \tag{10.24}$$

The current injected into the bus i is given by

$$I_i = Y_{i1}V_1 + Y_{i2}V_2 + \cdots + Y_{in}V_n = \sum_{k=1}^{n} Y_{ik} V_k$$

The bus current is given by

$$I_{\text{BUS}} = Y_{\text{BUS}} \cdot V_{\text{BUS}} \tag{10.25}$$

Hence, from Eqs. (10.22) and (10.23) for an n-bus system

$$I_i^* = \frac{P_i - jQ_i}{V_i^*} Y_{ii}V_i + \sum_{\substack{k=1 \\ k \neq i}}^{n} Y_{ik} V_k \tag{10.26}$$

and from Eq. (10.26)

$$V_i = -\frac{1}{Y_{ii}} \left[\frac{P_i - jQ_i}{V_i^*} - \sum_{\substack{k=1 \\ k \neq i}}^{n} Y_{ik} V_k \right] \tag{10.27}$$

Further

$$P_i + jQ_i = V_i \sum_{k=1}^{n} Y_{ik}^* V_k^* \tag{10.28}$$

In the polar form

$$P_i + jQ_i = \sum_{k=1}^{n} |V_i \quad V_k \quad Y_{ik}| \exp j\, (\delta_i \quad -\delta_k \quad -\theta_{ik}) \tag{10.29}$$

so that

$$P_i = \sum_{k=1}^{n} |V_i \quad V_k \quad Y_{ik}| \cos(\delta_i \quad -\delta_k \quad -\theta_{ik}) \tag{10.30}$$

and

$$Q_i = \sum_{k=1}^{n} \left| V_i \quad V_k \quad Y_{ik} \right| \sin \left(\delta_i \quad -\delta_k \quad -\theta_{ik} \right) \tag{10.31}$$

$$i = 1, 2, \ldots, n; \quad i \neq \text{slack bus}$$

The power flow Eqs. (10.30) and (10.31) are nonlinear and it is required to solve $2(n-1)$ such equations involving $|V_i|$, δ_i, P_i, and Q_i at each bus i for the load flow solution. Finally, the powers at the slack bus may be computed from which the losses and all other line flows can be ascertained. Y-matrix interactive methods are based on solution to power flow relations using their current mismatch at a bus given by

$$\Delta I_i = I_i - \sum_{k=1}^{n} Y_{ik} V_k \tag{10.32}$$

or using the voltage from

$$\Delta V_i = \frac{\Delta I_i}{Y_{ii}} \tag{10.33}$$

The convergence of the iterative methods depends on the diagonal dominance of the bus admittance matrix. The self-admittances of the buses are usually large, relative to the mutual admittances and thus, usually convergence is obtained. Junctions of very high and low series impedances and large capacitances obtained in cable circuits long, EHV lines, series, and shunt compensation are detrimental to convergence as these tend to weaken the diagonal dominance in the Y-matrix. The choice of slack bus can affect convergence considerably. In difficult cases, it is possible to obtain convergence by removing the least diagonally dominant row and column of Y. The salient features of the Y-matrix iterative methods are that the elements in the summation terms in Eqs. (10.26) or (10.27) are on the average only three, even for well-developed power systems. The sparsity of the Y-matrix and its symmetry reduces both the storage requirement and the computation time for iteration. For a large, well-conditioned system of n buses, the number of iterations required are of the order of n and total computing time varies approximately as n^2.

Instead of using Eq. (10.25), one can select the impedance matrix and rewrite the equation as

$$V = Y^{-1} I = Z \cdot I \tag{10.34}$$

The Z-matrix method is not usually very sensitive to the choice of the slack bus. It can easily be verified that the Z-matrix is not sparse. For problems that can be solved by both Z-matrix and Y-matrix methods, the former are rarely competitive with the Y-matrix methods.

10.7 GAUSS−SEIDEL METHOD

Gauss−Seidel iterative method is very simple in concept but may not yield convergence to the required solution. However, when the initial solution or starting point is very close to the actual solution, convergence is generally obtained. The following example illustrates the method.

Consider the equations:

$$2x + 3y = 22$$
$$3x + 4y = 31$$

The free solution to the above equations is $x = 5$ and $y = 4$.

If an interactive solution using the Gauss–Seidel method is required then let us assume a starting value for $x = 4.8$ which is nearer to the true value of 5 we obtain from the given equations

$$y = \frac{22 - 2x}{3} \quad \text{and} \quad x = \frac{34 - 4y}{3}$$

Iteration 1:	Let $x = 4.8$; $y = 4.13$
	with $y = 4.13$; $x = 6.2$
Iteration 2:	$x = 6.2$; $y = 3.2$
	$y = 3.2$; $x = 6.06$
Iteration 3:	$x = 6.06$; $y = 3.29$
	$y = 3.29$; $x = 5.94$
Iteration 4:	$x = 5.96$; $y = 3.37$
	$y = 3.37$; $x = 5.84$
Iteration 5:	$x = 5.84$; $y = 3.44$
	$y = 3.44$; $x = 5.74$
Iteration 6:	$x = 5.74$; $y = 3.5$
	$y = 3.5$; $x = 5.66$

The iterative solution slowly converges to the true solution. The convergence of the method depends upon the starting values for the iterative solution.

In many cases, the convergence may not be obtained at all. However, in case of power flow studies, as the bus voltages are not very far from the rated values and as all load flow studies are performed with per unit values assuming a flat voltage profile at all load buses of $(1 + j0)$ p.u. yields convergence in most of the cases with appropriate acceleration factors chosen.

10.8 GAUSS–SEIDEL ITERATIVE METHOD

In this method, voltages at all buses except at the slack bus are assumed. The voltage at the slack bus is specified and remains fixed at that value. The $(n - 1)$ bus voltage relations from Eq. (10.27):

$$V_i = \frac{1}{Y_{ii}} \left[\frac{P_i - jQ_i}{V_i^*} - \sum_{\substack{k=1 \\ k \neq 1}}^{n} Y_{ik} V_k \right] \tag{10.35}$$

$$i = 1, 2, \ldots n; \quad i \neq \text{slack bus}$$

are solved simultaneously for an improved solution.

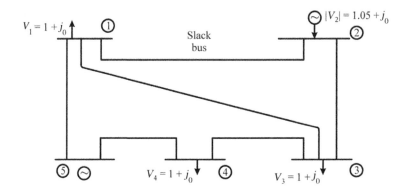

FIGURE 10.3

Five bus system for E10.3

Consider a five bus power system shown in Fig. 10.3.

	(1)	(2)	(3)	(4)	(5)
(1)	Y_{11}	Y_{12}	Y_{13}	0	Y_{15}
(2)	Y_{21}	Y_{22}	Y_{23}	0	0
(3)	Y_{31}	Y_{32}	Y_{33}	Y_{34}	0
(4)	0	0	Y_{43}	Y_{44}	Y_{45}
(5)	Y_{51}	0	0	Y_{54}	Y_{55}

Applying Eq. (10.35) for the five bus system, we obtain during $(k + 1)$th iteration

$$V_1^{k+1} = \frac{1}{Y_{11}} \left[\frac{P_1 - iQ_1}{\left(V_1^k\right)^*} - Y_{12}V_2^k - Y_{13}V_3^k - Y_{15}V_5^k \right] \tag{10.35a}$$

$$V_2^{k+1} = \frac{1}{Y_{22}} \left[\frac{P_2 - jQ_2}{V_2^*} - Y_{21}V_1^{k+1} - V_{23}V_3^k \right] \tag{10.35b}$$

$$V_3^{k+1} = \frac{1}{Y_{33}} \left[\frac{P_3 - jQ_3}{\left(V_3^k\right)^*} - Y_{31}V_1^{k+1} - Y_{32}V_2^{k+1} - Y_{34}V_4^k \right] \tag{10.35c}$$

and

$$V_4^{k+1} = \frac{1}{Y_{44}} \left[\frac{P_4 - jQ_4}{\left(V_4^k\right)^*} - Y_{43}V_3^{k+1} - Y_{45}V_5^k \right] \tag{10.35d}$$

Bus 5 is selected as slack bus or swing bus.

To start with all voltages at load buses are assumed $(1 + j_0)$ p.u., i.e.,

$$|V_1| = |V_3| = |V_4| = 1 \text{ p.u.}$$

and

$$\delta_1 = \delta_3 = \delta_4 = 0°$$

$|V_5|$ = voltage magnitude at swing bus is fixed at say 1.05 p.u. with $\delta_5 = 0°$ as reference.

Eqs. (10.35a), (10.35c), and (10.35d) are solved iteratively till convergence is reached. But Eq. (10.35b) is solved for Q_2 and its limits are checked at each iteration. If Q_2 reaches its limit (upper or lower), then the reactive generation at bus 3 is fixed at that Q_2, and bus 2 is also considered as a load bus in treatment and is solved for $|V_3|$. The equation for Q_2 is

$$Q_2 = I_m \left(V_i \sum_{k=1}^{4} Y_{ik}^* V_k^* \right) \tag{10.35e}$$

Substituting this value for Q_2, V_2^{k+1} is updated.

In order to accelerate the convergence, all newly computed values of bus voltages are substituted in Eq. (10.35). The bus voltage equation for the $(m + 1)$th iteration may then be written as

$$V_i^{(m+1)} = \frac{1}{Y_{ii}} \left[\frac{P_i - jQ_i}{V_i^{(m)*}} - \sum_{\substack{k=1 \\ k \neq 1}}^{i-1} Y_{ik} V_k^{(m+1)} - \sum_{k=i+1}^{n} Y_{ik} V_k^{(m)} \right] \tag{10.36}$$

The method converges slowly because of the loose mathematical coupling between the buses. The rate of convergence of the process can be increased by using acceleration factors to the solution obtained after each iteration.

10.8.1 ACCELERATION FACTOR

In Gauss–Seidal method of solution to the load flow problem, the iterative process is continued until the amount of correction to voltage at every bus is less than some prespecified precision index. However, the number of iterations required can be considerably reduced, if the correction to voltage at each bus is multiplied by some constant. This increases the amount of correction thus making the process of convergence faster. The multiplier that produces the improved convergence is called acceleration factor.

Let at any bus i the voltage at the end of $(k - 1)$th iteration be V_i^{k-1} and at the end of the kth iteration be V_i^k. Then, if α is the acceleration factor,

$$V_i^{(k)} \text{ (accelerated)} = V_i^{(k-1)} + \alpha(V_i^{(k)} - V_i^{(k-1)})$$

The accelerated value of $V_i^{(k)}$ is used for $(k + 1)$th iteration. Generally, the value of α is chosen as $1 \leq \alpha \leq 2$.

A fixed acceleration factor is sometimes used, using the relation:

$$\Delta V_i = \alpha \frac{\Delta S_i^*}{V_i^* Y_{ii}} \tag{10.37}$$

The use of acceleration factor amounts to a linear extrapolation of V_i. For a given system, it is quite often found that a near optimal choice of α exists. Some researchers have suggested even a complex value for α. But, it is found convenient to use real values for α (usually 1.5 or 1.6).

Alternatively, different acceleration factors may be used for real and imaginary parts of the voltage.

10.8.2 TREATMENT OF A PV BUS

The method of handling a PV bus requires rectangular coordinate representation for the voltages. Letting

$$V_i = v'_i + jv''_i \tag{10.38}$$

where v'_i and v''_i are the real and imaginary components of V_i the relationship.

$$v'^2_i + v''^2_i = |V_i|^2_{\text{scheduled}} \tag{10.39}$$

must be satisfied, so that the reactive bus power required to establish the scheduled bus voltage can be computed. The estimates of voltage components, $v'^{(m)}_i$ and $v''^{(m)}_i$ after m iterations must be adjusted to satisfy Eq. (10.39). The phase angle of the estimated bus voltage is

$$\delta^{(m)}_i = \tan^{-1}\left[\frac{v''^{(m)}_i}{v'^{(m)}_i}\right] \tag{10.40}$$

Assuming that the phase angles of the estimated and scheduled voltages are equal, then the adjusted estimates of $V'^{(m)}$ and $V''^{(m)}_i$ are

$$v'^{(m)}_{i(\text{new})} = |V_i|_{\text{scheduled}}\cos \delta^{(m)}_i \tag{10.41}$$

and

$$v''^{(m)}_{i(\text{new})} = |V_i|_{\text{scheduled}}\sin \delta^{(m)}_i \tag{10.42}$$

These values are used to calculate the reactive power $Q^{(m)}_i$. Using these reactive powers $Q^{(m)}_i$ and voltages $V^{(m)}_{i(\text{new})}$, a new estimate $V^{(m+1)}_i$ is calculated. The flow chart for computing the solution of load flow using the Gauss–Seidel method is given in Fig. 10.4.

While computing the reactive powers, the limits on the reactive source must be taken into consideration. If the calculated value of the reactive power is beyond limits, then its value is fixed at the limit that is violated and it is no longer possible to hold the desired magnitude of the bus voltage, the bus is treated as a PQ bus or load bus.

10.9 NEWTON–RAPHSON METHOD

The generated Newton–Raphson method is an interactive algorithm for solving a set of simultaneous nonlinear equations in an equal number of unknowns. Consider the set of nonlinear equations

$$f_i(x_1, x_2, \ldots, x_n) = y_i, \quad i = 1, 2, \ldots, n \tag{10.43}$$

with initial estimates for x_i

$$x_1^{(0)}, x_2^{(0)}, \ldots, x_n^{(0)}$$

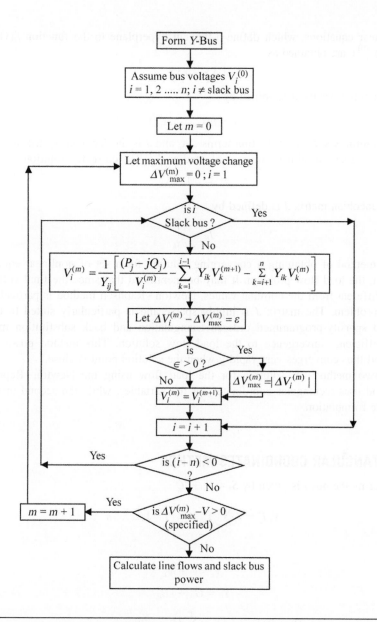

FIGURE 10.4

which are not far from the actual solution. Then using Taylor's series and neglecting the higher order terms, the corrected set of equations are

$$(x_1^{(0)} + \Delta x_1, x_2^{(0)} + \Delta x_2, \ldots, x_n^{(0)} + \Delta x_n) = y_i \tag{10.44}$$

where Δx_i are the corrections to x_i ($i = 1, 2, \ldots, n$).

A set of linear equations, which define a tangent hyperplane to the function $f_i(x)$ at the given iteration point $(x_i^{(0)})$, are obtained as

$$\Delta Y = J \Delta X \tag{10.45}$$

where ΔY is a column vector determined by

$$y_i - f_i(x_1^{(0)}, \ldots, x_n^{(0)})$$

ΔX is the column vector of correction terms Δx_i and J is the Jacobian matrix for the function f given by the first-order partial derivatives evaluated at $x_i^{(0)}$. The corrected solution is obtained as

$$x_i^{(1)} = x_i^{(0)} + \Delta x_i \tag{10.46}$$

The square Jacobian matrix J is defined by

$$J_{ik} = \frac{\partial f_i}{\partial x_k} \tag{10.47}$$

The above method of obtaining a converging solution for a set of nonlinear equations can be used for solving the load flow problem. It may be mentioned that since the final voltage solutions are not much different from the nominal values, Newton–Raphson method is particularly suited to the load flow problem. The matrix J is highly sparse and is particularly suited to the load flow application and sparsity-programmed ordered triangulation and back substitution methods result in quick and efficient convergence to the load flow solution. This method possesses quadratic convergence and thus converges very rapidly when the solution point is close.

There are two methods of solution for the load flow using the Newton–Raphson method. The first method uses rectangular coordinates for the variables, while the second method uses the polar coordinate formulation.

10.9.1 RECTANGULAR COORDINATES METHOD

The power entering the bus i is given by $S_i = P_i + jQ_i$

$$V_i I_i^* = V_i \sum_{k=1}^{n} Y_{ik}^* V_k^*, \quad i = 1, 2, \ldots, n \tag{10.48}$$

where

$$V_i = v_i' + jv_i''$$

$$Y_{ik} = G_{ik} + j\,B_{ik}$$

$$(P_i + jQ_i) = (v_i' + jv_i'') \sum_{k=1}^{n} (G_{ik} - jB_{ik})(v_k' - jv_k'') \tag{10.49}$$

Expanding the right side of the above equation and separating out the real and imaginary parts

$$P_i = v_i' \left[\sum \left(G_{ik}\, v_k' - B_{ik}\, v_k'' \right) + v_i'' \sum \left(G_{ik}\, v_k'' + B_{ik}\, v_k' \right) \right] \tag{10.50}$$

$$Q_i = -v_i' \left[\sum \left(G_{ik}\, v_k'' + B_{ik}\, v_k' \right) \right] + v_i'' \left[\left(G_{ik}\, v_k' - B_{ik}\, v_k'' \right) \right] \tag{10.51}$$

These are the two power relations at each bus and the linearized equations of the form (10.50) are written as (10.51)

$$
\begin{bmatrix} \Delta P_1 \\ \vdots \\ \Delta P_{n-1} \\ \Delta Q_1 \\ \vdots \\ \Delta Q_{n-1} \end{bmatrix} = \begin{bmatrix} \dfrac{\partial P_1}{\partial v_1'} & \cdots & \dfrac{\partial P_1}{\partial v_{n-1}'} & \dfrac{\partial P_1}{\partial v_1''} & \cdots & \dfrac{\partial P_1}{\partial v_{n-1}''} \\ \vdots & \cdots & \vdots & \vdots & \cdots & \vdots \\ \dfrac{\partial P_{n-1}}{\partial v_1'} & \cdots & \dfrac{\partial P_{n-1}}{\partial v_{n-1}'} & \dfrac{\partial P_{n-1}}{\partial v_n''} & \cdots & \dfrac{\partial P_{n-1}}{\partial v_{n-1}''} \\ \dfrac{\partial Q_1}{\partial v_1'} & \cdots & \dfrac{\partial Q_1}{\partial v_{n-1}'} & \dfrac{\partial Q_1}{\partial v_1''} & \cdots & \dfrac{\partial Q_1}{\partial v_{n-1}''} \\ \vdots & \cdots & \vdots & \vdots & \cdots & \vdots \\ \dfrac{\partial Q_{n-1}}{\partial v_1'} & \cdots & \dfrac{\partial Q_{n-1}}{\partial v_{n-1}'} & \dfrac{\partial Q_{n-1}}{\partial v_1''} & \cdots & \dfrac{\partial Q_{n-1}}{\partial v_{n-1}''} \end{bmatrix} \begin{bmatrix} \Delta v_1' \\ \vdots \\ \Delta v_{n-1}' \\ \Delta v_1'' \\ \vdots \\ \Delta v_{n-1}'' \end{bmatrix}
\tag{10.52}
$$

Matrix Eq. (10.52) can be solved for the unknowns $\Delta v_i'$ and $\Delta v_i''$ ($i = 1, 2, \ldots, n-1$), leaving the slack bus at the nth bus where the voltage is specified. Eq. (10.52) may be written compactly as

$$
\begin{bmatrix} \Delta P \\ \Delta Q \end{bmatrix} = \begin{bmatrix} H & N \\ M & L \end{bmatrix} \begin{bmatrix} \Delta v' \\ \Delta v'' \end{bmatrix}
\tag{10.53}
$$

where H, N, M, and L are the submatrices of the Jacobian. The elements of the Jacobian are obtained by differentiating Eqs. (10.50) and (10.51). The off-diagonal and diagonal elements of H matrix are given by

$$
\frac{\partial P_i}{\partial v_k'} = G_{ik}v_i' + B_{ik}v_i'', \quad i \neq k
\tag{10.54}
$$

$$
\frac{\partial P_i}{\partial v_i'} = \sum_{\substack{k=1 \\ k \neq i}}^{n} \left(G_{ik}v_k' - B_{ik}v_k'' \right) + 2v_i'G_{ii}
\tag{10.55}
$$

$$
i = 1, 2, \ldots, n, \quad k = 1, 2, \ldots, n, \ k \neq i
$$

The off-diagonal and diagonal elements of N are

$$
\frac{\partial P_i}{\partial v_k''} + G_{ik}v_i'' - B_{ik}v_i', \quad k \neq i
\tag{10.56}
$$

$$
\frac{\partial P_i}{\partial v_i''} = -2v_i''G_{ii} + \sum_{\substack{k=1 \\ k \neq i}}^{n} (G_{ik}v_k'' + B_{ik}v_k')
\tag{10.57}
$$

$$
i = 1, 2, \ldots, n, \quad k = 1, 2, \ldots, n, \ k \neq i
$$

The off-diagonal and diagonal elements of submatrix M are obtained as

$$
\frac{\partial Q_i}{\partial v_k'} + G_{ik}v_i'' - B_{ik}\partial v_i', \quad k \neq i, \ k \neq 1
\tag{10.58}
$$

$$\frac{\partial Q_i}{\partial v_i'} = -2B_{ii}v_i' - \sum_{\substack{k=1 \\ k \neq i}}^{n} (v_k'' G_{ik} + v_k' B_{ik}) \tag{10.59}$$

$$i = 1, 2, \ldots, n, \quad k = 1, 2, \ldots, n, \quad i \neq k$$

Finally, the off-diagonal and diagonal elements of L are given by

$$\frac{\partial Q_i}{\partial v_k''} = -v_i'' B_{ik} - v_i' G_{ik}, \quad i \neq k \tag{10.60}$$

$$\frac{\partial Q_i}{\partial v_i''} = -2B_{ii}v_i'' + \sum_{\substack{k=1 \\ k \neq i}}^{n} (G_{ik}v_k' - B_{ik}v_k'') \tag{10.61}$$

$$i = 1, 2, \ldots, n, \quad k = 1, 2, \ldots, n, \quad i \neq k$$

It can be noticed that

$$L_{ik} = H_{ik}$$

and

$$N_{ik} = M_{ik}$$

This property of symmetry of the elements reduces computer time and storage.

Treatment of Generator Buses

At all generator buses other than the swing bus, the voltage magnitudes are specified in addition to the real powers. At the ith generator bus

$$|V_i|^2 = v_i'^2 + v_i''^2 \tag{10.62}$$

Then, at all the generator nodes, the variable ΔQ_i will have to be replaced by

$$\Delta |V_i|^2$$

But,

$$|\Delta V_i|^2 = \frac{\partial(|\Delta_i|^2)}{\partial v_i'} \Delta V_i' + \frac{\partial(|\Delta_i|^2)}{\partial v_i''} \Delta V_i'' \tag{10.63}$$

$$= 2v_i'' \Delta v_i' + 2v_i'' \Delta v_i''$$

This is the only modification required to be introduced in Eq. (10.62).

10.9.2 THE POLAR COORDINATES METHOD

The equation for the complex power at node i in the polar form is given in Eq. (10.29) and the real and reactive powers at bus i are indicated in Eqs. (10.30) and (10.31). Reproducing them here once again for convenience

$$P_i = \sum_{k=1}^{n} |V_i \ V_k \ Y_{ik}| \cos(\delta_i - \delta_k - \theta_{ik})$$

and

$$Q_i = \sum_{k=1}^{n} |V_i \ V_k \ Y_{ik}| \sin(\delta_i - \delta_k - \theta_{ik})$$

The Jacobian is then formulated in terms of $|V|$ and δ instead of V_i' and V_i'' in this case. Eq. (10.53) then takes the form

$$\begin{bmatrix} \Delta P \\ \Delta Q \end{bmatrix} = \begin{bmatrix} H & N \\ M & L \end{bmatrix} \begin{bmatrix} \Delta\delta \\ \Delta|V| \end{bmatrix} \tag{10.64}$$

The off-diagonal and diagonal elements of the submatrices H, N, M, and L are determined by differentiating Eqs. (10.30) and (10.31) with respect to δ and $|V|$ as before. The off-diagonal and diagonal elements of H matrix are

$$\frac{\partial P_i}{\partial \delta_k} = |V_i \ V_k \ Y_{ik}| \sin(\delta_i - \delta_k - \theta_{ik}) \quad i \ne k \tag{10.65}$$

$$\frac{\partial P_i}{\partial \delta_i} = -\sum_{\substack{k=1 \\ k \ne 1}}^{n} |V_i V_k V_{ik}| \sin(\delta_i - \delta_k - \theta_{ik}) \tag{10.66}$$

The off-diagonal and diagonal elements of N matrix are

$$\frac{\partial P_i}{\partial |V_k|} = V_i \ Y_{ik} \cos(\delta_i - \delta_k - \theta_{ik}) \tag{10.67}$$

$$\frac{\partial P_i}{\partial |V_i|} = 2|V_i \ Y_{ii}| \cos\theta_{ii} + \sum_{\substack{k=1 \\ k \ne i}}^{n} |V_k \ Y_{ik}| \cos(\delta_i - \delta_k - \theta_{ik}) \tag{10.68}$$

The off-diagonal and diagonal elements of M matrix are

$$\frac{\partial Q_i}{\partial \delta_k} = -\left| V_i \quad V_k \quad Y_{ik} \right| \cos(\delta_i - \delta_k - \theta_{ik}) \tag{10.69}$$

$$\frac{\partial Q_i}{\partial \delta_i} = \sum_{\substack{k=1 \\ k \ne i}}^{n} \left| V_i \quad V_k \quad Y_{ik} \right| \cos(\delta_i - \delta_k - \theta_{ik}) \tag{10.70}$$

Finally, the off-diagonal and diagonal elements of L matrix are

$$\frac{\partial Q_i}{\partial |V_k|} = |V_i Y_{ik}| \sin(\delta_i - \delta_k - \theta_{ik}) \tag{10.71}$$

$$\frac{\partial Q_i}{\partial |V_i|} = -2|V_i Y_{ii}| \sin\theta_{ii} + \sum_{\substack{k=1 \\ k \ne i}}^{n} |V_k Y_{ik}| \sin(\delta_i - \delta_k - \theta_{ik}) \tag{10.72}$$

It is seen from the elements of the Jacobian in this case that the symmetry that existed in the rectangular coordinates case is no longer present now. By selecting the variable as $\Delta\delta$ and $\Delta|V|/|V|$ instead Eq. (10.64) will be in the form

$$\begin{bmatrix} \Delta P \\ \Delta Q \end{bmatrix} = \begin{bmatrix} H & N \\ M & L \end{bmatrix} \begin{bmatrix} \Delta\delta \\ \dfrac{\Delta|V|}{|V|} \end{bmatrix} \tag{10.73}$$

The terms of $H_{ik} = \partial P_i/\partial \delta_k = |V_i \; V_k \; Y_{ik}| \sin(\delta_i - \delta_k - \theta_{ik})$, $i \neq k$ $i = k$ (Eq. 10.67) remain unchanged with the modification $|\Delta V/V|$ introduced. However, the terms of L_{ik},

$$\frac{\partial Q_i}{\partial |V_k|} = |V_i \quad Y_{ik}| \sin(\delta_i - \delta_k - \theta_{ik}); \quad i \neq k$$

will get changed as

$$\frac{\partial Q_i}{\partial |V_k|} \cdot |V_k|$$

$$\frac{\partial Q}{\partial |V_k|}(\text{modified}) = |V_k||V_i \; Y_{ik}| \sin(\delta_i - \delta_k - \theta_{ik}) \tag{10.74}$$

$$= H_{ik}$$

Hence

$$L_{ik} = H_{ik}$$

In a similar manner

$$M_{ik} = \frac{\partial Q_i}{\partial \delta_k} = -|V_i \quad V_k \quad Y_{ik}| \cos(\delta_i - \delta_k - \theta_{ik})$$

remains the same even with $\Delta V/|V|$ variable but N_{ik} changes from

$$N_{ik} = |V_i \quad Y_{ik}| \cos(\delta_i - \delta_k - \theta_{ik})$$

into

$$N_{ik}(\text{modified}) = |V_k \quad V_i \quad Y_{ik}| \cos(\delta_i - \delta_k - \theta_{ik}) \; i \neq k \tag{10.75}$$

$$= - M_{ik}$$

Hence

$$N_{ik} = -M_{ik}$$

or, in other words, the symmetry is restored. The number of elements to be calculated for an n-dimensional Jacobian matrix are only $n + n^2/2$ instead of n^2, thus again saving computer time and storage. The flow chart for computer solution is given in Fig. 10.5.

Treatment of Generator Nodes
For a PV bus, the reactive power equations are replaced at the ith generator bus by

$$|V_i|^2 = v_i'^2 + v_i''^2$$

The elements of M are given by

$$M_{ik} = \frac{\partial(|V|)^2}{\partial \delta_k} = 0; \quad i \neq k$$

and

$$M_{ii} = \frac{\partial |V_i|^2}{\partial \delta_i} = 0$$

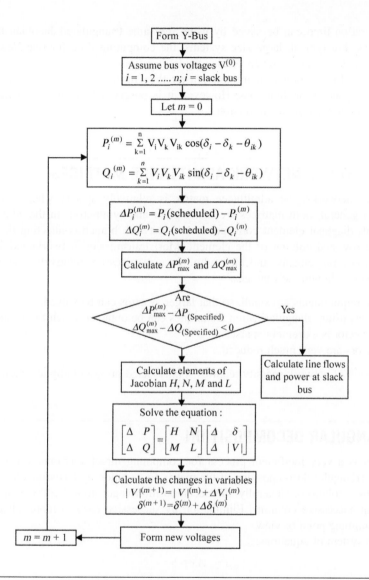

FIGURE 10.5

Then elements of L are given by

$$L_{ik} = \frac{\partial(|V_i|^2)}{\partial|V_k|}|V_k| = 0; \ \ i \neq k$$

and

$$L_{ii} = \frac{\partial(|V_i|)^2}{\partial|V_i|}|V_i| = 2|V_i|^2$$

Newtons method converges in 2–5 iterations from a flat start ($\{V\} = 1.0$ p.u. and $\delta = 0$) independent of system size. Previously stored solution can be used as starting values for rapid convergence. Iteration time can be saved by using the same triangulated Jacobian matrix for two or more iterations. For typical, large size systems, the computing time for one Newton–Raphson iteration is roughly equivalent to seven Gauss–Seidel iterations.

The rectangular formulation is marginally faster than the polar version because there are no time-consuming trigonometric functions. However, it is observed that the rectangular coordinates method is less reliable than the polar version.

10.10 SPARSITY OF NETWORK ADMITTANCE MATRICES

For many power networks, the admittance matrix is relatively sparse, whereas the impedance matrix is full. In general, both matrices are nonsingular and symmetric. In the admittance matrix, each non-zero off-diagonal element corresponds to a network branch connecting the pair of buses indicated by the row and column of the element. Most transmission networks exhibit irregularity in their connection arrangements, and their admittance matrices are relatively sparse. Such sparse systems possess the following advantages:

1. Their storage requirements are small, so that larger systems can be solved.
2. Direct solutions using triangularization techniques can be obtained much faster unless the independent vector is extremely sparse.
3. Round off errors are very much reduced.

The exploitation of network sparsity requires sophisticated programming techniques.

10.11 TRIANGULAR DECOMPOSITION

Matrix inversion is a very inefficient process for computing direct solutions, especially for large, sparse systems. Triangular decomposition of the matrix for solution by Gaussian elimination is more suited for load flow solutions. Generally, the decomposition is accomplished by elements below the main diagonal in successive columns. Elimination by successive rows is more advantageous from computer programming point of view.

Consider the system of equations:

$$Ax = b \tag{10.76}$$

where A is a nonsingular matrix, b is a known vector containing at least one nonzero element, and x is a column vector of unknowns.

To solve Eq. (10.76) by the triangular decomposition method, matrix A is augmented by b as shown

$$\begin{bmatrix} a_{11} & a_{12} & \cdots & a_{1n} & b_1 \\ a_{21} & a_{22} & \cdots & a_{2n} & b_2 \\ \vdots & \vdots & \vdots & \vdots & \vdots \\ a_{n1} & a_{n2} & \cdots & a_{nn} & b_n \end{bmatrix}$$

The elements of the first row in the augmented matrix are divided by a_{11} as indicated by the following step with superscripts denoting the stage of the computation:

$$a_{1j}^{(1)} = \left(\frac{1}{a_{11}}\right) a_{1j}, \quad j = 2, \ldots, n \tag{10.77}$$

$$b_1^{(1)} = \left(\frac{1}{a_{11}}\right) b_1 \tag{10.78}$$

In the next stage, a_{21} is eliminated from the second row using the relations:

$$a_{2j}^{(1)} = a_{2j} - a_{21} a_{1j}^{(1)}, \quad j = 2, \ldots, n \tag{10.79}$$

$$b_2^{(1)} = b_2 - a_{21} b_1^{(1)} \tag{10.80}$$

$$a_{2j}^{(2)} = \left(\frac{1}{a_{22}^{(2)}}\right) a_{2j}^{(1)}, \quad j = 3, \ldots, n$$

$$b_2^{(2)} = \left(\frac{1}{a_{22}^{(1)}}\right) b_2^{(1)} \tag{10.81}$$

The resulting matrix then becomes

$$\begin{bmatrix} 1 & 1_{12}^{(1)} & a_{13}^{(1)} & \cdots & a_{1n}^{(1)} & b_1^{(1)} \\ 0 & 1 & a_{23}^{(2)} & \cdots & a_{2n}^{(2)} & b_2^{(2)} \\ \vdots & \vdots & \vdots & \vdots & \vdots & \vdots \\ a_{n1} & a_{n2} & a_{n3} & \cdots & a_{nn} & b_n \end{bmatrix}$$

using the relations

$$b_{3j}^{(1)} = a_{3j} - a_{31} a_{1j}^{(1)}, \quad j = 2, \ldots, n \tag{10.82}$$

$$b_{(3)}^{1} = b_3 - a_{31} b_1^{(1)} \tag{10.83}$$

$$a_{3j}^{(2)} = a_{3j}^{(1)} - a_{32}^{(1)} a_{2j}^{(2)}, \quad j = 3, \ldots, n \tag{10.84}$$

$$b_3^{(2)} = b_3^{(1)} - a_{32}^{(1)} b_3^{(2)}, \quad j = 4, \ldots, n \tag{10.85}$$

$$a_3^{(3)} = \left(\frac{1}{a_{33}^{(2)}}\right) a_{3j}^{(1)}, \quad j = 4, \ldots, n \tag{10.86}$$

$$b_3^{(3)} = \left(\frac{1}{a_{33}^{(2)}}\right) b_3^{(2)} \tag{10.87}$$

The elements to the left of the diagonal in the third row are eliminated and further the diagonal element in the third row is made unity.

After n steps of computation for the nth-order system of Eq. (10.60), the augmented matrix will be obtained as

$$\begin{bmatrix} 1 & a_{12}^{(1)} & \cdots & a_{1n}^{(1)} & b_1^{(1)} \\ 0 & 1 & \cdots & a_{2n}^{(2)} & b_2^{(2)} \\ \vdots & \vdots & \vdots & \vdots & \vdots \\ 0 & 0 & \cdots & 1 & b_n^{(n)} \end{bmatrix}$$

By back substitution, the solution is obtained as

$$x_n = b_n^{(n)}$$
$$\underline{x_{n-1} = b_{n-1}^{(n-1)} - a_{n-1}^{(n-1)}, n \ldots x_n}$$
$$x_i = b_i^{(1)} - \sum_{j=i+1}^{n} a_{ij}^{(i)} x_j \qquad (10.90)$$

For matrix inversion of an nth-order matrix, the number of arithmetical operations required is n^3 while for the triangular decomposition it is approximately $\left(\frac{n^3}{3}\right)$.

10.12 OPTIMAL ORDERING

When the **A** matrix in Eq. (10.76) is sparse, it is necessary to see that the accumulation of non-zero elements in the upper triangle is minimized. This can be achieved by suitably ordering the equations, which is referred to as optimal ordering.

Consider the network system having five nodes as shown in Fig. 10.6.

The y-bus matrix of the network will have entries as follows:

$$
\begin{array}{c c c c c}
 & 1 & 2 & 3 & 4 \\
\hline
1 & \times & \times & \times & \times \\
2 & \times & \times & 0 & 0 \\
3 & \times & 0 & \times & 0 \\
4 & \times & 0 & 0 & \times
\end{array} = Y \qquad (10.91)
$$

After triangular decomposition, the matrix will be reduced to the form

$$
\begin{array}{c c c c c}
 & 1 & 2 & 3 & 4 \\
\hline
1 & 1 & \times & \times & \times \\
2 & 0 & 1 & \times & \times \\
3 & 0 & 0 & 1 & \times \\
4 & 0 & 0 & 0 & 1
\end{array} = Y \qquad (10.92)
$$

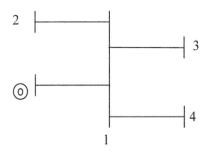

FIGURE 10.6

A five bus system.

By ordering the nodes as in Fig. 10.7, the bus admittance matrix will be of the form

$$
\begin{array}{c}
\quad 1\quad 2\quad 3\quad 4 \\
\begin{array}{c}1\\2\\3\\4\end{array}
\left|
\begin{array}{cccc}
\times & 0 & 0 & \times \\
0 & \times & 0 & \times \\
0 & 0 & \times & \times \\
\times & \times & \times & \times
\end{array}
\right| = Y
\end{array}
\tag{10.93}
$$

As a result of triangular decomposition, the **Y**-matrix will be reduced to

$$
\begin{array}{c}
\quad 1\quad 2\quad 3\quad 4 \\
\begin{array}{c}1\\2\\3\\4\end{array}
\left|
\begin{array}{cccc}
1 & 0 & 0 & \times \\
0 & 1 & 0 & \times \\
0 & 0 & 1 & \times \\
0 & 0 & 0 & 1
\end{array}
\right| = Y
\end{array}
\tag{10.94}
$$

Thus comparing the matrices in Eqs. (10.93) and (10.94), the nonzero off-diagonal entries are reduced from 6 to 3 by suitably numbering the nodes.

Tinney and Walker have suggested three methods for optimal ordering.

1. Number the rows according to the number of nonzero, off-diagonal elements before elimination. Thus rows with less number of off-diagonal elements are numbered first and the rows with large number last.
2. Number the rows so that at each step of elimination the next row to be eliminated is the one having fewest nonzero terms. This method required simulation of the elimination process to take into account the changes in the nonzero connections affected at each step.
3. Number the rows so that at each step of elimination, the next row to be eliminated is the one that will introduce fewest new nonzero elements. This requires the simulation of every feasible alternative at each step.

Scheme 1 is simple and fast. However, for power flow solutions, Scheme 2 has proved to be advantageous even with its additional computing time. If the number of iterations is large, Scheme 3 may prove to be advantageous.

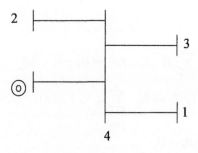

FIGURE 10.7

Renumbered five bus system.

10.13 DECOUPLED METHODS

All power systems exhibit in the steady state a strong interdependence between active powers and bus voltage angles and between reactive power and voltage magnitude. The coupling between real power and bus voltage magnitude and between reactive power and bus voltage phase angle are both relatively weak. This weak coupling is utilized in the development of the so-called decoupled methods. Recalling Eq. (10.78)

$$\begin{bmatrix} \Delta P \\ \Delta Q \end{bmatrix} = \begin{bmatrix} H & N \\ M & L \end{bmatrix} \begin{bmatrix} \Delta \delta \\ |V|/\Delta|V| \end{bmatrix}$$

by neglecting N and M submatrices as a first step, decoupling can be obtained so that

$$|\Delta P| = |H| \cdot |\Delta \delta| \tag{10.95}$$

and

$$|\Delta Q| = |L| \cdot |\Delta|V|/|V| \tag{10.96}$$

The decoupled method converges as reliably as the original Newton method from which it is derived. However, for very high accuracy the method requires more iterations because overall quadratic convergence is lost. The decoupled Newton method saves by a factor of 4 on the storage for the J-matrix and its triangulation. But, the overall saving is 35%–50% of storage when compared to the original Newton method. The computation per iteration is 10%–20% less than for the original Newton method.

10.14 FAST DECOUPLED METHODS

For security monitoring and outage-contingency evaluation studies, fast load flow solutions are required. A method developed for such an application is described in this section.

The elements of the submatrices H and L (Eqs. 10.65 and 10.71) are given by

$$H_{ik} = |V_i \quad V_k \quad Y_{ik}| \sin(\delta_i - \delta_k - \theta_{ik})$$
$$= |V_i \quad V_k \quad Y_{ik}| (\sin \delta_{ik} \cos \theta_{ik} - \cos \delta_{ik} \sin \theta_{ik})$$
$$= |V_i \quad V_k| (G_{ik} \sin \delta_{ik} - B_{ik} \cos \delta_{ik})$$

where

$$\delta_i - \delta_k = \delta_{ik}$$

$$H_{ii} = -\sum_{k=1}^{n} |V_i \quad V_k \quad Y_{ik}| \sin(\delta_i - \delta_k - \theta_{ik})$$
$$= -V_i^2 Y_{ii} \sin \theta_{ii} - \sum_{\substack{k=1 \\ k \neq i}}^{n} V_i V_k Y_{ik} \sin(\delta_i - \delta_k - \theta_{ik})$$
$$= V_i^2 B_{ii} - Q_i$$

In a similar manner

$$L_{ik} = \left| V_i \;\; V_k \;\; Y_{ik} \right| \sin(\delta_i - \delta_k - \theta_{ik})$$
$$= \left| V_i \;\; V_k \right| (B_{ik} \cos\theta_{ik} - G_{ik} \sin\theta_{ik})$$

$$L_{ii} = 2V_i \; Y_{ii} \sin\theta_{ii} + \sum_{\substack{k=1 \\ k \neq i}}^{n} V_k \; Y_{ik} \sin(\delta_i - \delta_k - \theta_{ik})$$

and with $\Delta|V|/|V|$ formulation on the right-hand side

$$L_{ii} = 2|V_i \; Y_{ii}|\sin\theta_{ii} + \sum_{\substack{k=1 \\ k \neq i}}^{n} V_i \; V_k \; Y_{ik} \sin(\delta_i - \delta_k - \theta_{ik})$$

$$= |V_i^2|B_{ii} + Q_i$$

Assuming that

$$\cos\delta_{ik} \cong 1$$
$$\sin\delta_{ik} \cong 0$$
$$G_{ii} \sin\delta_{ik} \leq B_{ik}$$

and

$$Q_i \leq B_{ii}|V_i|^2$$

and

$$\left.\begin{aligned} H_{ik} &= -|V_i \; V_k|B_{ik} \\ H_{ii} &= -|V_i|^2 \; B_{ii} \\ L_{ik} &= -|V_i \; V_k|B_{ik} \\ L_{ii} &= -|V_i|^2 \; B_{ii} \end{aligned}\right\}$$

Rewriting Eqs. (10.95) and (10.96)

$$|\Delta P| = |V|B'|V|\Delta\delta \tag{10.97}$$

$$|V| = |V|B''|V|\frac{\Delta|V|}{|V|} \tag{10.98}$$

Eq. (10.97) can be rewritten as

$$\frac{|\Delta P|}{|V|} = |V||B'|\Delta\delta$$

$$\frac{|\Delta P|}{|V|} \simeq [B'][\Delta\delta] \quad \text{setting}|V| \simeq 1 + j_0 \tag{10.99}$$

From Eq. (10.96)

$$\frac{|\Delta Q|}{|V|} = [B''][\Delta|V|] \tag{10.100}$$

Matrices B' and B'' represent constant approximations to the slopes of the tangent hyperplanes of the functions $\Delta P/|V|$ and $\Delta Q/|V|$, respectively. They are very close to the Jacobian submatrices H and L evaluated at system no-load.

Shunt reactances and off-nominal in-phase transformer taps which affect the Mvar flows are to be omitted from $[B']$ and for the same reason phase shifting elements are to be omitted from $[B'']$.

Both $[B']$ and $[B'']$ are real and "sparse" and need to be triangularized only once, at the beginning of the study since they contain network admittances only.

The method converges very reliably in two to five iterations with fairly good accuracy even for large systems. A good, approximate solution is obtained after the first or second iteration. The speed per iteration is roughly five times that of the original Newton method.

10.15 LOAD FLOW SOLUTION USING Z-BUS

10.15.1 BUS IMPEDANCE FORMATION

Any power network can be formed using the following possible methods of construction.

1. A line may be added to a reference point or bus
2. A bus may be added to any existing bus in the system other than the reference bus through a new line
3. A line may be added joining two existing buses in the system forming a loop.

The above three modes are illustrated in Fig. 10.8.

10.15.2 ADDITION OF A LINE TO THE REFERENCE BUS

If unit current is injected into bus k, no voltage will be produced at other buses of the systems.

$$Z_{ik} = Z_{ki} = 0, \quad i \neq k \tag{10.101}$$

The driving point impedance of the new bus is given by (Fig. 10.9)

$$Z_{kk} = Z_{\text{line}} \tag{10.102}$$

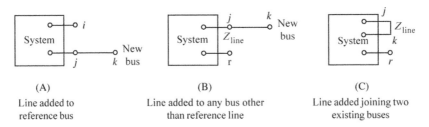

(A)	(B)	(C)
Line added to reference bus	Line added to any bus other than reference line	Line added joining two existing buses

FIGURE 10.8

Building of Z-bus.

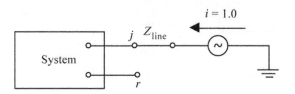

FIGURE 10.9

Addition of the line to reference line.

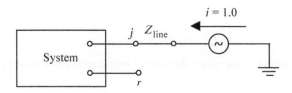

FIGURE 10.10

Addition of a radial line and new bus.

10.15.3 ADDITION OF A RADIAL LINE AND NEW BUS

Injection of unit current into the system through the new bus k produces voltages at all other buses of the system as shown in Fig. 10.10.

These voltages would of course be same as that would be produced if the current were injected instead at bus i as shown.

Therefore,

$$Z_{km} = Z_{im}$$

therefore,

$$Z_{mk} = Z_{mi}, \quad m \neq k \tag{10.103}$$

The dimension of the existing Z-bus matrix is increased by one. The off-diagonal elements of the new row and column are the same as the elements of the row and column of bus i of the existing system.

10.15.4 ADDITION OF A LOOP CLOSING TWO EXISTING BUSES IN THE SYSTEM

Since both the buses are existing buses in the system, the dimension of the bus impedance matrix will not increase in this case. However, the addition of the loop introduces a new axis which can be subsequently eliminated by Kron's reduction method.

The system in Fig. 10.11A can be represented alternatively as in Fig. 10.11B.

The link between i and k requires a loop voltage

$$V_{loop} = 1.0(Z_{ii} - Z_{2k} + Z_{kk} - Z_{ik} + Z_{line}) \tag{10.104}$$

for the circulation of unit current.

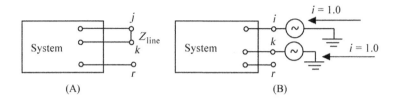

FIGURE 10.11

(A) Addition of a loop and (B) equivalent representation.

The loop impedance is

$$Z_{\text{loop}} = Z_{ii} + Z_{kk} - 2Z_{ik} + Z_{\text{line}} \qquad (10.105)$$

The dimension of $\mathbf{Z}$-matrix is increased due to the introduction of a new axis due to the loop 1

$$Z_{\ell\ell} = Z_{\text{loop}}$$

and

$$Z_{m-\ell} = Z_{mi} - Z_{mk}$$

$$Z_{\ell-m} = Z_{im} - Z_{km}; \quad m \neq \ell$$

The new loop axis can be eliminated now. Consider the matrix

$$\begin{bmatrix} Z_p & Z_q \\ Z_r & Z_s \end{bmatrix}$$

It can be proved easily that

$$Z'_p = Z_p - Z_q Z_s^{-1} Z_r \qquad (10.106)$$

Using Eq. (10.106), all the additional elements introduced by the loop can be eliminated. The method is illustrated in Example E.10.4.

10.15.5 GAUSS–SEIDEL METHOD USING *Z*-BUS FOR LOAD FLOW SOLUTION

An initial bus voltage vector is assumed as in the case of *Y*-bus method. Using these voltages, the bus currents are calculated using Eq. (10.25) or (10.26).

$$I = \frac{P_i - jQ_i}{V^*} - y_i v_i \qquad (10.107)$$

where y_i is the total shunt admittance at the bus i and $y_{ii}v_i$ is the shunt current flowing from bus i to ground.

A new bus voltage estimate is obtained for an *n*-bus system from the relation:

$$V_{\text{bus}} = Z_{\text{bus}} I_{\text{bus}} + V_R \qquad (10.108)$$

where V_R is the $(n-1) \times 1$-dimensional reference voltage vector containing in each element the slack bus voltage. It may be noted that since the slack bus is the reference bus, the dimension of the Z_{bus} is $(n-1 \times (n-1))$.

The voltages are updated from iteration to iteration using the relation:

$$V_i^{m+1} = V_S + \sum_{\substack{k=1 \\ k \neq S}}^{i-1} Z_{ik} I_k^{m+1} + \sum_{\substack{k=i \\ k \neq S}}^{n} Z_{ik} I_k^{(m)} \tag{10.109}$$

Then

$$V_k^{(m)} = \frac{P_k - jQ_k}{\left(V_k^{(m)}\right)^*} - y_k V_k^{(m+1)}$$

$$i = 1, 2, \ldots, n$$

$$S = \text{slack bus}$$

10.16 CONVERGENCE CHARACTERISTICS

The number of iterations required for convergence to solution depends considerably on the correction to voltage at each bus. If the correction ΔV_i at bus i is multiplied by a factor a, it is found that acceleration can be obtained to convergence rate. Then multiplier a is called acceleration factor. The difference between the newly computed voltage and the previous voltage at the bus is multiplied by an appropriate acceleration factor a. The value of a that generally improves the convergence is greater than are. In general $1 < a < 2$ and a typical value for $a = 1.5$ or 1.6 the use of acceleration factor amounts to a linear extrapolation of bus voltage V_i. For a given system, it is quite often found that a near optimal choice of a exists as suggested in the literature over a range of operating condition complex value is also suggested for a. Same suggested different a values for real and imaginary parts of the bus voltages.

The convergence of iterative methods depends upon the diagonal dominance of the bus admittance matrix. The self-admittances of the buses (diagonal terms) are usually large relative to the mutual admittances (off-diagonal terms). For this reason, convergence is obtained for power flow solution methods.

Junctions of high and low series impedances and large capacitances obtained in cable circuits, long EHV lines, series, and shunt. Compensation are detrimental to convergence as these tend to weaken the diagonal dominance in Y-bus matrix. The choice of swing bus may also affect convergence considerably. In difficult cases, it is possible to obtain convergence by removing the least diagonally dominant row and column of Y-bus. The salient features of Y-bus matrix iterative methods are that the elements in the summation term in equation 10.27 or 10.28 are on the average 2 or 3 only even for well-developed power systems. The sparsity of the Y matrix and its symmetry reduces both the storage requirement and the computation time for iteration.

For large well-conditioned system of n buses, the number of iterations required are of the order n and the total computing time varies approximately as n^2.

In contrast, the Newton–Raphson method gives convergence in three to four iterations. No acceleration factors are needed to the used. Being a gradient method solution is obtained and not must faster than any iterative method.

10.17 COMPARISON OF VARIOUS METHODS FOR POWER FLOW SOLUTION

The requirements of a good power flow method are high speed, low storage, and reliability for ill-conditioned problems. No single method meets all these requirements. It may be mentioned that for regular load flow studies NR method in polar coordinates and for special applications fast decoupled load flow solution methods have proned to be most useful than other methods. NR method is versatile, reliable, and accurate. Fast decoupled load flow method is fast and needs the least storage.

Convergence of iterative methods depends upon the dominance of the diagonal elements of the bus admittance matrix.

Advantages of the Gauss–Seidel method

1. The method is very simple in calculations and thus programing is easier.
2. The storage needed in the computer memory is relatively less.
3. In general, the method is applicable for smaller systems.

Disadvantages of the Gauss–Seidel method

1. The number of iterations needed is generally high and is also dependent on the acceleration factor selected.
2. For large systems, use of Gauss–Seidel method is practically prohibitive.
3. The time for convergence also increases dramatically with increase of number of buses.

Advantages of the Newton–Raphson method

1. The method is more accurate, faster, and reliable.
2. Requires less number of iteration for convergence. In fact, in three to four iterations good convergence is reached irrespective of the size of the system.
3. The number of iterations required is thus independent of the size of the system or the number of buses in the system.
4. The method is best suited for load flow solution to large size systems.
5. Decoupled and fast decoupled power flow solution can be obtained from the Newton–Raphson Polar Coordinates method. Hence, it also can serve as a base for security and contingency studies.

Disadvantages of the Newton–Raphson method

1. The memory needed is quite large for large size systems.
2. Calculations per iteration are also much larger than the Gauss–Seidel method.
3. Since, it is a gradient method, the method is quite involved and hence, programming is also comparatively difficult and complicated.

WORKED EXAMPLES

E.10.1. A three-bus power system is shown in Fig. E.10.1. The system parameters are given in Table E.10.1 and the load and generation data in Table E.10.2. The voltage at bus 2 is maintained at 1.03 p.u. The maximum and minimum reactive power limits of the

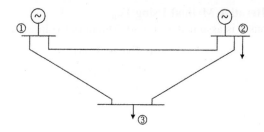

FIGURE E.10.1

A three-bus power system.

Table E.10.1 Impedance and Line Charging Admittances

Bus Code $i-k$	Impedance (p.u.) Z_{ik}	Line Charging Admittance (p.u.) y_i
1–2	$0.08 + j0.24$	0
1–3	$0.02 + j0.06$	0
2–3	$0.06 + j0.018$	0

Table E.10.2 Scheduled Generation, Loads, and Voltages

Bus No. i	Bus Voltage, V_i	Generation		Load	
		MW	Mvar	MW	Mvar
1	$1.05 + j0.0$	–	–	0	0
2	$1.03 + j0.0$	20	–	50	20
3	–	0	0	60	25

generation at bus 2 are 35 and 0 Mvar, respectively. Taking bus 1 as slack bus obtain the load flow solution using the Gauss–Seidel iterative method using Y_{Bus}

Solution:

The line admittance is obtained as

$$y_{12} = 1.25 - j3.75$$

$$y_{23} = 1.667 - j5.00$$

$$y_{13} = 5.00 - j15.00$$

The bus admittance matrix is formed using the procedure indicated in Section 2.1 as

$$Y_{Bus} = \begin{bmatrix} 6.25 & -j18.75 & -1.25 & +j3.75 & -5.0 & +j15.0 \\ -1.25 & +j3.73 & 2.9167 & -j8.75 & -j1.6667 & +j5.0 \\ -5.0 & +j15.0 & -1.6667 & +j5.0 & 6.6667 & -j20.0 \end{bmatrix}$$

Gauss–Seidel Iterative Method Using Y_{Bus}:

The voltage at bus 3 is assumed as $1 + j0$. The initial voltages are therefore

$$V_1^{(0)} = 1.05 + j0.0$$
$$V_2^{(0)} = 1.03 + j0.0$$
$$V_3^{(0)} = 1.00 + j0.0$$

Base MVA $= 100$

Iteration 1: It is required to calculate the reactive power Q_2 at bus 2, which is a P–V or voltage controlled bus

$$\delta_2^{(0)} = \tan^{-1}\left(\frac{e_2^a}{e_2'}\right) = 0$$

$$e_{2(\text{new})}' = |V_2|_{\text{sch}} \cos \delta_2 = (1.03)(1.0) = 1.03$$

$$e_{2(\text{new})}' = V_{2\text{sch}} \sin \delta_2^{(0)} = (1.03)(0.0) = 0.00$$

$$Q_2^{(0)} = \left[\left(e_{2(\text{new})}'\right)^2 B_{22} + \left(e_{2(\text{new})}''\right)^2 B_{22} \right]$$

$$+ \sum_{\substack{k=1 \\ k \neq 2}}^{3} \left[\left(e_{2(\text{new})}'' e_k' G_{2k} + e_k'' B_{2k}\right) - \left(e_{2(\text{new})}' e_k'' G_{2k} - e_k' B_{2k}\right) \right]$$

Substituting the values

$$Q_2^{(0)} = \left[(1.03)^2 8.75 + (0)^2 8.75\right] + 0(1.05)(-1.25) + 0.(-3.75) - 1.03\left[(0)(-1.25) - (1.05)(-3.75)\right]$$
$$+ (0)\left[(1)(-1.6667) + (0)(-5.0)\right] - 1.03\left[(0)(-1.6667) - (1)(-5)\right]$$
$$= 0.07725$$

Mvar generated at bus 2

$$= \text{Mvar injection into bus 2} + \text{load Mvar}$$
$$= 0.07725 + 0.2 = 0.27725 \text{ p.u.}$$
$$= 27.725 \text{ Mvar}$$

This is within the limits specified.

The voltage at bus i is

$$V_i^{(m+1)} = \frac{+1}{Y_{ii}}\left[\frac{P_1 - jQ_1}{V_i^{(m)*}} - \sum_{k=1}^{i-1} y_{ik} V_k^{(m+1)} - \sum_{k=i+1}^{n1} y_{ik} V_k^{(m)}\right]$$

$$V_2^{(1)} = \frac{1}{Y_{22}}\left[\frac{P_1 - jQ_2}{V_i^{(0)*}} - Y_{21}V_1 - Y_{23}V_3^{(0)}\right]$$

$$= \frac{1}{(2.9167 - j8.75)}$$

$$\left[\frac{-0.3 - 0.07725}{1.03 - j0.0} - (-1.25 + j3.75)(1.05 + j0.0) + (-1.6667 + j5.0)(1 + j0.0) \right]$$

$$V_2^{(1)} = 1.01915 - j0.032491$$
$$= 1.0196673 \angle -1.826°$$

An acceleration factor of 1.4 is used for both real and imaginary parts. The accelerated voltage is obtained using

$$v_2' = 1.03 + 1.4(1.01915 - 1.03) = 1.01481$$
$$v_2'' = 0.0 + 1.4(-0.032491 - 0.0) = -0.0454874$$

$$V_2^{(1)}(\text{accelerated}) = 1.01481 - j0.0454874 = 1.01583 \angle -2.56648°$$

The voltage at bus 3 is given by

$$V_2^{(1)} = \frac{1}{Y_{33}} \left[\frac{P_3 - jQ_3}{V_3^{(0)*}} - Y_{31}V_1 - Y_{32}V_2^{(1)} \right]$$

$$= \frac{1}{6.6667 - j20}$$

$$\left[\left(\frac{-0.6 + j0.25}{1 - j0} \right) - (-5 + j15)(1.05 + j0) - (-1.6667 + j5)(1.01481 - j0.0454874) \right]$$

$$= 1.02093 - j0.0351381$$

The accelerated value of $V_3^{(1)}$ obtained using

$$v_3' = 1.0 + 1.4(1.02093 - 1.0) = 1.029302$$
$$v_3'' = 0 + 1.4(-0.0351384 - 0) = -0.0491933$$

$$V_3^{(1)} = 1.029302 - j0.049933 = 1.03048 \angle -2.73624°$$

The voltages at the end of the first iteration are

$$V_1 = 1.05 + j0.0$$

$$V_2^{(1)} = 1.01481 - j0.0454874$$

$$V_3^{(1)} = 1.029302 - j0.0491933$$

Check for convergence: An accuracy of 0.001 is taken for convergence

$$\left[\Delta v_2'\right]^{(0)} = \left[v_2'\right]^{(1)} - \left[v_2'\right]^{(0)} = 1.01481 - 1.03 = -0.0152$$

$$\left[\Delta v_2''\right]^{(0)} = \left[v_2''\right]^{(1)} - \left[v_2''\right]^{(0)} = -0.0454874 - 0.0 = -0.0454874$$

$$\left[\Delta v_3''\right]^{(0)} = \left[v_3'\right]^{(1)} - \left[v_3'\right]^{(0)} = 1.029302 - 1.0 = 0.029302$$

$$\left[\Delta v_2''\right]^{(0)} = \left[\Delta v_2''\right]^{(1)} - \left[\Delta v_2''\right]^{(0)} = -0.0491933 - 0.0 = -0.0491933$$

The magnitudes of all the voltage changes are greater than 0.001.

Iteration 2: The reactive power Q_2 at bus 2 is calculated as before to give

$$\delta_2^{(1)} = \tan^{-1} \frac{[v_2'']^{(1)}}{[v_2']^{(1)}} = \tan^{-1} \left[\frac{-0.0454874}{1.01481} \right] = -2.56648°$$

$$[v_2']^{(1)} = |V_{2sch}| \cdot \cos \delta_2^{(1)} = 1.03 \cos(-2.56648°) = 1.02837$$

$$[v_2'']^{(1)} = |V_{2sch}| \cdot \sin \delta_2^{(1)} = 1.03 \sin(-2.56648°) = -0.046122$$

$$[V_{2new}]^{(1)} = 1.02897 - j0.046122$$

$$\begin{aligned} Q_2^{(1)} =\ & (1.02897)^2(8.75) + (-0.046122)^2(8.75) \\ & + (-0.046122)[1.05(-1.25) + (0)(-3.75)] \\ & - (1.02897)[(0)(-1.25) - (1.05)(-3.75)] \\ & + -(1.02897)[(-0.0491933)(-1.6667) - (1.029302)(-5)] \\ =\ & -0.0202933 \end{aligned}$$

Mvar to be generated at bus 2

$$= \text{Net Mvar injection into bus 2} + \text{load Mvar}$$
$$= -0.0202933 + 0.2 = 0.1797067 \text{ p.u.} = 17.97067 \text{ Mvar}$$

This is within the specified limits. The voltages are, therefore, the same as before

$$V_1 = 1.05 = j0.0$$

$$V_2^{(1)} = 1.02897 - j0.046122$$

$$V_3^{(1)} = 1.029302 - j0.0491933$$

The new voltage at bus 2 is obtained as

$$\begin{aligned} V_2^{(2)} =\ & \frac{1}{2.9167 - j8.75} \left[\frac{-0.3 + j0.0202933}{1.02827 + j0.046122} \right] \\ & - (-1.25 + j3.75)(1 - 05 + j0) \\ & - (-1.6667 + j5) \cdot (1.029302 - j0.0491933)] \\ =\ & 1.02486 - j0.0568268 \end{aligned}$$

The accelerated value of $V_2^{(2)}$ is obtained from

$$v_2' = 1.02897 + 1.4(1.02486 - 1.02897) = 1.023216$$

$$v_2'' = -0.046122 + 1.4(-0.0568268) - (-0.046122 = -0.0611087)$$

$$v_2^{(2)'} = 1.023216 - j0.0611087$$

The new voltage at bus 3 is calculated as

$$\begin{aligned} V_3^{(2)} =\ & \frac{1}{6.6667 - j20} \left[\frac{-6.6 + j0.25}{1.029302 + j0.0491933} \right] \\ & - (-5 + j15)(1.05 + j0.0) \\ & - (-1.6667 + j5.0) \cdot (1.023216 - j0.0611)] \\ =\ & 1.0226 - j0.0368715 \end{aligned}$$

The accelerated value of $V_2^{(2)}$ obtained from

$$v_3' = 1.029302 + 1.4(1.0226 - 1.029302) = 1.02$$

$$v_3'' = (-0.0491933) + 1.4(-0.0368715) + (0.0491933) = -0.03194278$$

$$V_3^{(2)} = 1.02 - j0.03194278$$

The voltages at the end of the second iteration are

$$V_1 = 1.05 = j0.0$$

$$V_2^{(2)} = 1.023216 - j0.0611087$$

$$V_3^{(2)} = 1.02 - j0.03194278$$

The procedure is repeated till convergence is obtained at the end of the sixth iteration. The results are tabulated in Table E.10.3.
Line flow from bus 1 to bus 2

$$S_{12} = V_1(V_1^* - V_2^*)Y_{12}^* = 0.228975 + j0.017396$$

Line flow from bus 2 to bus 1

$$S_{21} = V_2(V_2^* - V_1^*)Y_{21}^* = -0.22518 - j0.0059178$$

Similarly, the other line flows can be computed and are tabulated in Table E.10.4. The slack bus power obtained by adding the flows in the lines terminating at the slack bus is

$$P_1 + jQ_1 = 0.228975 + j0.017396 + 0.684006 + j0.225$$
$$= (0.912981 + j0.242396)$$

Table E.10.3 Bus Voltage

Iteration	Bus 1	Bus 2	Bus 3
0	$1.05 + j0$	$1.03 + j0$	$1.0 + j0$
1	$1.05 + j0$	$1.01481 - j0.04548$	$1.029302 - j0.049193$
2	$1.05 + j0$	$1.023216 - j0.0611087$	$1.02 - j0.0319428$
3	$1.05 + j0$	$1.033476 - j0.0481383$	$1.027448 - j0.03508$
4	$1.05 + j0$	$1.0227564 - j0.051329$	$1.0124428 - j0.0341309$
5	$1.05 + j0$	$1.027726 - j0.0539141$	$1.0281748 - j0.0363943$
6	$1.05 + j0$	$1.029892 - j0.05062$	$1.020301 - j0.0338074$
7	$1.05 + j0$	$1.028478 - j0.0510117$	$1.02412 - j0.034802$

Table E.10.4 Line Flows

Line	P	Power Flow	Q
1-2	$+0.228975$		0.017396
2-1	-0.225183		0.0059178
1-3	0.68396		0.224
3-1	-0.674565		-0.195845
2-3	-0.074129		0.0554
3-2	0.07461		-0.054

Newton–Raphson Polar Coordinates Method

The bus admittance matrix is written in polar form as

$$Y_{\text{Bus}} = \begin{bmatrix} 19.7642 \angle -71.6° & 3.95285 \angle -108.4° & 15.8114 \angle -108.4° \\ 3.95285 \angle -108.4° & 9.22331 \angle -71.6° & 5.27046 \angle -108.4° \\ 15.8114 \angle -108.4° & 5.27046 \angle -108.4° & 21.0819 \angle -71.6° \end{bmatrix}$$

Note that

$$\angle Y_{ii} = -71.6°$$

and

$$\angle Y_{ik} = -180° - 71.6° = 108.4°$$

The initial bus voltages are

$$V_1 = 1.05 \angle 0°$$

$$V_2^{(0)} = 1.03 \angle 0°$$

$$V_2^{(0)} = 1.0 \angle 0°$$

The real and reactive powers at bus 2 are calculated as follows:

$$P_2 = |V_2 V_1 Y_{21}| \cos(\delta_2^{(0)} - \delta_1 - \theta_{21}) + |V_2^2 \, Y_{22}| \cos(-\theta_{22})$$

$$+ |V_2 V_3 Y_{23}| \cos(\delta_2^{(0)} - \delta_3^{(0)} - \theta_{23})$$

$$= (1.03)(1.05)(3.95285)\cos(108.4°) + (1.03)^2(9.22331)\cos(-108.4°)$$

$$= (1.03)^2(9.22331)\cos(71.6°) + (1.03)(1.0)(5.27046)\cos(-108.4°)$$

$$= 0.02575$$

$$Q_2 = |V_2 V_1 Y_{21}| \sin(\delta_2^{(0)} - \delta_1 - \theta_{21}) + |V_2^2 \, Y_{22}| \sin(-\theta_{22})$$

$$+ |V_2 V_3 Y_{23}| \sin(\delta_2^{(0)} - \delta_3^{(0)} - \theta_{23})$$

$$= (1.03)(1.05)(3.95285)\sin(-108.4°) + (1.03)^2(9.22331)\sin(71.6°)$$

$$+ (1.03)(1.0)(5.27046)\sin(108.4°)$$

$$= 0.07725$$

Generation of p.u. Mvar at bus 2

$$= 0.2 + 0.07725$$
$$= 0.27725 = 27.725 \text{ Mvar}$$

This is within the limits specified. The real and reactive powers at bus 3 are calculated in a similar way.

$$P_3 = |V_3^{(0)} V_1 Y_{31}| \cos(\delta_3^{(0)} - \delta_1 - \theta_{31}) + |V_3^{(0)} V_2 Y_{32}| \cos(\delta_3^{(0)} - \delta_1 - \theta_{32})$$

$$+ |V_3^{(0)2} \, Y_{33}| \cos(-\theta_{33})$$

$$= (1.0)(1.05)(15.8114) \cos(-108.4°) + (1.0)(1.03)(5.27046) \cos(-108.4°)$$

$$+ (1.0)^2(21.0819) \cos(71.6°)$$

$$= -0.3$$

$$Q_3 = \left| V_3^{(0)} V_1 Y_{31} \right| \sin \left(\delta_3^{(0)} - \delta_1 - \theta_{31} \right) + \left| V_3^{(0)} V_2 Y_{32} \right| \sin \left(\delta_3^{(0)} - \delta_2 - \theta_{32} \right)$$

$$+ \left| V_3^{(0)2} Y_{33} \right| \sin \left(- \theta_{33} \right)$$

$$= (1.0)1.05(15.8114)\sin(- 108.4°) + (1.0)(1.03)(5.27046)\sin(-108.4°)$$

$$+ (1.0)^2(21.0891)\sin(71.6°)$$

$$= - 0.9$$

The difference between scheduled and calculated powers is

$$\Delta P_2^{(0)} = - 0.3 - 0.02575 = - 0.32575$$
$$\Delta P_3^{(0)} = - 0.6 - (- 0.3) = - 0.3$$
$$\Delta Q_2^{(0)} = - 0.25 - (- 0.9) = - 0.65$$

It may be noted that ΔQ_2 has not been computed since bus 2 is voltage controlled bus. Since

$$\left| \Delta P_2^{(0)} \right|, \quad \left| \Delta P_3^{(0)} \right| \quad \text{and} \quad \left| \Delta Q_3^{(0)} \right|$$

are greater than the specified limit of 0.01, the next iteration is computed.

Iteration 1: Elements of the Jacobian are calculated as follows:

$$\frac{\partial P_2}{\partial \delta_2} = - |V_2 V_1 Y_{21}| \sin \left(\delta_2^{(0)} - \delta_1^{(0)} - \theta_{21} \right) + \left| V_2 V_3^{(0)} Y_{23} \right| \sin \left(\delta_2^{(0)} - \delta_3^{(0)} - \theta_{23} \right)$$

$$= - (1.03)(1.05)(3.95285) \sin (108.4°) + (1.03)(1.0)(5.27046) \sin (-108.4°)$$

$$= 9.2056266$$

$$\frac{\partial P_2}{\partial \delta_3} = \left| V_2 V_3^{(0)} Y_{23} \right| \sin \left(\delta_2^{(0)} - \delta_3^{(0)} - \theta_{23} \right)$$

$$= (1.03)(1.0)(5.27046) \sin (-108.4°) = -5.15$$

$$\frac{\partial P_3}{\partial \delta_2} = \left| V_3^{(0)} V_1 Y_{31} \right| \sin \left(\delta_3^{(0)} - \delta_2^{(0)} - \theta_{32} \right)$$

$$= (0.0)(1.03)(5.27046) \sin (-108.4°)$$

$$= - 5.15$$

$$\frac{\partial P_3}{\partial \delta_3} = \left| V_3^{(0)} V_1 Y_{31} \right| \sin \left(\delta_3^{(0)} - \delta_1 - \theta_{31} \right) + \left| V_3^{(0)} V_2 Y_{32} \right| \sin \left(\delta_3^{(0)} - \delta_2^{(0)} - \theta_{32} \right)$$

$$= - (1.0)(1.05)(15.8114)\sin(- 108.4°) - 5.15$$

$$= 20.9$$

$$\frac{\partial P_2}{\partial v_3} = |V_2 Y_{23}|\cos(\delta_2^{(0)} - \delta_3^{(0)} - \theta_{23})$$

$$= (1.03)(5.27046)\cos(108.4°)$$

$$= - 1.7166724$$

$$\frac{\partial P_3}{\partial V_3} = 2|V_3 \ Y_{33}|\cos\theta_{33} + |V_1 \ Y_{31}|\cos(\delta_3^{(0)} - \delta_2^{(0)} - \theta_{32}) + |V_2 \ Y_{32}|\cos(\delta_3^{(0)} - \delta_2^{(0)} - \theta_{32})$$

$$= 2(1.0)(21.0819)\cos(71.6°) + (1.05)(15.8114)\cos(-108.4°)$$
$$+ (1.03)(5.27046)\cos(-108.4°)$$
$$= 6.366604$$

$$\frac{\partial Q_3}{\partial \delta_2} = -\left|V_3^{(0)}V_1 \ Y_{32}\right|\cos\left(\delta_3^{(0)} - \delta_2^{(0)} - \theta_{32}\right)$$
$$= (1.0)(1.03)(5.27046)\cos(-108.4°)$$
$$= 1.7166724$$

$$\frac{\partial Q_3}{\partial \delta_3} = \left|V_3^{(0)}V_1 \ Y_{32}\right|\cos\left(\delta_3^{(0)} - \delta_1 - \theta_{31}\right) + \left|V_3^{(0)} \ V_2 \ Y_{32}\right|\cos\left(\delta_2^{(0)} - \delta_3^{(0)} - \theta_{32}\right)$$

$$= (1.0)(1.05)(15.8114)\cos(-108.4°) - 1.7166724$$
$$= -6.9667$$

$$\frac{\partial Q_3}{\partial V_3} = 2V_3^{(0)} \ Y_{33} \sin(-\theta_{33}) + |V_1 \ Y_{31}| \sin(\delta_3^{(0)} - \delta_1 - \theta_{31}) + |V_2 \ Y_{32}| \sin(\delta_3^{(0)} - \delta_2^{(0)} - \theta_{32})$$
$$= 2(1.0)(21.0819) \sin(71.6°) + (1.05)(15.8114) \sin(-108.4°) + (1.03)(5.27046) \sin(-108.4°)$$
$$= 19.1$$

From Eq. (10.69)

$$\begin{bmatrix} -0.32575 \\ -0.3 \\ 0.65 \end{bmatrix} = \begin{bmatrix} 9.20563 & -5.15 & -1.71667 \\ -5.15 & 20.9 & 6.36660 \\ 1.71667 & -6.9967 & 19.1 \end{bmatrix} \cdot \begin{bmatrix} |\Delta\delta_2| \\ \Delta\delta_3 \\ \Delta|V_3| \end{bmatrix}$$

Following the method of triangulation and back substitutions

$$\begin{bmatrix} -0.35386 \\ -0.3 \\ -0.035386 \end{bmatrix} = \begin{bmatrix} 1 & -0.55944 & -0.18648 \\ -5.15 & 20.9 & 6.36660 \\ +1.71667 & -6.9667 & 19.1 \end{bmatrix} \cdot \begin{bmatrix} \Delta\delta_2 \\ \Delta\delta_3 \\ \Delta|V_3| \end{bmatrix}$$

$$\begin{bmatrix} -0.35386 \\ -0.482237 \\ +0.710746 \end{bmatrix} = \begin{bmatrix} 1 & -0.55944 & -0.18648 \\ 0 & 18.02 & 5.40623 \\ 0 & -6.006326 & 19.42012 \end{bmatrix} \cdot \begin{bmatrix} \Delta\delta_2 \\ \Delta\delta_3 \\ \Delta|V_3| \end{bmatrix}$$

Finally,

$$\begin{bmatrix} -0.35386 \\ -0.0267613 \\ 0.55 \end{bmatrix} = \begin{bmatrix} 1 & -0.55944 & -0.18648 \\ 0 & 1 & 0.3 \\ 0 & 0 & 21.22202 \end{bmatrix} \cdot \begin{bmatrix} \Delta\delta_2 \\ \Delta\delta_3 \\ \Delta|V_3| \end{bmatrix}$$

Thus

$$\Delta|V_3| = (0.55)/(21.22202) = 0.025917$$

$$\Delta\delta_3 = -0.0267613 - (0.3)(0.025917)$$
$$= -0.0345364 \text{ rad}$$
$$= -1.98°$$

$$\Delta\delta_2 = -0.035286 - (-0.55944)(-0.034536) - (-0.18648)(0.025917)$$
$$= -0.049874 \text{ rad}$$
$$= -2.8575°$$

At the end of the first iteration, the bus voltages are

$$V_1 = 1.05 \angle 0°$$
$$V_2 = 1.03 \angle 2.85757°$$
$$V_3 = 1.025917 \angle -1.9788°$$

The real and reactive powers at bus 2 are computed as

$$P_2^{(1)} = (1.03)(1.05)(3.95285)\left[\cos(-2.8575) - 0(-108.4°)\right]$$
$$+ (1.03)^2(1.025917)(5.27046) \cos\left[(-2.8575) - (-1.9788) - 108.4°\right]$$
$$= -0.30009$$

$$Q_2^{(1)} = (1.03)(1.05)(3.95285)\left[\sin(-2.8575) - 0(-108.4°)\right]$$
$$+ (1.03)^2(9.22331) \sin\left[(-2.85757) - (-1.9788) - 108.4°\right]$$
$$= 0.043853$$

Generation of reactive power at bus 2

$$= 0.2 + 0.043853 = 0.243856 \text{ p.u. Mvar}$$
$$= 24.3856 \text{ Mvar}$$

This is within the specified limits.
The real and reactive powers at bus 3 are computed as

$$P_3^{(1)} = (1.025917)(1.05)(15.8117) \cos\left[(-1.09788) - 0 - 108.4°\right]$$
$$+ (1.025917)(1.03)(5.27046) \cos\left[(-1.0988) - (-2.8575) - 108.4\right]$$
$$+ (1.025917)^2(21.0819) \cos(71.6°)$$
$$= -0.60407$$

$$Q_3^{(1)} = (1.025917)(1.05)(15.8114) \sin\left[(-1.977) - 108.4°\right]$$
$$+ (1.025917)(1.03)(5.27046) \sin\left[(-1.9788) - (-2.8575) - 108.4°\right]$$
$$+ (1.025917)^2(21.0819) \sin(71.6°)$$
$$= -0.224$$

The differences between scheduled powers and calculated powers are

$$\Delta P_2^{(1)} = -0.3 - (-0.30009) = 0.00009$$

$$\Delta P_3^{(1)} = -0.6 - (-0.60407) = 0.00407$$

$$\Delta Q_3^{(1)} = -0.25 - (-0.2224) = -0.0276$$

Even though the first two differences are within the limits the last one $Q_2^{(1)}$ is greater than the specified limit 0.01. The next iteration is carried out in a similar manner. At the end of the second iteration, even ΔQ_3 also is found to be within the specified tolerance. The results are tabulated in Tables E.10.5 and E.10.6.

E.10.2. **Consider the bus system shown in** Fig. E.10.2.

The following is the data:

Table E.10.5 Bus Voltages

Iteration	Bus 1	Bus 2	Bus 3
0	$1.05 \angle 0°$	$1.03 \angle 0°$	$1. \angle 0°$
1	$1.05 \angle 0°$	$1.03 \angle -2.85757$	$1.025917 \angle -1.9788$
2	$1.05 \angle 0°$	$1.03 \angle -2.8517$	$1.02476 \angle -1.947$

Table E.10.6 Line Flows

Line	P	Power Flow	Q
1−2	0.2297		0.016533
2−1	−0.22332		−0.0049313
1−3	0.68396		0.224
3−1	−0.674565		−0.0195845
2−3	−0.074126		0.0554
3−2	0.07461		−0.054

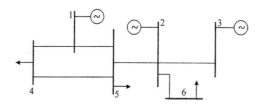

FIGURE E.10.2

A six bus power system.

Line Impedance (p.u.)	Real		Imaginary	
1−4	0.57000	E-1	0.845	E-1
1−5	1.33000	E-2	3.600	E-2
2−3	3.19999	E-2	1.750	E-1
2−5	1.73000	E-2	0.560	E-1
2−6	3.00000	E-2	1.500	E-1
4−5	1.94000	E-2	0.625	E-1

Scheduled generation and bus voltages:

Bus Code P	Assumed Bus Voltage	Generation		Load	
		MW p.u.	Mvar p.u.	MW p.u.	Mvar p.u.
1	$1.05 + j0.0$ (specified)	−	−	−	−
2	−	1.2	0.05	−	−
3	−	1.2	0.05	−	−
4	−	−	−	1.4	0.05
5	−	−	−	0.8	0.03
6	−	−	−	0.7	0.02

1. Taking bus 1 as slack bus and using an accelerating factor of 1.4, perform load flow by the Gauss−Seidel method. Take precision index as 0.0001.
2. Solve the problem also using the Newton−Raphson polar coordinate method.

Solution:
The bus admittance matrix is obtained as

	Bus Code	Admittance (p.u.)
$P-Q$	Real	Imaginary
1−1	14.516310	− 32.57515
1−4	− 5.486446	8.13342
1−5	− 9.029870	24.44174
2−2	7.329113	− 28.24106
2−3	− 1.011091	5.529494
2−5	− 5.035970	16.301400
2−6	− 1.282051	6.410257
3−2	− 1.011091	5.529404
3−3	1.011091	− 5.529404
4−1	− 5.486446	8.133420
4−4	10.016390	− 22.727320
4−5	− 4.529948	14.593900
5−1	− 9.029870	24.441740
5−2	− 5.035970	16.301400
5−4	− 4.529948	14.593900

(Continued)

Continued		
	Bus Code	**Admittance (p.u.)**
P–Q	**Real**	**Imaginary**
5–5	18.595790	− 55.337050
6–2	− 1.282051	6.410257
6–6	1.282051	− 6.410254

All the bus voltages, $V^{(0)}$, are assumed to be $1 + j0$ except the specified voltage at bus 1 which is kept fixed at $1.05 + j0$. The voltage equations for the first Gauss−Seidel iteration are

$$V_2^{(1)} = \frac{1}{Y_2}\left[\frac{P_2 - jQ_2}{V_2^{(0)*}} - Y_{23}V_3^{(0)} - Y_{25}V_5^{(0)} - Y_{26}V_6^{(0)}\right]$$

$$V_3^{(1)} = \frac{1}{Y_{33}}\left[\frac{P_3 - jQ_3}{V_3^{(0)*}} - Y_{32}V_2^{(1)}\right]$$

$$V_4^{(1)} = \frac{1}{Y_{44}}\left[\frac{P_4 - jQ_4}{V_4^{(0)*}} - Y_{41}V_1 - Y_{45}V_5^{(0)}\right]$$

$$V_5^{(1)} = \frac{1}{Y_{55}}\left[\frac{P_5 - jQ_5}{V_5^{(0)*}} - Y_{51}V_l - Y_{51}V_2^{(1)} - Y_{54}V_4^{(1)}\right]$$

$$V_6^{(1)} = \frac{1}{Y_{66}}\left[\frac{P_6 - jQ_6}{V_6^{(0)*}} - Y_{62}V_2^{(1)}\right]$$

Substituting the values, the equations for solution are

$$V_2^{(1)} = \left(\frac{1}{7.329113} - j28.24100\right) \times \left[\frac{1.2 - j0.05}{1 - j0}\right]$$

$$- (- 1.011091 + j5.529404) \times (1 + j0) - (- 5.03597 + j16.3014)(1 + j0)$$

$$- (1 - 282051 + j16.3014)(1 + j0)$$

$$= 1.016786 + j0.0557924$$

$$V_3^{(1)} = \left(\frac{1}{1.011091} - j5.52424\right) \times \left[\frac{1.2 - j0.05}{1 - j0}\right]$$

$$- (- 1.011091 + j5.529404) \times (1.016786 + j0.0557924)$$

$$= 1.089511 + j0.3885233$$

$$V_4^{(1)} = \left(\frac{1}{10.01639} - j22.72732\right) \times \left[\frac{- 1.4 + j0.005}{1 - j0}\right]$$

$$- (- 5.486446 + j8.133342) \times (1.05 + j0)$$

$$- (- 4.529948 + j14.5939)(1 + j0)$$

$$= 0.992808 - j0.0658069$$

$$V_5^{(1)} = \left(\frac{1}{18.59579} - j55.33705 \right) \times \left[\frac{-0.8 + j0.03}{1 - j0} \right]$$

$$- (-9.02987 + j24.44174) \times (1.05 + j0)$$

$$- (-5.03597 + j16.3014)(1.016786 + j0.0557929)$$

$$- (-4.529948 + j14.5939)(0.992808 - j0.0658069)$$

$$= 1.028669 - j0.01879179$$

$$V_6^{(1)} = \left(\frac{1}{1.282051} - j6.410257 \right) \times \left[\frac{-0.7 + j0.02}{1 - j0} \right]$$

$$- (-1.282051 - j6.410257) \times (1.016786 + j0.0557924)$$

$$= 0.989904 - j0.0669962$$

The results of these iterations are given in Table E.10.7.
In the polar form, all the voltages at the end of the 14th iteration are given in Table E.10.8.

Table E.10.7

It. No.	Bus 2	Bus 3	Bus 4	Bus 5	Bus 6
0	1 + j0.0	1 + j0.0	1 + j0.0	1 + j0.0	1 + j0.0
1	1.016789 + j0.0557924	1.089511 + j0.3885233	0.992808 − j0.0658069	1.02669 − j0.01879179	0.989901 − j0.0669962
2	1.05306 + j0.1018735	1.014855 + j0.2323309	1.013552 − j0.0577213	1.042189 + j0.0177322	1.041933 + j0.0192121
3	1.043568 + j0.089733	1.054321 + j0.3276035	1.021136 − j0.0352727	1.034181 + j0.00258192	1.014571 − j0.02625271
4	1.047155 + j0.101896	1.02297 + j0.02763564	1.012207 − j0.0500558	1.035391 + j0.00526437	1.02209 + j0.00643566
5	1.040005 + j0.093791	1.03515 + j0.3050814	1.61576 − j0.04258692	0.033319 + j0.003697056	1.014416 − j0.01319787
6	1.04212 + j0.0978431	1.027151 + j0.2901358	1.013044 − j0.04646546	10.33985 + j0.004504417	1.01821 − j0.001752973
7	1.040509 + j0.0963405	1.031063 + j0.2994083	1.014418 − j0.0453101	1.033845 + j0.00430454	1.016182 − j0.00770669
8	1.041414 + j0.097518	1.028816 + j0.294465	1.013687 − j0.0456101	1.033845 + j0.004558826	1.017353 − j0.0048398
9	1.040914 + j0.097002	1.030042 + j0.2973287	1 − 014148 − j0.04487629	1.033711 + j0.004413647	1.016743 − j0.0060342
10	1.041203 + j0.0972818	1.02935 + j0.2973287	1.013881 − j0.04511174	1.03381 + j0.004495542	1.017089 − j0.00498989
11	1.041036 + j0.097164	1.029739 + j0.296598	1.01403 − j0.04498312	1.03374 + j0.004439559	1.016877 − j0.00558081
12	1.041127 + j0.0971998	1.029518 + j0.2960784	1.013943 − j0.04506212	1.033761 + j0.00447096	1.016997 − j0.00524855
13	1.041075 + j0.0971451	1.029642 + j0.2963715	1.019331 − j0.04501488	1.033749 + j0.004454002	1.016927 − j0.00543323
14	1.041104 + j0.0971777	1.02571 + j0.2962084	1.0013965 − j0.04504223	1.033756 + j0.004463713	1.016967 − j0.00053283

Table E.10.8

Bus	Voltage Magnitude (p.u.)	Phase Angle (°)
1	1.05	0
2	1.045629	5.3326
3	1.071334	16.05058
4	1.014964	− 2.543515
5	1.033765	2.473992
6	1.016981	− 3.001928

E.10.3. For the given sample power system find load flow solution using the N−R polar coordinates method, decoupled method, and fast decoupled method.

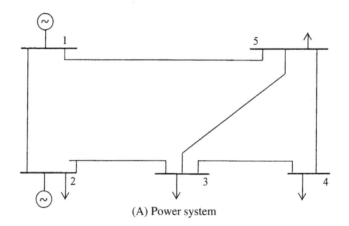

(A) Power system

Bus Code	Line Impedance Z_{pq}	Line Charging
1–2	0.02 + j0.24	j0.02
2–3	0.04 + j0.02	j0.02
3–5	0.15 + j0.04	j0.025
3–4	0.02 + j0.06	j0.01
4–5	0.02 + j0.04	j0.01
5–1	0.08 + j0.02	j0.2

(B) Line data

FIGURE E.10.3

Five bus system for E.10.3. (A) Power system, (B) Line data, (C) generation and load data, and (D) bus admittance matrix

Bus Code (Slack)	Generation		Load	
	Mw	Mvar	MW	Mvar
1	0	0	0	0
2	50	25	15	10
3	0	0	45	20
4	0	0	40	15
5	0	0	50	25

(C) Generation and load data

	1	2	3	4	5
1	$11.724 - j24.27$	$-10 + j20$	$0 + j0$	$0 + j0$	$-1.724 + j4.31$
2	$-10 + j20$	$10.962 - j24.768$	$-0.962 + j4.808$	$0 + j0$	$0 + j0$
3	$0 + j0$	$-0.962 + j4.808$	$6.783 - j21.944$	$-5 + j15$	$-0.822 + j2.192$
4	$0 + j0$	$0 + j0$	$-5 + j15$	$15 - j34.98$	$-10 + j20$
5	$-1.724 + j4.31$	$0 + j0$	$-0.82 + j2.192$	$-10 + j20$	$12.546 - j26.447$

(D) Bus admittance matrix

FIGURE E.10.3

(Continued)

Solution:
The residual or mismatch vector for iteration no. 1 is
  $dp[2] = 0.04944$
  $dp[3] = -0.041583$
  $dp[4] = -0.067349$
  $dp[5] = -0.047486$
  $dQ[2] = -0.038605$
  $dQ[3] = -0.046259$
  $dQ[4] = -0.003703$
  $dQ[5] = -0.058334.$

The new voltage vector after iteration 1 is
Bus no. 1: E: 1.000000; F: 0.000000
Bus no. 2: E: 1.984591; F: -0.008285
Bus no. 3: E: 0.882096; F: -0.142226
Bus no. 4: E: 0.86991; F: -0.153423
Bus no. 5: E: 0.875810; F: $-0.142707.$

The residual or mismatch vector for iteration no. 2 is
$dp[2] = 0.002406$
$dp[3] = -0.001177$

$dp[4] = -0.004219$
$dp[5] = -0.000953$
$dQ[2] = -0.001087$
$dQ[3] = -0.002261$
$dQ[4] = -0.000502$
$dQ[5] = -0.002888.$

The new voltage vector after iteration 2 is
Bus no. 1: E: 1.000000; F: 0.000000
Bus no. 2: E: 0.984357; F: -0.008219
Bus no. 3: E: 0.880951; F: -0.142953
Bus no. 4: E: 0.868709; F: -0.154322
Bus no. 5: E: 0.874651; F: $-0.143439.$

The residual or mismatch vector for iteration no. 3 is
$dp[2] = 0.000005$
$dp[3] = -0.000001$
$dp[4] = -0.000013$
$dp[5] = -0.000001$
$dQ[2] = -0.000002$
$dQ[3] = -0.000005$
$dQ[4] = -0.000003$
$dQ[5] = -0.000007.$

The final load flow solution (for allowable error .0001):

Bus no. 1	Slack $P = 1.089093$	$Q = 0.556063$	$E = 1.000000$	$F = 0.000000$
Bus no. 2	pq $P = 0.349995$	$Q = 0.150002$	$E = 0.984357$	$F = -0.008219$
Bus no. 3	pq $P = -0.449999$	$Q = -0.199995$	$E = 0.880951$	$F = -0.1429531$
Bus no. 4	pq $P = -0.399987$	$Q = -0.150003$	$E = 0.868709$	$F = -0.154322$
Bus no. 5	pq $P = -0.500001$	$Q = -0.249993$	$E = 0.874651$	$F = -0.143439$

Decoupled load flow solution (polar coordinate method)
The residual or mismatch vector for iteration no. 0 is
$dp[2] = 0.350000$
$dp[3] = -0.450000$
$dp[4] = -0.400000$
$dp[5] = -0.500000$
$dQ[2] = -0.190000$
$dQ[3] = -0.145000$
$dQ[4] = -0.130000$
$dQ[5] = -0.195000.$

The new voltage vector after iteration 0:
Bus no. 1: E: 1.000000; F: 0.000000
Bus no. 2: E: 0.997385; F: −0.014700
Bus no. 3: E: 0.947017; F: −0.148655
Bus no. 4: E: 0.941403; F: −0.161282
Bus no. 5: E: 0.943803; F: −0.150753.

The residual or mismatch vector for iteration no. 1 is
$dp[2] = 0.005323$
$dp[3] = −0.008207$
$dp[4] = −0.004139$
$dp[5] = −0.019702$
$dQ[2] = −0.067713$
$dQ[3] = −0.112987$
$dQ[4] = −0.159696$
$dQ[5] = −0.210557.$

The new voltage vector after iteration 1:
Bus no. 1: E: 1.000000; F: 0.000000
Bus no. 2: E: 0.982082; F: −0.013556
Bus no. 3: E: 0.882750; F: −0.143760
Bus no. 4: E: 0.870666; F: −0.154900
Bus no. 5: E: 0.876161; F: −0.143484.

The residual or mismatch vector for iteration no. 2 is
$dp[2] = 0.149314$
$dp[3] = −0.017905$
$dp[4] = −0.002305$
$dp[5] = −0.006964$
$dQ[2] = −0.009525$
$dQ[3] = −0.009927$
$dQ[4] = −0.012938$
$dQ[5] = 0.007721.$

The new voltage vector after iteration 2:
Bus no. 1: E: 1.000000; F: 0.000000
Bus no. 2: E: 0.981985; F: −0.007091
Bus no. 3: E: 0.880269; F: −0.142767
Bus no. 4: E: 0.868132; F: −0.154172
Bus no. 5: E: 0.874339; F: −0.143109.

The residual or mismatch vector for iteration no. 3 is
$dp[2] = 0.000138$
$dp[3] = 0.001304$

$dp[4] = 0.004522$

$dp[5] = -0.006315$

$dQ[2] = 0.066286$

$dQ[3] = 0.006182$

$dQ[4] = -0.001652$

$dQ[5] = -0.002233.$

The new voltage vector after iteration 3:

Bus no. 1: E: 1.000000; F: 0.000000

Bus no. 2: E: 0.984866; F: -0.007075

Bus no. 3: E: 0.881111; F: -0.142710

Bus no. 4: E: 0.868848; F: -0.154159

Bus no. 5: E: 0.874862; F: $-0.143429.$

The residual or mismatch vector for iteration no. 4 is

$dp[2] = -0.031844$

$dp[3] = 0.002894$

$dp[4] = -0.000570$

$dp[5] = 0.001807$

$dQ[2] = -0.000046$

$dQ[3] = 0.000463$

$dQ[4] = 0.002409$

$dQ[5] = -0.003361.$

The new voltage vector after iteration 4:

Bus no. 1: E: 1.000000; F: 0.000000

Bus no. 2: E: 0.984866; F: -0.008460

Bus no. 3: E: 0.881121; F: -0.142985

Bus no. 4: E: 0.868849; F: -0.1546330

Bus no. 5: E: 0.874717; F: $-0.143484.$

The residual or mismatch vector for iteration no. 5 is

$dp[2] = 0.006789$

$dp[3] = -0.000528$

$dp[4] = -0.000217$

$dp[5] = -0.0000561$

$dQ[2] = -0.000059$

$dQ[3] = -0.000059$

$dQ[4] = -0.000635$

$dQ[5] = -0.000721.$

The new voltage vector after iteration 5:

Bus no. 1: E: 1.000000; F: 0.000000

Bus no. 2: E: 0.984246; F: -0.008169

Bus no. 3: E: 0.880907; F: -0.142947
Bus no. 4: E: 0.868671; F: -0.154323
Bus no. 5: E: 0.874633; F: -0.143431.

The residual or mismatch vector for iteration no. 6 is
$dp[2] = 0.000056$
$dp[3] = 0.000010$
$dp[4] = 0.000305$
$dp[5] = -0.000320$
$dQ[2] = 0.003032$
$dQ[3] = -0.000186$
$dQ[4] = -0.000160$
$dQ[5] = -0.000267$.

The new voltage vector after iteration 6:
Bus no. 1: E: 1.000000; F: 0.000000
Bus no. 2: E: 0.984379; F: -0.008165
Bus no. 3: E: 0.880954; F: -0.142941
Bus no. 4: E: 0.868710; F: -0.154314
Bus no. 5: E: 0.874655; F: -0.143441.

The residual or mismatch vector for iteration no. 7 is
$dp[2] = -0.001466$
$dp[3] = 0.000106$
$dp[4] = -0.000073$
$dp[5] = 0.000156$
$dQ[2] = 0.000033$
$dQ[3] = 0.000005$
$dQ[4] = 0.000152$
$dQ[5] = -0.000166$.

The new voltage vector after iteration 7:
Bus no. 1: E: 1.000000; F: 0.000000
Bus no. 2: E: 0.954381; F: -0.008230
Bus no. 3: E: 0.880958; F: -0.142957
Bus no. 4: E: 0.868714; F: -0.154325
Bus no. 5: E: 0.874651; F: -0.143442.

The residual or mismatch vector for iteration no. 8 is
$dp[2] = -0.000022$
$dp[3] = 0.000001$
$dp[4] = -0.000072$
$dp[5] = -0.000074$
$dQ[2] = -0.000656$

$dQ[3] = 0.000037$
$dQ[4] = -0.000048$
$dQ[5] = -0.000074$.

The new voltage vector after iteration 8:
Bus no. 1: E: 1.000000; F: 0.000000
Bus no. 2: E: 0.984352; F: -0.008231
Bus no. 3: E: 0.880947; F: -0.142958
Bus no. 4: E: 0.868706; F: -0.154327
Bus no. 5: E: 0.874647; F: -0.143440.

The residual or mismatch vector for iteration no. 9 is
$dp[2] = 0.000318$
$dp[3] = -0.000022$
$dp[4] = 0.000023$
$dp[5] = -0.000041$
$dQ[2] = -0.000012$
$dQ[3] = -0.000000$
$dQ[4] = 0.000036$
$dQ[5] = -0.000038$.

The new voltage vector after iteration 9:
Bus no. 1: E: 1.000000; F: 0.000000
Bus no. 2: E: 0.984352; F: -0.008217
Bus no. 3: E: 0.880946; F: -0.142954
Bus no. 4: E: 0.868705; F: -0.154324
Bus no. 5: E: 0.874648; F: -0.143440.

The residual or mismatch vector for iteration no. 10 is
$dp[2] = 0.000001$
$dp[3] = -0.000001$
$dp[4] = 0.000017$
$dp[5] = -0.000017$
$dQ[2] = 0.000143$
$dQ[3] = -0.000008$
$dQ[4] = 0.000014$
$dQ[5] = -0.000020$

The new voltage vector after iteration 10:
Bus no. 1: E: 1.000000; F: 0.000000
Bus no. 2: E: 0.984658; F: -0.008216
Bus no. 3: E: 0.880949; F: -0.142954
Bus no. 4: E: 0.868707; F: -0.154324
Bus no. 5: E: 0.874648; F: -0.143440.

The residual or mismatch vector for iteration no. 11 is

$dp[2] = -0.000069$

$dp[3] = 0.000005$

$dp[4] = -0.000006$

$dp[5] = 0.000011$

$dQ[2] = 0.000004$

$dQ[3] = -0.000000$

$dQ[4] = 0.000008$

$dQ[5] = -0.000009.$

The final load flow solution after 11 iterations
(for allowable error .0001)

The final load flow solution (for allowable error .0001):

Bus no. 1	Slack $P = 1.089043$	$Q = 0.556088$	$E = 1.000000$	$F = 0.000000$
Bus no. 2	pq $P = 0.350069$	$Q = 0.150002$	$E = 0.984658$	$F = -0.008216$
Bus no. 3	pq $P = -0.450005$	$Q = -0.199995$	$E = 0.880949$	$F = -0.142954$
Bus no. 4	pq $P = -0.399994$	$Q = -0.150003$	$E = 0.868707$	$F = -0.154324$
Bus no. 5	pq $P = -0.500011$	$Q = -0.249991$	$E = 0.874648$	$F = -0.143440$

Fast Decoupled Load Flow Solution (Polar Coordinate Method)

The residual or mismatch vector for iteration no. 0 is

$dp[2] = 0.350000$

$dp[3] = -0.450000$

$dp[4] = 0.400000$

$dp[5] = -0.500000$

$dQ[2] = 0.190000$

$dQ[3] = -0.145000$

$dQ[4] = 0.130000$

$dQ[5] = -0.195000.$

The new voltage vector after iteration 0:

Bus no. 1: E: 1.000000; F: 0.000000

Bus no. 2: E: 0.997563; F: -0.015222

Bus no. 3: E: 0.947912; F: -0.151220

Bus no. 4: E: 0.942331; F: -0.163946

Bus no. 5: E: 0.944696; F: $-0.153327.$

The residual or mismatch vector for iteration no. 1 is

$dp[2] = 0.004466$

$dp[3] = -0.000751$

$dp[4] = 0.007299$

$dp[5] = -0.012407$

$dQ[2] = 0.072548$

$dQ[3] = -0.118299$
$dQ[4] = 0.162227$
$dQ[5] = -0.218309$.

The new voltage vector after iteration 1:
Bus no. 1: E: 1.000000; F: 0.000000
Bus no. 2: E: 0.981909; F: -0.013636
Bus no. 3: E: 0.882397; F: -0.143602
Bus no. 4: E: 0.869896; F: -0.154684
Bus no. 5: E: 0.875752; F: -0.143312

The residual or mismatch vector for iteration no. 2 is
$dp[2] = 0.153661$
$dp[3] = -0.020063$
$dp[4] = 0.005460$
$dp[5] = -0.009505$
$dQ[2] = 0.011198$
$dQ[3] = -0.014792$
$dQ[4] = -0.000732$
$dQ[5] = -0.002874$.

The new voltage vector after iteration 2:
Bus no. 1: E: 1.000000; F: 0.000000
Bus no. 2: E: 0.982004; F: -0.007026
Bus no. 3: E: 0.880515; F: -0.142597
Bus no. 4: E: 0.868400; F: -0.153884
Bus no. 5: E: 0.874588; F: -0.143038

The residual or mismatch vector for iteration no. 3 is
$dp[2] = -0.000850$
$dp[3] = -0.002093$
$dp[4] = 0.000155$
$dp[5] = -0.003219$
$dQ[2] = 0.067612$
$dQ[3] = -0.007004$
$dQ[4] = -0.003236$
$dQ[5] = -0.004296$.

The new voltage vector after iteration 3:
Bus no. 1: E: 1.000000; F: 0.000000
Bus no. 2: E: 0.984926; F: -0.007086
Bus no. 3: E: 0.881246; F: -0.142740
Bus no. 4: E: 0.869014; F: -0.154193
Bus no. 5: E: 0.874928; F: -0.143458.

The residual or mismatch vector for iteration no. 4 is

$dp[2] = -0.032384$

$dp[3] = 0.003011$

$dp[4] = -0.001336$

$dp[5] = -0.002671$

$dQ[2] = -0.000966$

$dQ[3] = -0.000430$

$dQ[4] = -0.000232$

$dQ[5] = -0.001698.$

The new voltage vector after iteration 4:

Bus no. 1: E: 1.000000; F: 0.000000

Bus no. 2: E: 0.984862; F: -0.008488

Bus no. 3: E: 0.881119; F: -0.143053

Bus no. 4: E: 0.868847; F: -0.154405

Bus no. 5: E: 0.874717; F: $-0.143501.$

The residual or mismatch vector for iteration no. 5 is

$dp[2] = 0.000433$

$dp[3] = 0.000006$

$dp[4] = -0.000288$

$dp[5] = 0.000450$

$dQ[2] = -0.014315$

$dQ[3] = -0.000936$

$dQ[4] = -0.000909$

$dQ[5] = -0.001265.$

The new voltage vector after iteration 6:

Bus no. 1: E: 1.000000; F: 0.000000

Bus no. 2: E: 0.984230; F: -0.008463

Bus no. 3: E: 0.881246; F: -0.143008

Bus no. 4: E: 0.869014; F: -0.154357

Bus no. 5: E: 0.874607; F: $-0.143433.$

The residual or mismatch vector for iteration no. 6 is

$dp[2] = 0.006981$

$dp[3] = -0.000528$

$dp[4] = 0.000384$

$dp[5] = -0.000792$

$dQ[2] = 0.000331$

$dQ[3] = 0.000039$

$dQ[4] = -0.000155$

$dQ[5] = 0.000247.$

The residual or mismatch vector for iteration no. 7 is
$dp[2] = -0.000144$
$dp[3] = -0.000050$
$dp[4] = 0.000080$
$dp[5] = -0.000068$
$dQ[2] = 0.003107$
$dQ[3] = -0.000162$
$dQ[4] = -0.000255$
$dQ[5] = -0.000375.$

The new voltage vector after iteration 7:
Bus no. 1: E: 1.000000; F: 0.000000
Bus no. 2: E: 0.984386; F: -0.008166
Bus no. 3: E: 0.880963; F: -0.142943
Bus no. 4: E: 0.868718; F: -0.154316
Bus no. 5: E: 0.874656; F: $-0.143442.$

The residual or mismatch vector for iteration no. 8 is
$dp[2] = -0.001523$
$dp[3] = -0.000105$
$dp[4] = -0.000115$
$dp[5] = -0.000215$
$dQ[2] = 0.000098$
$dQ[3] = -0.000024$
$dQ[4] = -0.000037$
$dQ[5] = -0.000038.$

The new voltage vector after iteration 8:
Bus no. 1: E: 1.000000; F: 0.000000
Bus no. 2: E: 0.984380; F: -0.008233
Bus no. 3: E: 0.880957; F: -0.142961
Bus no. 4: E: 0.868714; F: -0.154329
Bus no. 5: E: 0.874651; F: $-0.143442.$

The residual or mismatch vector for iteration no. 9 is
$dp[2] = -0.000045$
$dp[3] = 0.000015$
$dp[4] = -0.000017$
$dp[5] = 0.000008$
$dQ[2] = 0.000679$
$dQ[3] = 0.000031$
$dQ[4] = -0.000072$
$dQ[5] = -0.000105.$

The new voltage vector after iteration 9:
Bus no. 1: E: 1.000000; F: 0.000000
Bus no. 2: E: 0.984350; F: -0.008230
Bus no. 3: E: 0.880945; F: -0.142958
Bus no. 4: E: 0.868704; F: -0.154326
Bus no. 5: E: 0.874646; F: -0.143440.

The residual or mismatch vector for iteration no. 10 is
$dp[2] = 0.000334$
$dp[3] = -0.000022$
$dp[4] = 0.000033$
$dp[5] = -0.000056$
$dQ[2] = 0.000028$
$dQ[3] = 0.000007$
$dQ[4] = -0.000007$
$dQ[5] = 0.000005$.

The new voltage vector after iteration 10:
Bus no. 1: E: 1.000000; F: 0.000000
Bus no. 2: E: 0.984352; F: -0.008216
Bus no. 3: E: 0.880946; F: -0.142953
Bus no. 4: E: 0.898705; F: -0.154323
Bus no. 5: E: 0.874648; F: -0.143440.

The residual or mismatch vector for iteration no. 11 is
$dp[2] = -0.000013$
$dp[3] = -0.000004$
$dp[4] = 0.000003$
$dp[5] = -0.000000$
$dQ[2] = 0.000149$
$dQ[3] = -0.000007$
$dQ[4] = 0.000020$
$dQ[5] = -0.000027$.

The new voltage vector after iteration 11:
Bus no. 1: E: 1.000000; F: 0.000000
Bus no. 2: E: 0.984358; F: -0.008216
Bus no. 3: E: 0.880949; F: -0.142954
Bus no. 4: E: 0.868707; F: -0.154324
Bus no. 5: E: 0.874648; F: -0.143440.

The residual or mismatch vector for iteration no. 12 is

$dp[2] = -0.000074$

$dp[3] = 0.000005$

$dp[4] = -0.000009$

$dp[5] = -0.000014$

$dQ[2] = 0.000008$

$dQ[3] = -0.000002$

$dQ[4] = -0.000001$

$dQ[5] = -0.000000$.

The load flow solution

Bus no. 1	Slack $P = 1.089040$	$Q = 0.556076$	$E = 1.000000$	$F = 0.000000$
Bus no. 2	pq $P = 0.350074$	$Q = 0.150008$	$E = 0.984358$	$F = -0.008216$
Bus no. 3	pq $P = -0.450005$	$Q = -0.199995$	$E = 0.880949$	$F = -0.142954$
Bus no. 4	pq $P = -0.399991$	$Q = -0.150001$	$E = 0.868707$	$F = -0.154324$
Bus no. 5	pq $P = -0.500014$	$Q = -0.250000$	$E = 0.874648$	$F = -0.143440$

E.10.4. Obtain the load flow solution to the system given in example E.10.1 using Z-Bus. Use the Gauss–Seidel method. Take accuracy for convergence as 0.0001.

Solution:

The bus impedance matrix is formed as indicated in Section 10.15. The slack bus is taken as the reference bus. In this example, as in Example 10.1 bus 1 is chosen as the slack bus.

1. Add element 1–2. This is addition of a new bus to the reference bus

$$Z_{\text{BUS}} = \begin{array}{cc} & (2) \\ (2) & \boxed{0.05 + j0.24} \end{array}$$

2. Add element 1–3. This is also addition of a new bus to the reference bus

$$Z_{\text{BUS}} = \begin{array}{c|c|c} & (2) & (3) \\ \hline (2) & 0.08 + j0.24 & 0.0 + j0.0 \\ \hline (3) & 0.0 + j0.0 & 0.02 + j0.06 \end{array}$$

3. Add element 2–3. This is the addition of a link between two existing buses 2 and 3.

$$Z_{2-\text{loop}} = Z_{\text{loop}-2} = Z_{22} - Z_{23} = 0.08 + j0.24$$

$$Z_{3-\text{loop}} = Z_{\text{loop}-3} = Z_{32} - Z_{33} = -(0.02 + j0.06)$$

$$Z_{\text{loop}-\text{loop}} = Z_{22} + Z_{33} - 2Z_{23} + Z_{23,23}$$
$$= (0.08 + j0.24) + (0.02 + j0.06)(0.06 + j0.18)$$
$$= 0.16 + j0.48$$

$$Z_{\text{BUS}} = \begin{array}{c} (2) \\ (3) \\ (\ell) \end{array} \begin{array}{|c|c|c|} \hline \text{(2)} & \text{(3)} & \text{(}\ell\text{)} \\ \hline 0.08 + j0.024 & 0 + j0 & 0.08 + j0.24 \\ \hline 0.0 + j0.0 & 0.02 + j0.06 & -(0.02 + j0.06) \\ \hline 0.08 + j0.24 & -(0.02 + j0.006) & 0.16 + j0.48 \\ \hline \end{array}$$

The loop is now eliminated

$$Z'_{22} = Z_{22} - \frac{Z_{2-\text{loop}}\, Z_{\text{loop}-2}}{Z_{\text{loop}-\text{loop}}}$$

$$= (0.08 + j0.24) - \frac{(0.8 + j0.24)^2}{0.16 + j0.48}$$

$$= 0.04 + j0.12$$

$$Z'_{23} = Z'_{32} = \left[Z_{23} - \frac{Z_{2-\text{loop}}\, Z_{\text{loop}-3}}{Z_{\text{loop}-\text{loop}}} \right]$$

$$= (0.0 + j0.0) - \frac{(0.8 + j0.24)(-0.02 - j0.06)}{0.16 + j0.48}$$

$$= 0.01 + j0.03$$

Similarly $Z'_{33} = 0.0175 + j0.0526$
The Z-Bus matrix is thus

$$Z_{\text{Bus}} = \left[\begin{array}{c|c} 0.04 + j0.12 & 0.01 + j0.03 \\ \hline 0.01 + j0.03 & 0.017 + j0.0525 \end{array} \right]$$

$$= \left[\begin{array}{c|c} 0.1265 \angle 71.565^\circ & 0.031623 \angle 71.565^\circ \\ \hline 0.031623 \angle 71.565^\circ & 0.05534 \angle 71.565^\circ \end{array} \right]$$

The voltages at buses 2 and 3 are assumed to be

$$V_2^{(0)} = 1.03 + j0.0$$

$$V_3^{(0)} = 1.0 + j0.0$$

Assuming that the reactive power injected into bus 2 is zero:

$$Q_2 = 0.0$$

The bus currents $I_2^{(0)}$ and $I_3^{(0)}$ are computed as

$$I_2^{(0)} = \frac{-0.3 + j0.0}{1.03 - j0.0} = -0.29126 - j0.0 = 0.29126 \angle 180^\circ$$

$$I_3^{(0)} = \frac{-0.6 + j0.25}{1.0 + j0.0} = -0.6 - j0.25 = 0.65 \angle 157.38^\circ$$

Iteration 1: The voltage at bus 2 is computed as

$$V_2^{(1)} = V_2^{(1)} = V_1 + Z_{22} I_2^{(0)} + Z_{23} I^{(0)}$$
$$= 1.05 \angle 0° + (0.1265 \angle 71.565°(0.29126 \angle 180°$$
$$+ (0.031623 \angle 71.565°)(0.65 \angle 157.38°)$$
$$= 1.02485 - j0.05045$$
$$= 1.02609 \angle -2.8182$$

The new bus current $I_2^{(0)}$ is now calculated as

$$\Delta I_2^{(0)} = \frac{V_2^{(1)}}{Z_{22}} \left[\frac{|V_{sch}|}{|V_2^{(1)}|} - 1 \right]$$

$$= \frac{1.02609 \angle -2.8182}{0.1265 \angle 71.565°} \times \left(\frac{1.03}{1.02609} - 1 \right) = 0.0309084 \angle -74.3832°$$

$$Q^{(0)} = \mathrm{Im}[V_2^{(1)} \Delta I_2^{(0)*}]$$
$$= \mathrm{Im} \left[1.02609 \angle -2.8182° \right] (0.0309084 \angle 74.383°)$$
$$= 0.03$$

$$Q_2^{(1)} = Q_2^{(0)} + \Delta Q_2^{(0)} = 0.0 + 0.03 = 0.03$$

$$I_2^{(1)} = \frac{-0.3 - j0.3}{1.02609 \angle -2.8182°} = 0.29383 \angle 182.8918°$$

Voltage at bus 3 is now calculated as

$$V_3^{(1)} = V_1 + Z_{32} I_2^{(1)} + Z_{33} I_3^{(0)}$$
$$= 1.05 \angle 0.0° + (0.031623 \angle 71.565°)(0.29383 \angle 182.832°)$$
$$+ (0.05534 \angle 71.565°)(0.65 \angle 157.38°)$$
$$= (1.02389 - j0.036077) = 1.0245 \angle -2.018°$$

$$I_3^{(1)} = \frac{0.65 \angle 157.38°}{1.0245 \angle 2.018°} = 0.634437 \angle 155.36°$$

The voltages at the end of the first iteration are:

$$V_1 = 1.05 \angle 0°$$
$$V_2^{(1)} = 1.02609 \angle -2.8182°$$
$$V_3^{(1)} = 1.0245 \angle -2.018°$$

The differences in voltages are

$$\Delta V_2^{(1)} = (1.02485 - j0.05045) - (1.03 + j0.0)$$
$$= -0.00515 - j0.05045$$

$$\Delta V_3^{(1)} = (1.02389 - j0.036077) - (1.0 + j0.0)$$
$$= (0.02389 - j0.036077)$$

Both the real and imaginary parts are greater than the specified limit 0.001.

Iteration 2:

$$V_2^{(2)} = V_1 + Z_{22}I_2^{(1)(1)} + Z_{23}I_3^{(1)(1)}$$

$$= 1.02 \angle 0° + (0.1265 \angle 71.565°)(0.29383 \angle 182.892°)$$

$$+ (0.031623 \angle 71.565°)(0.63447 \angle 155.36°)$$

$$= 1.02634 - j0.050465 - 1.02758 \angle - 2.81495°$$

$$\Delta I_2^{(1)} = \frac{1.02758 \angle - 2.81495°}{1.1265 \angle - 71.565°} \left[\frac{1.03}{1.02758} - 1 \right]$$

$$= 0.01923 \angle - 74.38°$$

$$\Delta Q_2^{(1)} = \text{Im}\left[V_2^{(2)}(\Delta I_2^{(1)})^* \right]$$

$$= \text{Im}(1.02758 \angle - 2.81495°)(0.01913 \angle 74.38°)$$

$$= 0.0186487$$

$$Q_2^{(2)} = Q_2^{(1)} + \Delta Q_2^{(1)}$$

$$= 0.03 + 0.0186487 = 0.0486487$$

$$I_2^{(2)} = \frac{-0.3 - j0.0486487}{1.02758 \angle 2.81495°}$$

$$= 0.295763 \angle 186.393°$$

$$V_3^{(2)} = 1.05 \angle 0° + (0.31623 \angle 71.565°(0.295763 \angle 186.4°$$

$$+ 0.05534 \angle 71.565°)(0.634437 \angle 155.36°)$$

$$I_3^{(2)} = \frac{0.65 \angle 157.38°}{1.02466 \angle 1.9459°} = 0.6343567 \angle 155.434°$$

$$\Delta V_2^{(1)} = (1.02634 - j0.050465) - (1.02485 - j0.05041) = 0.00149 - j0.000015$$

$$\Delta V_3^{(1)} = (1.024 - j0.034793) - (1.02389 - j0.036077) = 0.00011 + j0.00128$$

As the accuracy is still not enough, another iteration is required.
Iteration 3:

$$V_2^{(3)} = 1.05 \angle 0° + (0.1265 \angle 71.565°)(0.295763 \angle 186.4°)$$

$$+ (0.031623 \angle 71.565°)(0.63487 \angle 155.434°)$$

$$= 1.0285187 - j0.051262$$

$$= 1.0298 \angle - 2.853°$$

$$I_2^{(2)} = \frac{1.0298 \angle - 2.853°}{0.1265 \angle 71.565°} \left[\frac{1.03}{1.0298} - 1 \right] = 0.001581 \angle 74.418°$$

$$\Delta Q_2^{(2)} = 0.00154456$$

$$Q_2^{(3)} = 0.0486487 + 0.001544 = 0.0502$$

$$I_2^{(3)} = \frac{-0.3 - j0.0502}{0.0298 \angle 2.853°} = 0.29537 \angle 186.647°$$

$$V_3^{(3)} = 1.05 \angle 0° + (0.031623 \angle 71.565°) + (0.29537 \angle 186.647°)$$
$$+ (0.05534 \angle 71.565°)(0.634357 \angle 155.434°)$$
$$= 1.024152 - j0.034817 = 1.02474 \angle -1.9471°$$

$$I_3^{(3)} = \frac{-0.65 - \angle 157.38°}{1.02474 \angle 1.9471°} = 0.6343 \angle 155.433°$$

$$\Delta V_2^{(2)} = (1.0285187 - j0.051262) - (1.02634 - j0.050465)$$
$$= 0.0021787 - 0.000787$$

$$\Delta V_3^{(2)} = (1.024152 - j0.034817) - (1.024 - j0.034793)$$
$$= 0.000152 - j0.00002$$

Iteration 4:

$$V_2^{(4)} = 1.02996 \angle -2.852°$$
$$\Delta I_2^{(3)} = 0.0003159 \angle -74.417°$$
$$\Delta Q_2^{(3)} = 0.0000867$$

$$Q_2^{(4)} = 0.0505$$
$$I_2^{(4)} = 0.29537 \angle 186.7°$$
$$V_2^{(4)} = 1.02416 - j0.034816 = 1.02475 \angle -1.947°$$
$$\Delta V_2^{(3)} = 0.000108 + j0.000016$$

$$\Delta V_3^{(3)} = 0.00058 + j0.000001$$

The final voltages are

$$V_1 = 1.05 + j0.0$$
$$V_2 = 1.02996 \angle -2.852°$$
$$V_3 = 1.02475 \angle -1.947°$$

The line flows may be calculated further if required.

PROBLEMS

P.10.1. Obtain a load flow solution for the system shown in Fig. P.10.1 use
1. Gauss—Seidel method
2. N—R Polar Coordinates Method

Bus Code $p-q$	Impedance Z_{pq}	Line Charges Ypq/s
1–2	0.02 + j0.2	0.0
2–3	0.01 + j0.025	0.0
3–4	0.02 + j0.4	0.0
3–5	0.02 + 0.05	0.0
4–5	0.015 + j0.04	0.0
1–5	0.015 + j0.04	0.0

Values are given in p.u. on a base of 100 Mva.
The scheduled powers are as follows:

Bus Code $p-q$	Impedance Z_{pq}	Line Charges Ypq/s
1–2	0.02 + j0.2	0.0
2–3	0.01 + j0.025	0.0
3–4	0.02 + j0.4	0.0
3–5	0.02 + 0.05	0.0
4–5	0.015 + j0.04	0.0
1–5	0.015 + j0.04	0.0

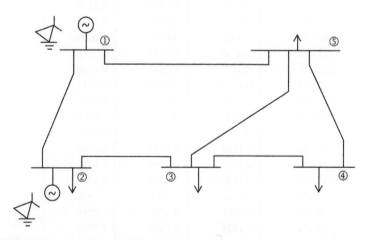

FIGURE P.10.1

Five bus system (A) with line data and (B) for P.10.3.

	Generation		Load	
Bus Code (P)	Mw	Mvar	MW	Mvar
1 (slack bus)	0	0	0	0
2	80	35	25	15
3	0	0	0	0

(*Continued*)

Continued				
	Generation		Load	
Bus Code (*P*)	**Mw**	**Mvar**	**MW**	**Mvar**
4	0	0	45	15
5	0	0	55	20

Take voltage at bus 1 as $1 \angle 0°$ p.u.

P.10.2. Repeat Problem P.10.1 with line charging capacitance $Y_{pq}/2 = j0.025$ for each line.

P.10.3. Obtain the decoupled and fast decoupled load flow solution for the system in P.10.1 and compare the results with the exact solution.

P.10.4. For the 51 bus system shown in Fig. P.10.1, the system data is given as follows in p.u. Perform load flow analysis for the system

Line Data	**Resistance**	**Reactance**	**Capacitance**
2−3	0.0287	0.0747	0.0322
3−4	0.0028	0.0036	0.0015
3−6	0.0614	0.1400	0.0558
3−7	0.0247	0.0560	0.0397
7−8	0.0098	0.0224	0.0091
8−9	0.0190	0.0431	0.0174
9−10	0.0182	0.0413	0.0167
10−11	0.0205	0.0468	0.0190
11−12	0.0660	0.0150	0.0060
12−13	0.0455	0.0642	0.0058
13−14	0.1182	0.2360	0.0213
14−15	0.0214	0.2743	0.0267
15−16	0.1336	0.0525	0.0059
16−17	0.0580	0.3532	0.0367
17−18	0.1550	0.1532	0.0168
18−19	0.1550	0.3639	0.0350
19−20	0.1640	0.3815	0.0371
20−21	0.1136	0.3060	0.0300
20−23	0.0781	0.2000	0.0210
23−24	0.1033	0.2606	0.0282
12−25	0.0866	0.2847	0.0283
25−26	0.0159	0.0508	0.0060
26−27	0.0872	0.2870	0.0296
27−28	0.0136	0.0436	0.0045
28−29	0.0136	0.0436	0.0045
29−30	0.0125	0.0400	0.0041
30−31	0.0136	0.0436	0.0045
27−31	0.0136	0.0436	0.0045

Line Data	Resistance	Reactance	Capacitance
Continued			
30−32	0.0533	0.1636	0.0712
32−33	0.0311	0.1000	0.0420
32−34	0.0471	0.1511	0.0650
30−51	0.0667	0.1765	0.0734
51−33	0.0230	0.0622	0.0256
35−50	0.0240	0.1326	0.0954
35−36	0.0266	0.1418	0.1146
39−49	0.0168	0.0899	0.0726
36−38	0.0252	0.1336	0.1078
38−1	0.0200	0.1107	0.0794
38−47	0.0202	0.1076	0.0869
47−43	0.0250	0.1336	0.1078
42−43	0.0298	0.1584	0.1281
40−41	0.0254	0.1400	0.1008
41−43	0.0326	0.1807	0.1297
43−45	0.0236	0.1252	0.1011
43−44	0.0129	0.0715	0.0513
45−46	0.0054	0.0292	0.0236
44−1	0.0330	0.1818	0.1306
46−1	0.0343	0.2087	0.1686
1−49	0.0110	0.0597	0.1752
49−50	0.0071	0.0400	0.0272
37−38	0.0014	0.0077	0.0246
47−39	0.0203	0.1093	0.0879
48−2	0.0426	0.1100	0.0460
3−35	0.0000	0.0500	0.0000
7−36	0.0000	0.0450	0.0000
11−37	0.0000	0.0500	0.0000
14−47	0.0000	0.0900	0.0000
16−39	0.0000	0.0900	0.0000
18−40	0.0000	0.0400	0.0000
20−42	0.0000	0.0800	0.0000
24−43	0.0000	0.0900	0.0000
27−45	0.0000	0.0900	0.0000
26−44	0.0000	0.0500	0.0000
30−46	0.0000	0.0450	0.0000
1−34	0.0000	0.0630	0.0000
21−2	0.0000	0.2500	0.0000
4−5	0.0000	0.2085	0.0000
19−41	0.0000	0.0800	0.0000

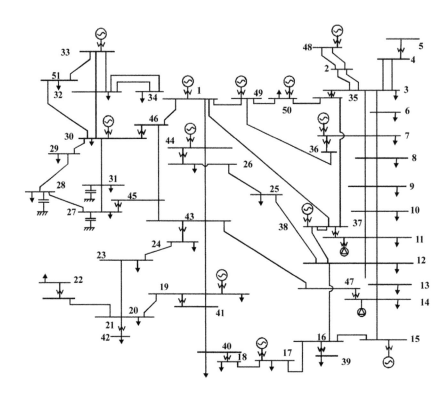

Bus $P-Q$	TAP
3–35	1.0450
7–36	1.0450
11–37	1.0500
14–47	1.0600
16–39	1.0600
18–40	1.0900
19–41	1.0750
20–42	1.0600
24–43	1.0750
30–46	1.0750
1–34	1.0875
21–22	1.0600
5–4	1.0800
27–45	1.0600
26–44	1.0750

Bus Data—Voltage and Scheduled Powers

Bus No.	Voltage Magnitude (p.u.)	Voltage Phase Angle	Real Power (p.u.)	Reactive Power (p.u.)
1	1.0800	0.0000	0.0000	0.0000
2	1.0000	0.0000	− 0.5000	− 0.2000
3	1.0000	0.0000	− 0.9000	− 0.5000
4	1.0000	0.0000	0.0000	0.0000
5	1.0000	0.0000	− 0.1190	0.0000
6	1.0000	0.0000	− 0.1900	− 0.1000
7	1.0000	0.0000	− 0.3300	− 0.1800
8	1.0000	0.0000	− 0.4400	− 0.2400
9	1.0000	0.0000	− 0.2200	− 0.1200
10	1.0000	0.0000	− 0.2100	− 0.1200
11	1.0000	0.0000	− 0.3400	− 0.0500
12	1.0000	0.0000	− 0.2400	− 0.1360
13	1.0000	0.0000	− 0.1900	− 0.1100
14	1.0000	0.0000	− 0.1900	− 0.0400
15	1.0000	0.0000	0.2400	0.0000
16	1.0000	0.0000	− 0.5400	− 0.3000
17	1.0000	0.0000	− 0.4600	− 0.2100
18	1.0000	0.0000	− 0.3700	− 0.2200
19	1.0000	0.0000	− 0.3100	− 0.0200
20	1.0000	0.0000	− 0.3400	− 0.1600
21	1.0000	0.0000	0.0000	0.0000
22	1.0000	0.0000	− 0.1700	− 0.0800
23	1.0000	0.0000	− 0.4200	− 0.2300
24	1.0000	0.0000	− 0.0800	− 0.0200
25	1.0000	0.0000	− 0.1100	− 0.0600
26	1.0000	0.0000	− 0.2800	− 0.1400
27	1.0000	0.0000	− 0.7600	− 0.2500
28	1.0000	0.0000	− 0.8000	− 0.3600
29	1.0000	0.0000	− 0.2500	− 0.1300
30	1.0000	0.0000	− 0.4700	0.0000
31	1.0000	0.0000	− 0.4200	− 0.1800
32	1.0000	0.0000	− 0.3000	− 0.1700
33	1.0000	0.0000	0.5000	0.0000
34	1.0000	0.0000	− 0.5800	− 0.2600
35	1.0000	0.0000	0.0000	0.0000
36	1.0000	0.0000	0.0000	0.0000
37	1.0000	0.0000	0.0000	0.0000
38	1.0000	0.0000	1.7000	0.0000
39	1.0000	0.0000	0.0000	0.0000
40	1.0000	0.0000	0.0000	0.0000

Bus No.	Voltage Magnitude (p.u.)	Voltage Phase Angle	Real Power (p.u.)	Reactive Power (p.u.)
41	1.0000	0.0000	0.0000	0.0000
42	1.0000	0.0000	0.0000	0.0000
43	1.0000	0.0000	0.0000	0.0000
44	1.0000	0.0000	1.7500	0.0000
45	1.0000	0.0000	0.0000	0.0000
46	1.0000	0.0000	0.0000	0.0000
47	1.0000	0.0000	0.0000	0.0000
48	1.0000	0.0000	0.5500	0.0000
49	1.0000	0.0000	3.5000	0.0000
50	1.0000	0.0000	1.2000	0.0000
51	1.0000	0.0000	− 0.5000	− 0.3000

Bus No.	Voltage at VCB	Reactive Power Limit
15	1.0300	0.1800
30	1.0000	0.0400
33	1.0000	0.4800
38	1.0600	0.9000
44	1.0500	0.4500
48	1.0600	0.2000
49	1.0700	0.5600
50	1.0700	1.500

P.10.5. The data for a 13 machine, 71 bus, 94 line system is given. Obtain the load flow solution.
Data:

No. of buses	71
No. of lines	94
Base power (MVA)	200
No. of machines	13
No. of shunt loads	23

Bus no	Generation		Load	Power
1	–	–	0.0	0.0
2	0.0	0.0	0.0	0.0
3	506.0	150.0	0.0	0.0
4	0.0	0.0	0.0	0.0
5	0.0	0.0	0.0	0.0
6	100.0	32.0	0.0	0.0

Continued				
Bus no	**Generation**		**Load**	**Power**
7	0.0	0.0	12.8	8.3
8	300.0	125.0	0.0	0.0
9	0.0	0.0	185.0	130.0
10	0.0	0.0	80.0	50.0
11	0.0	0.0	155.0	96.0
12	0.0	0.0	0.0	0.0
13	0.0	0.0	100.0	62.0
14	0.0	0.0	0.0	0.0
15	180.0	110.0	0.0	0.0
16	0.0	0.0	73.0	45.5
17	0.0	0.0	36.0	22.4
18	0.0	0.0	16.0	9.0
19	0.0	0.0	32.0	19.8
20	0.0	0.0	27.0	16.8
21	0.0	0.0	32.0	19.8
22	0.0	0.0	0.0	0.0
23	0.0	0.0	75.0	46.6
24	0.0	0.0	0.0	0.0
25	0.0	0.0	133.0	82.5
26	0.0	0.0	0.0	0.0
27	300.0	75.0	0.0	0.0
28	0.0	0.0	30.0	20.0
29	260.0	70.0	0.0	0.0
30	0.0	0.0	120.0	0.0
31	0.0	0.0	160.0	74.5
32	0.0	0.0	0.0	99.4
33	0.0	0.0	0.0	0.0
34	0.0	0.0	112.0	69.5
35	0.0	0.0	0.0	0.0
36	0.0	0.0	50.0	32.0
37	0.0	0.0	147.0	92.0
38	0.0	0.0	93.5	88.0
39	25.0	30.0	0.0	0.0
40	0.0	0.0	0.0	0.0
41	0.0	0.0	225.0	123.0
42	0.0	0.0	0.0	0.0
43	0.0	0.0	0.0	0.0
44	180.0	55.0	0.0	0.0
45	0.0	0.0	0.0	0.0
46	0.0	0.0	78.0	38.6

(*Continued*)

Continued				
Bus no	**Generation**		**Load**	**Power**
47	0.0	0.0	234.0	145.0
48	340.0	250.0	0.0	0.0
49	0.0	0.0	295.0	183.0
50	0.0	0.0	40.0	24.6
51	0.0	0.0	227.0	142.0
52	0.0	0.0	0.0	0.0
53	0.0	0.0	0.0	0.0
54	0.0	0.0	108.0	68.0
55	0.0	0.0	25.5	48.0
56	0.0	0.0	0.0	0.0
57	0.0	0.0	55.6	35.6
58	0.0	0.0	42.0	27.0
59	0.0	0.0	57.0	27.4
60	0.0	0.0	0.0	0.0
61	0.0	0.0	0.0	0.0
62	0.0	0.0	40.0	27.0
63	0.0	0.0	33.2	20.6
64	300.0	75.0	0.0	0.0
65	0.0	0.0	0.0	0.0
66	96.0	25.0	0.0	0.0
67	0.0	0.0	14.0	6.5
68	90.0	25.0	0.0	0.0
69	0.0	0.0	0.0	0.0
70	0.0	0.0	11.4	7.0
71	0.0	0.0	0.0	0.0

Line Data						
Line No.	**From Bus**	**To Bus**	**Line Impedance**		**1/2 Y charge**	**Turns Ratio**
1	9	8	0.0000	0.0570	0.0000	1.05
2	9	7	0.3200	0.0780	0.0090	1.00
3	9	5	0.0660	0.1600	0.0047	1.00
4	9	10	0.0520	0.1270	0.0140	1.00
5	10	11	0.0660	0.1610	0.0180	1.00
6	7	10	0.2700	0.0700	0.0070	1.00
7	12	11	0.0000	0.0530	0.0000	0.95
8	11	13	0.0600	0.1480	0.0300	1.00
9	14	13	0.0000	0.0800	0.0000	1.00
10	13	16	0.9700	0.2380	0.0270	1.00

Line Data *Continued*

Line No.	From Bus	To Bus	Line Impedance		1/2 Y charge	Turns Ratio
11	17	15	0.0000	0.0920	0.0000	1.05
12	7	6	0.0000	0.2220	0.0000	1.05
13	7	4	0.0000	0.0800	0.0000	1.00
14	4	3	0.0000	0.0330	0.0000	1.05
15	4	5	0.0000	0.1600	0.0000	1.00
16	4	12	0.0160	0.0790	0.0710	1.00
17	12	14	0.0160	0.0790	0.0710	1.00
18	17	16	0.0000	0.0800	0.0000	0.95
19	2	4	0.0000	0.0620	0.0000	1.00
20	4	26	0.0190	0.0950	0.1930	0.00
21	2	1	0.0000	0.0340	0.0000	1.05
22	31	26	0.0340	0.1670	0.1500	1.00
23	26	25	0.0000	0.0800	0.0000	0.95
24	25	23	0.2400	0.5200	0.1300	1.00
25	22	23	0.0000	0.0800	0.0000	0.95
26	24	22	0.0000	0.0840	0.0000	0.95
27	22	17	0.0480	0.2500	0.0505	1.00
28	2	24	0.0100	0.1020	0.3353	1.00
29	23	21	0.0366	0.1412	0.0140	1.00
30	21	20	0.7200	0.1860	0.0050	1.00
31	20	19	0.1460	0.3740	0.0100	1.00
32	19	18	0.0590	0.1500	0.0040	1.00
33	18	16	0.0300	0.0755	0.0080	1.00
34	28	27	0.0000	0.0810	0.0000	1.05
35	30	29	0.0000	0.0610	0.0000	1.05
36	32	31	0.0000	0.0930	0.0000	0.95
37	31	30	0.0000	0.0800	0.0000	0.95
38	28	32	0.0051	0.0510	0.6706	1.00
39	3	33	0.0130	0.0640	0.0580	1.00
40	31	47	0.0110	0.0790	0.1770	1.00
41	2	32	0.0158	0.1570	0.5100	1.00
42	33	34	0.0000	0.0800	0.0000	0.95
43	35	33	0.0000	0.0840	0.0000	0.95
44	35	24	0.0062	0.0612	0.2120	1.00
45	34	36	0.0790	0.2010	0.0220	1.00
46	36	37	0.1690	0.4310	0.0110	1.00
47	37	38	0.0840	0.1880	0.0210	1.00
48	40	39	0.0000	0.3800	0.0000	1.05
49	40	38	0.0890	0.2170	0.0250	1.00
50	38	41	0.1090	0.1960	0.2200	1.00

(Continued)

Line Data *Continued*

Line No.	From Bus	To Bus	Line Impedance		1/2 Y charge	Turns Ratio
51	41	51	0.2350	0.6000	0.0160	1.00
52	42	41	0.0000	0.0530	0.0000	0.95
53	45	42	0.0000	0.0840	0.0000	0.95
54	47	49	0.2100	0.1030	0.9200	1.00
55	49	48	0.0000	0.0460	0.0000	1.05
56	49	50	0.0170	0.0840	0.0760	1.00
57	49	42	0.0370	0.1950	0.0390	1.00
58	50	51	0.0000	0.0530	0.0000	0.95
59	52	50	0.0000	0.0840	0.0000	0.95
60	50	55	0.0290	0.1520	0.0300	1.00
61	50	53	0.0100	0.0520	0.0390	1.00
62	53	54	0.0000	0.0800	0.0000	0.95
63	57	54	0.0220	0.0540	0.0060	1.00
64	55	56	0.0160	0.0850	0.0170	1.00
65	56	57	0.0000	0.0800	0.0000	1.00
66	57	59	0.0280	0.0720	0.0070	1.00
67	59	58	0.0480	0.1240	0.0120	1.00
68	60	59	0.0000	0.0800	0.0000	1.00
69	53	60	0.0360	0.1840	0.3700	1.00
70	45	44	0.0000	0.1200	0.0000	1.05
71	45	46	0.0370	0.0900	0.0100	1.00
72	46	41	0.0830	0.1540	0.0170	1.00
73	46	59	0.1070	0.1970	0.0210	1.00
74	60	61	0.0160	0.0830	0.0160	1.00
75	61	62	0.0000	0.0800	0.0000	0.95
76	58	62	0.0420	0.1080	0.0020	1.00
77	62	63	0.0350	0.0890	0.0090	1.00
78	69	68	0.0000	0.2220	0.0000	1.05
79	69	61	0.0230	0.1160	0.1040	1.00
80	67	66	0.0000	0.1880	0.0000	1.05
81	65	64	0.0000	0.0630	0.0000	1.05
82	65	56	0.0280	0.1440	0.0290	1.00
83	65	61	0.0230	0.1140	0.0240	1.00
84	65	67	0.0240	0.0600	0.0950	1.00
85	67	63	0.0390	0.0990	0.0100	1.00
86	61	42	0.0230	0.2293	0.0695	1.00
87	57	67	0.0550	0.2910	0.0070	1.00
88	45	70	0.1840	0.4680	0.0120	1.00
89	70	38	0.1650	0.4220	0.0110	1.00
90	33	71	0.0570	0.2960	0.0590	1.00
91	71	37	0.0000	0.0800	0.0000	0.95

Line Data *Continued*

Line No.	From Bus	To Bus	Line Impedance		1/2 Y charge	Turns Ratio
92	45	41	0.1530	0.3880	0.1000	1.00
93	35	43	0.0131	0.1306	0.4293	1.00
94	52	52	0.0164	0.1632	0.5360	1.00

Shunt Load Data

S. No.	Bus No.	Shunt	Load
1	2	0.00	− 0.4275
2	13	0.00	0.1500
3	20	0.00	0.0800
4	24	0.00	− 0.2700
5	28	0.00	− 0.3375
6	31	0.00	0.2000
7	32	0.00	− 0.8700
8	34	0.00	0.2250
9	35	0.00	− 0.3220
10	36	0.00	0.1000
11	37	0.00	0.3500
12	38	0.00	0.2000
13	41	0.00	0.2000
14	43	0.00	− 0.2170
15	46	0.00	0.1000
16	47	0.00	0.3000
17	50	0.00	0.1000
18	51	0.00	0.1750
19	52	0.00	− 0.2700
20	54	0.00	0.1500
21	57	0.00	0.1000
22	59	0.00	0.0750
23	21	0.00	0.0500

QUESTIONS

10.1. Explain the importance of load flow studies.

10.2. Discuss briefly the bus classification.

10.3. What is the need for a slack bus or reference bus? Explain.

10.4. Explain the Gauss−Seidel method of load flow solution.

10.5. Discuss the method of Newton−Raphson method in general and explain its applicability for power flow solution.

10.6. Explain the Polar Coordinates method of Newton−Raphson load flow solution.

10.7. Give the Cartesian coordinates method or rectangular coordinates method of Newton–Raphson load flow solution.

10.8. Give the flow chart for Q. no. 6.

10.9. Give the flow chart for Q. no. 7.

10.10. Explain sparsity and its application in power flow studies.

10.11. How are generator buses are P, V buses treated in load flow studies?

10.12. Give the algorithm for decoupled load flow studies.

10.13. Explain the fast decoupled load flow method.

10.14. Compare the Gauss–Seidel and Newton–Raphson methods for power flow solution.

10.15. Compare the Newton–Raphson method, decoupled load flow method, and fast decoupled load flow method.

SHORT CIRCUIT ANALYSIS

Electrical networks and machines are subject to various types of faults while in operation. During the fault period, the current flowing is determined by the internal e.m.f.s of the machines in the network, and by the impedances of the network and machines. However, the impedances of machines may change their values from those that exist immediately after the fault occurrence to different values during the fault till the fault is cleared. The network impedance may also change, if the fault is cleared by switching operations. It is, therefore, necessary to calculate the short circuit current at different instants when faults occur. For such fault analysis studies and in general for power system analysis, it is very convenient to use per unit system and percentage values. In the following, the various models for analysis are explained.

11.1 PER UNIT QUANTITIES

The per unit value of any quantity is the ratio of the actual value in any units to the chosen base quantity of the same dimensions expressed as a decimal.

$$\text{Per unit quantity} = \frac{\text{actual value in any units}}{\text{base or reference value in the same units}}$$

In power systems, the basic quantities of importance are voltage, current, impedance, and power. For all per unit calculations, a base kVA or MVA and a base kV are to be chosen. Once the base values or reference values are chosen, the other quantities can be obtained as follows.

Selecting the total or three-phase kVA as base kVA, for a three-phase system

$$\text{Base current in amperes} = \frac{\text{base kVA}}{\sqrt{3}\,[\text{base kV (line-to-line)}]}$$

$$\text{Base impedance in ohms} = \left[\frac{\text{base kV (line-to-line)}^2 \times 1000}{\sqrt{3}\,[(\text{base kVA})/3]} \right]$$

$$\text{Base impedance in ohms} = \frac{\text{base kV (line-to-line)}^2}{\text{base MVA}}$$

Hence,

$$\text{Base impedance in ohm} = \frac{\text{base kV (line-to-line)}^2 \times 1000}{\text{base kVA}}$$

where base kVA and base MVA are the total or three-phase values.

Power Systems Analysis. DOI: http://dx.doi.org/10.1016/B978-0-08-101111-9.00011-2

If phase values are used

$$\text{Base current in amperes} = \frac{\text{base kVA}}{\text{base kV}}$$

$$\text{Base impedance in ohm} = \frac{\text{base voltage}}{\text{base current}}$$

$$= \frac{(\text{base kV})^2 \times 1000}{\text{base kVA per phase}}$$

$$\text{Base impedance in ohm} = \frac{(\text{base kV})^2}{\text{base MVA per phase}}$$

In all the above relations, the power factor is assumed unity, so that

$$\text{base power KW} = \text{base kVA}$$

Now,

$$\text{Per unit impedance} = \frac{(\text{actual impedance in ohm}) \times \text{kVA}}{(\text{base kV})^2 \times 1000}$$

Sometimes, it may be required to use the relation:

$$\text{Actual impedance in ohm} = \frac{(\text{per unit impedance in ohms}) (\text{base kV})^2}{\text{base kVA}} \times 1000$$

Very often the values are in different base values. In order to convert the per unit impedance from given base to another base, the following relation can be derived easily.

Per unit impedance on new base

$$Z_{new} \text{ p.u.} = Z_{given} \text{p.u.} \left(\frac{\text{new kVA base}}{\text{given kVA base}} \right) \left(\frac{\text{given kV base}}{\text{new kV base}} \right)^2$$

11.2 ADVANTAGES OF PER UNIT SYSTEM

1. While performing calculations, referring quantities from one side of the transformer to the other side serious errors may be committed. This can be avoided by using per unit system.
2. Voltages, currents, and impedances expressed in per unit do not change when they are referred from one side of transformer to the other side. This is a great advantage.
3. Per unit impedances of electrical equipment of similar type usually lie within a narrow range, when the equipment ratings are used as base values.
4. Transformer connections do not affect the per unit values.
5. Manufacturers usually specify the impedances of machines and transformers in per unit or percent of name plate ratings.

11.3 THREE-PHASE SHORT CIRCUITS

In the analysis of symmetrical three-phase short circuits, the following assumptions are generally made.

1. Transformers are represented by their leakage reactances. The magnetizing current and core losses are neglected. Resistances, shunt admittances are not considered. Star-delta phase shifts are also neglected.
2. Transmission lines are represented by series reactances. Resistances and shunt admittances are neglected.
3. Synchronous machines are represented by constant voltage sources behind subtransient reactances. Armature resistances, saliency, and saturation are neglected.
4. All nonrotating impedance loads are neglected.
5. Induction motors are represented just as synchronous machines with constant voltage source behind a reactance. Smaller motor loads are generally neglected.

Per unit impedances of transformers: Consider a single-phase transformer with primary and secondary voltages and currents denoted by V_1, V_2 and I_1, I_2, respectively.
We have

$$\frac{V_1}{V_2} = \frac{I_2}{I_1}$$

Base impedance for primary $= \dfrac{V_1}{I_1}$

Base impedance for secondary $= \dfrac{V_2}{I_2}$

Per unit impedance referred to primary $= \dfrac{Z_1}{(V_1/I_1)} = \dfrac{I_1 Z_1}{V_1}$

Per unit impedance referred to secondary $= \dfrac{I_2 Z_2}{V_2}$

Again, actual impedance referred to secondary $= Z_1 \left(\dfrac{V_2}{V_1}\right)^2$

Per unit impedance referred to secondary

$$= \frac{Z_1 (V_2/V_1)^2}{(V_2/I_2)} = Z_1 \cdot \frac{V_2^2}{V_1^2} \cdot \frac{I_2}{V_2} = \frac{Z_1(V_2 I_2)}{V_1^2} = Z_1 \frac{(V_1 I_1)}{V_1^2} = \frac{Z_1 I_1}{V_1}$$

$$= \text{Per unit impedance referred to primary}$$

Thus the per unit impedance referred remains the same for a transformer on either side.

11.4 REACTANCE DIAGRAMS

In power system analysis, it is necessary to draw an equivalent circuit for the system. This is an impedance diagrams. However, in several studies, including short circuit analysis it is sufficient to

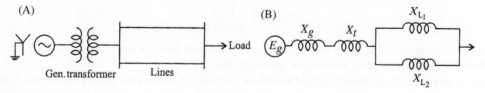

FIGURE 11.1

Reactance diagram. (A) Power system. (B) Equivalent single line reactance diagram.

consider only reactances neglecting resistances. Hence, we draw reactance diagrams. For three-phase balanced systems, it is simpler to represent the system by a single line diagram without losing the identity of the three-phase system. Thus single line reactance diagrams can be drawn for calculation.

This is illustrated by the system shown in Fig. 11.1A and B and by its single line reactance diagram.

11.5 PERCENTAGE VALUES

The reactances of generators, transformers, and reactors are generally expressed in percentage values to permit quick short circuit calculation.

Percentage reactance is defined as

$$\%X = \frac{IX}{V} \times 100$$

where

$I = $ full load current
$V = $ phase voltage
$X = $ reactance in ohms per phase.

Short circuit current I_{SC} in a circuit then can be expressed as

$$I_{SC} = \frac{V}{X} = \frac{V \cdot I}{V \cdot (\%X)} \times 100$$

$$= \frac{I \cdot 100}{\%X}$$

Percentage reactance can expressed in terms of kVA and kV as follows.
From equation:

$$X = \frac{(\%X) \cdot V}{I.100} = \frac{(\%X)V^2}{100 \cdot V \cdot I} = \frac{(\%X)\dfrac{V}{1000} \cdot \dfrac{V}{1000} \times 1000}{100 \cdot \dfrac{V}{1000} \cdot I}$$

$$= \frac{(\%X)\,(KV)^2\,10}{KVA}$$

Alternatively

$$(\%X) = X \cdot \frac{kVA}{10\,(kV)^2}$$

As has been stated already in short circuit analysis since the reactance X is generally greater than three times the resistance, resistances are neglected.

But, in case percentage resistance and therefore, percentage impedance values are required then, in a similar manner we can define

$$\%R = \frac{IR}{V} \times 100$$

and

$$\%Z = \frac{IZ}{V} \times 100 \text{ with usual notation.}$$

The percentage values of R and Z also do not change with the side of the transformer or either side of the transformer they remain constant. The ohmic values of R, X, and Z change from one side to the other side of the transformer.

When a fault occurs the potential falls to a value determined by the fault impedance. Short circuit current is expressed in term of short circuit kVA based on the normal system voltage at the point of fault.

11.6 SHORT CIRCUIT kVA

It is defined as the product of normal system voltage and short circuit current at the point of fault expressed in kVA.

Let

V = normal phase voltage in volts
I = full load current in amperes at base kVA
$\%X$ = percentage reactance of the system expressed on base kVA.

The short circuit current

$$I_{SC} = I \cdot \frac{100}{\%X}$$

The three-phase or total short circuit kVA

$$= \frac{3 \cdot V \, I_{SC}}{1000} = \frac{3 \cdot V \cdot I \, 100}{(\%X) \, 1000} = \frac{3V \, I}{1000} \cdot \frac{100}{\%X}$$

Therefore Short circuit kVA = base kVA $\times \dfrac{100}{(\%X)}$

In a power system or even in a single power station, different equipments may have different ratings. Calculations are required to be performed where different components or units are rated differently. The percentage values specified on the name plates will be with respect to their name plate ratings. Hence, it is necessary to select a common base kVA or MVA and also a base kV. The following are some of the guidelines for the selection of base values.

1. Rating of the largest plant or unit for base MVA or kVA
2. The total capacity of a plant or system for base MVA or kVA
3. Any arbitrary value.

$$(\%X)_{\text{On new base}} = \left(\frac{\text{Base kVA}}{\text{Unit kVA}}\right)(\%X \text{ at unit kVA})$$

If a transformer has 8% reactance on 50 kVA base, its value at 100 kVA base will be

$$(\%X)_{100 \text{ kVA}} = \left(\frac{100}{50}\right) \times 8 = 16\%$$

Similarly the reactance values change with voltage base as per the relation

$$X_2 = \left(\frac{V_2}{V_1}\right)^2 \cdot X_1$$

where
X_1 = reactance at voltage V_1 and
X_2 = reactance at voltage V_2.
For short circuit analysis, it is often convenient to draw the reactance diagrams indicating the values in per unit.

11.7 IMPORTANCE OF SHORT CIRCUIT CURRENTS

Knowledge of short circuit current values is necessary for the following reasons:

1. Fault currents which are several times larger than the normal operating currents produce large electromagnetic forces and torques which may adversely affect the stator end windings. The forces on the end windings depend on both the d.c. and a.c. components of stator currents.
2. The electrodynamic forces on the stator end windings may result in the displacement of the coils against one another. This may result in loosening of the support or damage to the insulation of the windings.
3. Following a short circuit, it is always recommended that the mechanical bracing of the end windings is to be checked for any possible loosening.
4. The electrical and mechanical forces that develop due to a sudden three-phase short circuit are generally severe when the machine is operating under loaded condition.
5. As the fault is cleared within three cycles generally the heating efforts are not considerable.

Short circuits may occur in power systems due to system overvoltages caused by lightning or switching surges or due to equipment insulation failure or even due to insulator contamination. Sometimes even mechanical causes may create short circuits. Other well-known reasons include line-to-line, line-to-ground, or line-to-line faults on overhead lines. The resultant short circuit has to be interrupted within few cycles by the circuit breaker.

It is absolutely necessary to select a circuit breaker that is capable of operating successfully when maximum fault current flows at the circuit voltage that prevails at that instant. An insight can be gained when we consider an R–L circuit connected to an alternating voltage source, the circuit being switched on through a switch.

11.8 ANALYSIS OF R–L CIRCUIT

Consider the circuit in Fig. 11.2.
Let $e = E_{max} \sin(\omega t + \alpha)$ when the switch S is closed at $t = 0^+$

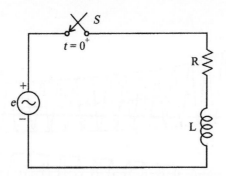

FIGURE 11.2

RL circuit with switch.

$$e = E_{\max} \sin(\omega t + \alpha) = R + L\frac{di}{dt}$$

α is determined by the magnitude of voltage when the circuit is closed.
The general solution is

$$i = \frac{E_{\max}}{|Z|}\left[\sin(\omega t + \alpha - \theta) - e^{-Rt/L}\sin(\alpha - \theta)\right]$$

where

$$|Z| = \sqrt{R^2 + \omega^2 L^2}$$

and

$$\theta = \tan^{-1}\frac{\omega L}{R}$$

The current contains two components:

$$\text{a.c. component} = \frac{E_{\max}}{|Z|}\sin(\omega t + \alpha - \theta)$$

and

$$\text{d.c. component} = \frac{E_{\max}}{|Z|}e^{-Rt/L}\sin(\alpha - \theta)$$

If the switch is closed when $\alpha - \theta = \pi$ or when $\alpha - \theta = 0$, the d.c. component vanishes.
The d.c. component is a maximum when $\alpha - \theta = \pm\frac{\pi}{2}$.

11.9 THREE-PHASE SHORT CIRCUIT ON UNLOADED SYNCHRONOUS GENERATOR

If a three-phase short circuit occurs at the terminals of a salient pole synchronous generator, we obtain typical oscillograms as shown in Fig. 11.3 for the short circuit currents of the three phases.

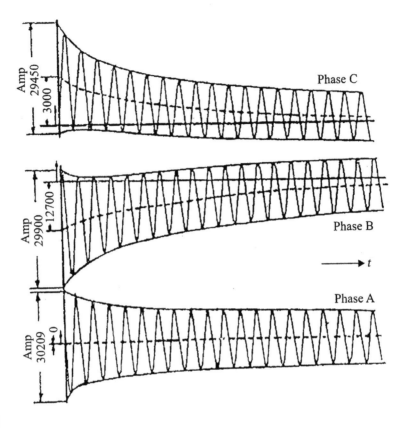

FIGURE 11.3

Oscillograms of the armature currents after a short circuit.

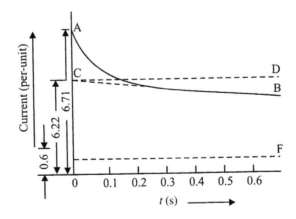

FIGURE 11.4

Alternating component of the short circuit armature current.

Fig. 11.4 shows the alternating component of the short circuit current when the d.c. component is eliminated. The fast changing subtransient component and the slowly changing transient components are shown at A and C. Fig. 11.5 shows the electrical torque. The changing field current is shown in Fig. 11.6.

From the oscillogram of a.c. component, the quantities x''_d, x''_q, x'_d, and x'_q can be determined.

If V is the line to neutral prefault voltage, then the a.c. component $i_{a.c.} = V/x''_q = I''$, the r.m.s. subtransient short circuit. Its duration is determined by T''_d, the subtransient direct axis time constant. The value of $i_{a.c.}$ decreases to V/x'_d when $t > T''_d$.

With T'_d as the direct axis transient time constant when $t > T'_d$

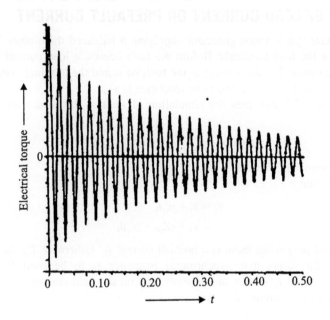

FIGURE 11.5

Electrical torque on three-phase terminal short circuit.

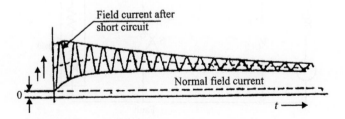

FIGURE 11.6

Oscillogram of the field current after a short circuit.

$$i_{\text{a.c.}} = \frac{V}{x_d}$$

The maximum d.c. off-set component that occurs in any phase at $\alpha = 0$ is

$$i_{\text{d.c. max}}(t) = \sqrt{2}\,\frac{V}{x''_d}\,e^{-t/TA}$$

where T_A is the armature time constant.

11.10 EFFECT OF LOAD CURRENT OR PREFAULT CURRENT

Consider a three-phase synchronous generator supplying a balanced three-phase load. Let a three-phase fault occurs at the load terminals. Before the fault occurs, a load current I_L is flowing into the load from the generator. Let the voltage at the fault be v_f and the terminal voltage of the generator be v_t. Under fault conditions, the generator reactance is x''_d.

The circuit in Fig. 11.7 indicates the simulation of fault at the load terminals by a parallel switch S.

$$E''_g = V_t + jx''_d I_L = V_f + (X_{\text{ext}} + jx''_d)I_L$$

where E''_g is the subtransient internal voltage.

For the transient state

$$E'_g = V_t + jx'_d I_L$$
$$= V_f + (Z_{\text{ext}} + jx'_d)I_L$$

E''_g or E'_g are used only when there is a prefault current I_L. Otherwise E_g, the steady-state voltage in series with the direct axis synchronous reactance is to be used for all calculations. E_g remains the same for all I_L values and depends only on the field current. Every time, of course, a new E''_g is required to be computed.

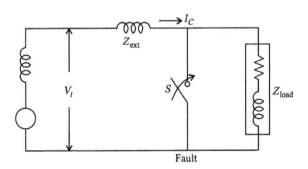

FIGURE 11.7

Fault simulation for synchronous machine.

11.11 REACTORS

Whenever faults occur in power system, large currents flow. Especially, if the fault is a dead short circuit at the terminals or bus bars enormous currents flow damaging the equipment and its components. To limit the flow of large currents under these circumstances, current limiting reactors are used. These reactors are large coils constructed for high self-inductance.

They are also located that the effect of the fault does not affect other parts of the system and is thus localized. From time to time, new generating units are added to an existing system to augment the capacity. When this happens, the fault current level increases and it may become necessary to change the switch gear. With proper use of reactors, addition of generating units does not necessitate changes in existing switch gear.

11.11.1 CONSTRUCTION OF REACTORS

These reactors are built with nonmagnetic core so that the saturation of core with consequent reduction in inductance and increased short circuit currents is avoided. Alternatively, it is possible to use iron core with air gaps included in the magnetic core so that saturation is avoided.

11.11.2 CLASSIFICATION OF REACTORS

There are three types of reactors: (1) generator reactors, (2) feeder reactors, and (3) bus bar reactors.

The above classification is based on the location of the reactors. Reactors may be connected in series with the generator in series with each feeder or to the bus bars.

1. *Generator reactors*: The reactors are located in series with each of the generators as shown in Fig. 11.8 so that current flowing into a fault F from the generator is limited.

   Disadvantages:
   a. In the event of a fault occurring on a feeder, the voltage at the remaining healthy feeders also may lose synchronism requiring resynchronization later.

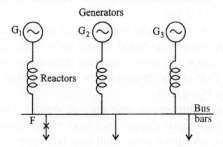

FIGURE 11.8

Generator reactors.

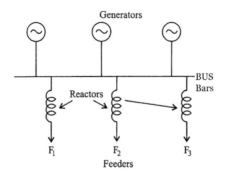

FIGURE 11.9

Feeder reactors.

b. There is a constant voltage drop in the reactors and also power loss, even during normal operation. Since modern generators are designed to withstand dead short circuit at their terminals, generator reactors are nowadays not used except for old units in operation.

2. *Feeder reactors*: In this method of protection, each feeder is equipped with a series reactor as shown in Fig. 11.9.

In the event of a fault on any feeder the fault current drawn is restricted by the reactor. *Disadvantages*:

a. Voltage drop and power loss still occurs in the reactor for a feeder fault. However, the voltage drop occurs only in that particular feeder reactor.

b. Feeder reactors do not offer any protection for bus bar faults. Nevertheless, bus bar faults occur very rarely.

As series reactors inherently create voltage drop, system voltage regulation will be impaired. Hence they are to be used only in special case such as for short feeders of large cross-section.

3. *Bus bar reactors*: In both the above methods, the reactors carry full load current under normal operation. The consequent disadvantage of constant voltage drops and power loss can be avoided by dividing the bus bars into sections and interconnect the sections through protective reactors. There are two ways of doing this.

a. *Ring system*: In this method, each feeder is fed by one generator. Very little power flows across the reactors during normal operation. Hence the voltage drop and power loss are negligible. If a fault occurs on any feeder, only the generator to which the feeder is connected will feed the fault and other generators are required to feed the fault through the reactor. This is shown in Fig. 11.10.

b. *Tie bar system*: This is an improvement over the ring system. This is shown in Fig. 11.11. Current fed into a fault has to pass through two reactors in series between sections.

Another advantage is that additional generation may be connected to the system without requiring changes in the existing reactors.

The only disadvantage is that this system requires an additional bus bar system, the tie bar.

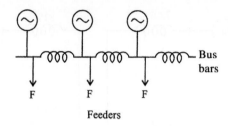

FIGURE 11.10

Ring system.

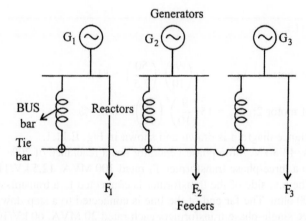

FIGURE 11.11

Tie bar system.

WORKED EXAMPLES

E.11.1. Two generators rated at 10 MVA, 11 kV and 15 MVA, 11 kV, respectively, are connected in parallel to a bus. The bus bars feed two motors rated 7.5 MVA and 10 MVA, respectively. The rated voltage of the motors is 9 kV. The reactance of each generator is 12% and that of each motor is 15% on their own ratings. Assume 50 MVA, 10 kV base and draw the reactance diagram.

Solution:

The reactances of the generators and motors are calculated on 50 MVA, 10 kV base values.

Reactance of generator 1: $X_{G_1} = 12 \cdot \left(\frac{11}{10}\right)^2 \cdot \left(\frac{50}{10}\right) = 72.6\%$

Reactance of generator 2: $X_{G_2} = 12 \left(\frac{11}{10}\right)^2 \cdot \left(\frac{50}{10}\right) = 48.4\%$

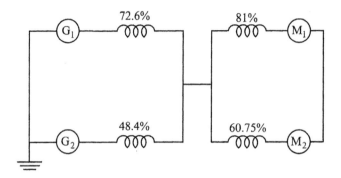

72.6%
81%
48.4%
60.75%

FIGURE E.11.1

Reactance diagram for E.11.1.

$$\text{Reactance of motor 1: } X_{M_1} = 15 \cdot \left(\frac{9}{10}\right)^2 \left(\frac{50}{7.5}\right) = 81\%$$

$$\text{Reactance of motor 2: } X_{M_2} = 15 \left(\frac{9}{10}\right)^2 \left(\frac{50}{10}\right) = 60.75\%$$

The reactance diagram is drawn and shown in Fig. E.11.1.

E.11.2. A 100-MVA, 13.8-kV, three-phase generator has a reactance of 20%. The generator is connected to a three-phase transformer T_1 rated 100 MVA, 12.5 kV/110 kV with 10% reactance. The h.v. side of the transformer is connected to a transmission line of reactance 100 ohm. The far end of the line is connected to a step-down transformer T_2, made of three single-phase transformers each rated 30 MVA, 60 kV/10 kV with 10% reactance the generator supplies two motors connected on the l.v. side T_2 as shown in Fig. E.11.2. The motors are rated at 25 MVA and 50 MVA both at 10 kV with 15% reactance. Draw the reactance diagram showing all the values in per unit. Take generator rating as base.

Solution:

Base MVA = 100

Base kV = 13.8

$$\text{Base kV for the line} = 13.8 \times \frac{110}{12.5} = 121.44$$

$$\text{Line-to-line voltage ratio of } T_2 = \frac{\sqrt{3} \times 66 \ kV}{10 \ kV} = \frac{114.31}{10}$$

$$\text{Base voltage for motors} = \frac{121.44 \times 10}{114.31} = 10.62 \ kV$$

$\%X$ for generators = 20% = 0.2 p.u.

$$\%X \text{ for transformer } T_1 = 10 \times \left(\frac{12.5}{13.8}\right)^2 \times \frac{100}{100} = 8.2\%$$

$\%X$ for transformer T_2 on $\sqrt{3} \times 66$: 10 kV and 3×30 MVA base = 10%

$\%X$ for T_2 on 100 MVA, and 121.44 kV:10.62 kV is

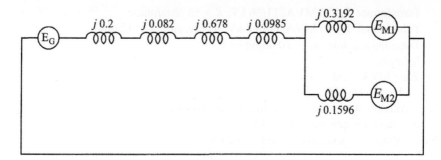

FIGURE E.11.2

Reduced reactance diagram.

$$\%X\ T_2 = 10 \times \left(\frac{10}{10.62}\right)^2 \times \left(\frac{100}{90}\right) = 9.85\% = 0.0985\ \text{p.u.}$$

$$\text{Base reactance for line} = \left(\frac{121.44}{100}\right)^2 = 147.47\ \text{ohm}$$

$$\text{Reactance of line} = \frac{100}{147.47} = 0.678\ \text{p.u.}$$

$$\text{Reactance of motor M}_1 = 10 \times \left(\frac{10}{10.62}\right)^2 \left(\frac{90}{25}\right) = 31.92\%$$

$$= 0.3192\ \text{p.u.}$$

$$\text{Reactance of motor M}_2 = 10 \times \left(\frac{10}{10.62}\right)^2 \left(\frac{90}{50}\right) = 15.96\%$$

The reactance diagram is shown in Fig. E.11.2.

E.11.3. Obtain the per unit representation for the three-phase power system shown in Fig. E.11.3.

Generator 1: 50 MVA, 10.5 kV; $X = 1.8$ ohm
Generator 2: 25 MVA, 6.6 kV; $X = 1.2$ ohm
Generator 3: 35 MVA, 6.6 kV; $X = 0.6$ ohm

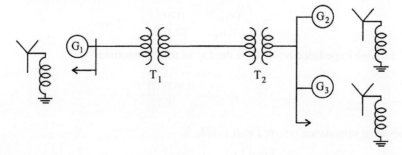

FIGURE E.11.3

Three-phase power system.

Transformer T_1: 30 MVA, 11/66 kV, $X = 15$ ohm/phase
Transformer T_2: 25 MVA, 66/6.2 kV, as h.v. side $X = 12$ ohm
Transmission line: $X_L = 20$ ohm/phase.
Solution:
Let base MVA = 50
Base kV = 66 ($L-L$)
Base voltage on transmission as line 1 p.u. (66 kV)
Base voltage for generator 1: 11 kV
Base voltage for generators 2 and 3: 6.2 kV

p.u. reactance of transmission line $= \dfrac{20 \times 50}{66^2} = 0.229$ p.u.

p.u. reactance of transformer $T_1 = \dfrac{15 \times 50}{66^2} = 0.172$ p.u.

p.u. reactance of transformer $T_2 = \dfrac{12 \times 50}{66^2} = 0.1377$ p.u.

p.u. reactance of generator $1 = \dfrac{1.8 \times 50}{(11)^2} = 0.7438$ p.u.

p.u. reactance of generator $2 = \dfrac{1.2 \times 50}{(6.2)^2} = 1.56$ p.u.

p.u. reactance of generator $3 = \dfrac{0.6 \times 50}{(6.2)^2} = 0.78$ p.u.

E.11.4. A single-phase two winding transformer is rated 20 kVA, 480/120 V at 50 Hz. The equivalent leakage impedance of the transformer referred to l.v. side is 0.0525 78.13 ohm using transformer ratings as base values. Determine the per unit leakage impedance referred to the h.v. side and l.v. side.
Solution:
Let base kVA = 20
Base voltage on h.v. side = 480 V
Base voltage on l.v. side = 120 V
The leakage impedance on the l.v. side of the transformer

$$Z_{12} = \frac{(V_{\text{base}})^2}{VA_{\text{base}}} = \frac{(120)^2}{20,000} = 0.72 \text{ ohm}$$

p.u. leakage impedance referred to the l.v. of the transformer

$$Z_{p.u.\ 2} = \frac{0.0525 \,\lfloor 78.13^\circ}{0.72}$$

Equivalent impedance referred to h.v. side is

$$\left(\frac{400}{120}\right)^2 [(0.0525)\, 70.13^\circ] = (0.84)\, 78.13^\circ$$

The base impedance on the h.v. side of the transformer is $\dfrac{(480)^2}{20,000} = 11.52$ ohm
p.u. leakage impedance referred to h.v. side

$$= \frac{(0.84)\,78.13°}{11.52} = (0.0729)\,78.13° \text{ p.u.}$$

E.11.5. A single-phase transformer is rated at 110/440 V, 3 kVA. Its leakage reactance measured on 110 V side is 0.05 ohm. Determine the leakage impedance referred to 440 V side.

Solution:

Base impedance on 110 V side $= \dfrac{(0.11)^2 \times 1000}{3} = 4.033$ ohm

Per unit reactance on 110 V side $= \dfrac{0.05}{4.033} = 0.01239$ p.u.

Leakage reactance referred to 440 V side $= (0.05)\left(\dfrac{440}{110}\right)^2 = 0.8$ ohm

Base impedance referred to 440 V side $= \dfrac{0.8}{64.53} = 0.01239$ p.u.

E.11.6. Consider the system shown in Fig. E.11.4. Selecting 10,000 kVA and 110 kV as base values, find the p.u. impedance of the 200 ohm load referred to 110 kV side and 11 kV side.

Solution:

Base voltage at $\rho = 11$ kV

Base voltage at $R = \dfrac{110}{2} = 55$ kV

Base impedance at $R = \dfrac{55^2 \times 1000}{10,000} = 302.5$ ohm

p.u. impedance at $R = \dfrac{200 \text{ ohm}}{302.5 \text{ ohm}} = 0.661$ ohm

Base impedance at $\phi = \dfrac{110^2 \times 1000}{10,000} = 1210$ ohm

Load impedance referred to $\phi = 200 \times 2^2 = 800$ ohm

p.u. impedance of load referred to $\phi = \dfrac{800}{1210} = 0.661$

Similarly base impedance at $P = \dfrac{11^2 \times 1000}{10,000} = 121.1$ ohm

Impedance of load referred to $P = 200 \times 2^2 \times 0.1^2 = 8$ ohm

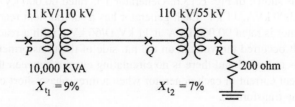

FIGURE E.11.4

Step-up and step-down transformer system.

p.u. impedance of load at $P = \dfrac{8}{12.1} = 0.661$ ohm

E.11.7. Three transformers each rated 30 MVA at 38.1/3.81 kV are connected in star-delta with a balanced load of three 0.5-ohm, star-connected resistors. Selecting a base of 900 MVA 66 kV for the h.v. side of the transformer find the base values for the l.v. side.

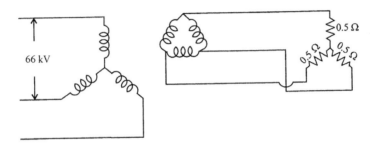

FIGURE E.11.5

System for E.11.7.

    Solution:

Base impedance on l.v. side $= \dfrac{(\text{base kV}_{L\text{-}L})^2}{\text{base MVA}} = \dfrac{(3.81)^2}{90} = 0.1613$ ohm

p.u. load resistance on l.v. side $= \dfrac{0.5}{0.1613} = 3.099$ p.u.

Base impedance on h.v. side $= \dfrac{(66)^2}{90} = 48.4 \ ohm$

Load resistance referred to h.v. side $= 0.5 \times \left(\dfrac{66}{3.81}\right)^2 = 150$ ohm

p.u. load resistance referred to h.v. side $= \dfrac{150}{48.4} = 3.099$ p.u.

The per unit load resistance remains the same.

E.11.8. Two generators are connected in parallel to the l.v. side of a three-phase delta-star transformer as shown in Fig. E.13.6. Generator 1 is rated 60,000 kVA, 11 kV. Generator 2 is rated 30,000 kVA, 11 kV. Each generator has a subtransient reactance of $x''_d = 25\%$. The transformer is rated 90,000 kVA at 11 kV D/66 kV g with a reactance of 10%. Before a fault occurred the voltage on the h.t. side of the transformer is 63 kV. The transformer is unloaded and there is no circulating current between the generators. Find the subtransient current in each generator when a three-phase short circuit occurs on the h.t. side of the transformer.

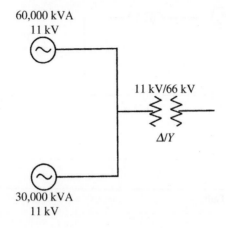

60,000 kVA
11 kV

11 kV/66 kV

Δ/Y

30,000 kVA
11 kV

FIGURE E.11.6

System for E.11.8.

    Solution:

Let the line voltage on the h.v. side be the base kV = 66 kV

Let the base kVA = 90,000 kVA

Generator 1: $x''_d = 0.25 \times \dfrac{90,000}{60,000} = 0.375$ p.u.

For generator 2: $x''_d = \dfrac{90,000}{30,000} = 0.75$ p.u.

The internal voltage for generator 1

$$E_{G_1} = \frac{0.63}{0.66} = 0.955 \text{ p.u.}$$

The internal voltage for generator 2

$$E_{G_2} = \frac{0.63}{0.66} = 0.955 \text{ p.u.}$$

The reactance diagram is shown in Fig. E.11.7 when switch S is closed, the fault condition is simulated. As there is no circulating current between the generators, the equivalent reactance of the parallel circuit is $\dfrac{0.375 \times 0.75}{0.375 + 0.75} = 0.25$ p.u.

The subtransient current $I'' = \dfrac{0.955}{(j0.25 + j0.10)} = j2.7285$ p.u.

The voltage as the delta side of the transformer is $(-j2.7285)\,(j0.10) = 0.27205$ p.u.

I''_1 = the subtransient current flowing into fault from generator

$$I''_1 = \frac{0.955 - 0.2785}{j0.375} = 1.819 \text{ p.u.}$$

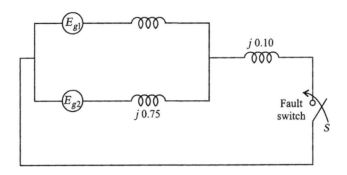

FIGURE E.11.7

Reduced system for simulation of fault.

Similarly,

$$I''_2 = \frac{0.955 - 0.27285}{j0.75} = -j1.819 \text{ p.u.}$$

The actual fault currents supplied in amperes are

$$I''_1 = \frac{1.819 \times 90,000}{\sqrt{3} \times 11} = 8592.78 \text{ A}$$

$$I''_2 = \frac{0.909 \times 90,000}{\sqrt{3} \times 11} = 4294.37 \text{ A}$$

E.11.9. R station with two generators feeds through transformers a transmission system operating at 132 kV. The far end of the transmission system consisting of 200-km-long double circuit line is connected to load from bus B. If a three-phase fault occurs at bus B, determine the total fault current and fault current supplied by each generator.

Select 75 MVA and 11 kV on l.v. side and 132 kV on h.v. side as base values.

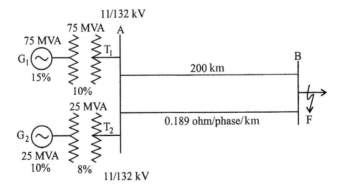

FIGURE E.11.8

System for E.11.9.

Solution:

p.u. x of generator $1 = j0.15$ p.u.

$$2 = j = 0.10\frac{75}{25}$$

p.u. x of generator

$$= j0.3 \text{ p.u.}$$

p.u. x of transformer $T_1 = j0.1$

p.u. x of transformer $T_2 = j0.08 \times \dfrac{75}{25} = j0.24$

p.u. x of each line $= \dfrac{j0.180 \times 200 \times 75}{132 \times 132} = j0.1549$

The equivalent reactance diagram is shown in Fig. E.11.9A−C.

Fig. E.11.9A−C can be reduced further into

$$Z_{eq} = j0.17 + j0.07745 = j0.248336$$

Total fault current $\dfrac{1\angle 0°}{j\,0.248336} = -j\,4.0268$ p.u.

Base current for 132 kV circuit $= \dfrac{75 \times 1000}{\sqrt{3} \times 132} = 328$ A

Hence actual fault current $= -j4.0268 \times 328 = 1321\text{A}\angle -90°$

Base current for 11 kV side of the transformer $= \dfrac{75 \times 1000}{\sqrt{3} \times 11} = 3936.6$ A

Actual fault current supplied from 11 kV side $= 3936.6 \times 4.0248 = 15851.9\text{A}\angle -90°$

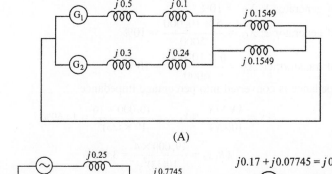

(A)

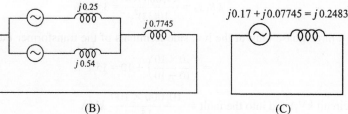

(B)   (C)

FIGURE E.11.9

Reduced equivalent reactance diagram.

$$\text{Fault current supplied by generator 1} = \frac{1585139 \angle - 90° \times j0.54}{j0.54 + j0.25} = -j10835.476 \text{ A}$$

$$\text{Fault current supplied by generator 2} = \frac{15851.9 \times j0.25}{j0.79} = 5016.424 \text{ A} \angle - 90°$$

E.11.10. A 33-kV line has a resistance of 4 ohm and reactance of 16 ohm, respectively. The line is connected to a generating station bus bars through a 6000-kVA step-up transformer which has a reactance of 6%. The station has two generators rated 10,000 kVA with 10% reactance and 5000 kVA with 5% reactance. Calculate the fault current and short circuit kVA when a three-phase fault occurs at the h.v. terminals of the transformers and at the load end of the line.

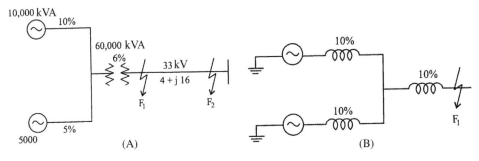

FIGURE E.11.10

System and reduced system for fault simulation.

Solution:

Let 10,000 kVA be the base kVA

Reactance of generator 1: $X_{G_1} = 10\%$

Reactance of generator 2: $X_{G_2} = \dfrac{5 \times 10,000}{5000} = 10\%$

Reactance of transformer: $X_T = \dfrac{6 \times 10,000}{6000} = 10\%$

The line impedance is converted into percentage impedance

$$\%X = \frac{\text{kVA} \cdot X}{10(\text{kV})^2}; \quad \%X_{\text{Line}} = \frac{10,000 \times 16}{10 \times (33)^2} = 14.69\%$$

$$\%R_{\text{Line}} = \frac{19,000 \times 4}{10(33)^2} = 3.672\%$$

1. For a three-phase fault at the h.v. side terminals of the transformer fault impedance

$$= \left(\frac{10 \times 10}{10 + 10} \right) + 10 = 15\%$$

$$\text{Short circuit kVA fed into the fault} = \frac{10,000 \times 100}{15} \text{ kVA}$$

$$= 66666.67 \text{ kVA}$$

$$= 66.67 \text{ MVA}$$

For a fault at F_2 the load end of the line the total reactance to the fault

$$= 15 + 14.69$$
$$= 29.69\%$$

Total resistance to fault $= 3.672\%$

Total impedance to fault $= \sqrt{3.672^2 + 29.69^2}$

$$= 29.916\%$$

Short circuit kVA into fault $= \dfrac{100}{29.916} \times 10,000$

$$= 33433.63 \text{ kVA}$$
$$= 33.433 \text{ MVA}$$

E.11.11. Fig. E.11.11A shows a power system where load at bus 5 is fed by generators at bus 1 and bus 4. The generators are rated at 100 MVA; 11 kV with subtransient reactance of 25%. The transformers are rated each at 100 MVA, 11/112 kV and have a leakage reactance of 8%. The lines have an inductance of 1 mH/phase/km. Line L_1 is 100 km long while lines L_2 and L_3 are each of 50 km in length. Find the fault current and MVA for a three-phase fault at bus 5.

Solution:

Let base MVA = 100 MVA

Base voltage for l.v. side = 11 kV

Base voltage for h.v. side = 112 kV

Base impedance for h.v. side of transformer

$$= \frac{112 \times 112}{100} = 125.44 \text{ ohm}$$

Base impedance for l.v. side of transformer

$$= \frac{11 \times 11}{100} = 1.21 \text{ ohm}$$

Reactance of line $L_1 = 2 \times p \times 50 \times 1 \times 10^{-3} \times 100 = 31.4$ ohm

Per unit reactance of line $L_1 = \dfrac{31.4}{125.44} = 0.25$ p.u.

p.u. impedance of line $L_2 = \dfrac{2\pi \times 50 \times 1 \times 10^{-3} \times 50}{125.44} = 0.125$ p.u.

p.u. impedance of line $L_3 = 0.125$ p.u.

The reactance diagram is shown in Fig. 11.11B.

By performing conversion of delta into star at A, B, and C, the star impedances are

$$Z_1 = \frac{j0.25 \times j0.125}{j0.25 + j0.125 + j0.125} = j0.0625$$

$$Z_2 = \frac{j0.25 \times j0.125}{j0.5} = j0.0625$$

and

$$Z_3 = \frac{j0.125 \times j0.125}{j0.5} = j0.03125$$

The reactance diagram obtained is shown in E.11.11C.
This can be further reduced into Fig. E.11.11D.
Finally this can be put first into Fig. E.11.11E and later into Fig. E.11.11F.

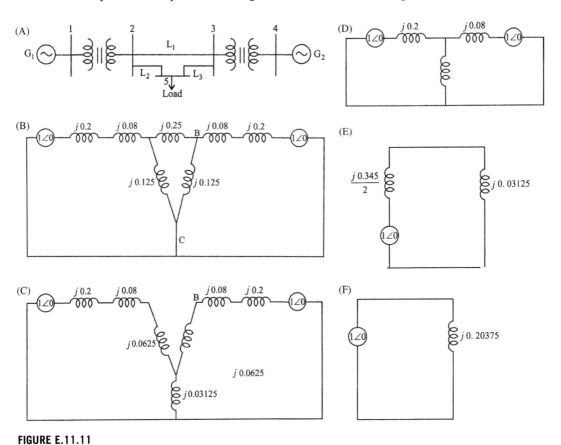

FIGURE E.11.11

System for E.11.11 and its stepwise reduction.

$$\text{Fault MVA} = \frac{1}{0.20375} = 4.90797 \text{ p.u.}$$

$$= 100 \text{ MVA} \times 4.90797 = 490.797 \text{ MVA}$$

$$\text{Fault current} = \frac{1}{j0.20375} = 4.90797 \text{ p.u.}$$

$$\text{Base current} = \frac{100 \times 10^6}{\sqrt{3} \times 112 \times 10^3} = 515.5 \text{ amp}$$

$$\text{Fault current} = 4.90797 \times 515.5$$
$$= 2530 \text{ amp}$$

E.11.12. Two motors having transient reactances 0.3 p.u. and subtransient reactances 0.2 p.u. based on their own ratings of 6 MVA, 6.8 kV are supplied by a transformer rated 15 MVA, 112 kV/6.6 kV and its reactance is 0.18 p.u. A three-phase short circuit occurs at the terminals of one of the motors. Calculate (1) the subtransient fault current, (2) subtransient current in circuit breaker A, (3) the momentary circuit rating of the breaker, and (4) if the circuit breaker has a breaking time of four cycles calculate the current to be interrupted by the circuit breaker A.

Solution:

Let base MVA = 15

Base kV for l.v. side = 6.6 kV

Base kV for h.v side = 112 kV

For each motor $x''_d = 0.2 \times \dfrac{15}{6} = 0.5$ p.u.

For each motor $x''_d = 0.3 \times \dfrac{15}{6} = 0.75$ p.u.

The reactance diagram is shown in Fig. E.11.12B.

Under fault condition, the reactance diagram can be further simplified into Fig. E.11.12C.

$$\text{Impedance to fault} = \cfrac{1}{\cfrac{1}{j0.18} + \cfrac{1}{j0.5} + \cfrac{1}{j0.5}}$$

$$\text{Subtransient fault current} = \frac{1\angle 0°}{j0.1047} = -j9.55 \text{ p.u.}$$

$$\text{Base current} = \frac{15 \times 10^6}{\sqrt{3} \times 6.6 \times 10^3} = 1312.19 \text{ A}$$

$$\text{Subtransient fault current} = 1312.19 \times (-j9.55)$$
$$= 12531.99 \text{ amp (lagging)}$$

1. Total fault current from the infinite bus.

$$\frac{-1\angle 0°}{j0.18} = -j5.55 \text{ p.u.}$$

Fault current from each motor $= \frac{1\angle 0°}{j0.5} = -j2$ p.u.

Fault current into breaker A is sum of the two currents from the infinite bus and from motor 1

$$= -j5.55 + (-j2) = -j7.55 \text{ p.u.}$$

Total fault current into breaker $= -j7.55 \times 1312.19$
$$= 9907 \text{ amp}$$

2. Manentary fault current taking into the d.c. off-set component is approximately

$$1.6 \times 9907 = 15851.25 \text{ A}$$

3. For the transient condition, i.e., after four cycles the motor reactance changes to 0.3 p.u.

The reactance diagram for the transient state is shown in Fig. E.11.12D.

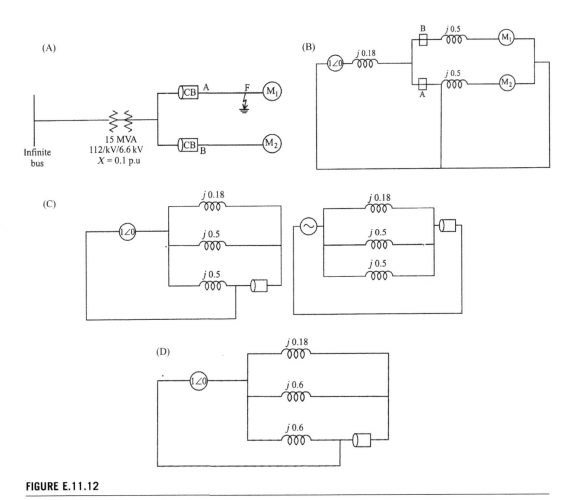

FIGURE E.11.12

System for E.11.12 and stepwise reduction. *CB*, circuit breaker.

The fault impedance is $\dfrac{1}{\dfrac{1}{j0.15} + \dfrac{1}{j0.6} + \dfrac{1}{j0.6}} = j\,0.1125$ p.u.

The fault current $= \dfrac{1\angle 0°}{j0.1125} = j8.89$ p.u.

Transient fault current $\begin{aligned} &= -j8.89 \times 1312.19 \\ &= 11665.37 \text{ A} \end{aligned}$

If the d.c. off-set current is to be considered, it may be increased by a factor of say 1.1.

So that the transient fault current $= 11665.37 \times 1.1$

$$= 12831.9 \text{ amp}$$

E.11.13. Consider the power system shown in Fig. E.11.13A.

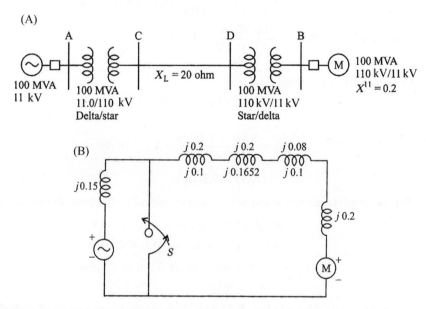

FIGURE E.11.13

System for E.11.13 and its stepwise reduction.

The synchronous generator is operating at its rated MVA at 0.95 lagging power factor and at rated voltage. A three-phase short circuit occurs at bus A calculate the per unit value of (1) subtransient fault current, (2) subtransient generator and motor currents. Neglect prefault current. Also compute (3) the subtransient generator and motor currents including the effect of prefault currents.

$$\text{Base line impedance} = \frac{(110)^2}{100} = 121 \text{ ohm}$$

$$\text{Line reactance in per unit} = \frac{20}{121} = 0.1652 \text{ p.u.}$$

The reactance diagram including the effect of the fault by switch S is shown in Fig. E.11.13B.

Looking into the network from the fault using the Thevenin's theorem

$$Z_{th} = jX_{th} = j\left(\frac{0.15 \times 0.565}{0.15 + 0.565}\right) = j\,0.1185$$

1. The subtransient fault current

$$I''_m = \frac{0.565}{0.565 + 0.15} I''_f = \frac{0.565 \times j8.4388}{0.7125} = j6.668$$

2. The motor subtransient current

$$I''_m = \frac{0.15}{0.715} I''_f = \frac{0.15}{0.715} \times 8.4388 = j1770 \text{ p.u.}$$

3. Generator base current $= \dfrac{100 \text{ MVA}}{\sqrt{3} \times 11 \text{ kV}} = 5.248 \text{ kA}$

Generator prefault current $= \dfrac{100}{\sqrt{3} \times 11} \left[\cos^{-1} 0.95 \right]$

$$= 5.248 \angle -18°.19 \text{ kA}$$

$$I_{\text{load}} = \frac{5.248 \angle -18°.19}{5.248} = 1 \angle -18°.19$$

$$= (0.95 - j0.311) \text{p.u.}$$

The subtransient generator and motor currents including the prefault currents are

$$I''_g = j6.668 + 0.95 - j0.311 = -j6.981 + 0.95$$
$$= (0.95 - j6.981) \text{p.u.} = 7.045 - 82.250 \text{ p.u.}$$

$$I''_m = -j1.77 - 0.95 + j0.311 = -0.95 - j1.459$$
$$= 1.74 \angle -56.93°$$

E.11.14. Consider the system shown in Fig. E.11.14A. The percentage reactance of each alternator is expressed on its own capacity determine the short circuit current that will flow into a dead three-phase short circuit at *F*.

Solution:

Let base kVA = 25,000 and base kV = 11

$$\%X \text{ of generator } 1 = \frac{25,000}{10,000} \times 40 = 100\%$$

$$\%X \text{ of generator } 2 = \frac{25,000}{15,000} \times 60 = 100\%$$

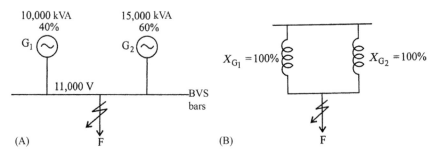

FIGURE E.11.14

System for E.11.14 and its stepwise reduction.

Line current at 25,000 kVA and 11 kV $= \dfrac{25,000}{\sqrt{3} \times 11} \times \dfrac{10^3}{10^3} = 1312.19$ A

The reactance diagram is shown in Fig. E.11.14B.

The net percentage reactance up to the fault $= \dfrac{100 \times 100}{100 + 100} = 50\%$

Short circuit current $= \dfrac{I \times 100}{\%X} = \dfrac{1312.19 \times 100}{50} = 2624.30$A

E.11.15. A three-phase, 25-MVA, 11-kV alternator has internal reactance of 6%. Find the external reactance per phase to be connected in series with the alternator so that steady-state short circuit current does not exceed six times the full load current.

Solution:

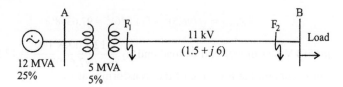

FIGURE E.11.15

System for E.11.15.

$$\text{Full load current} = \frac{25 \times 10^6}{\sqrt{3} \times 11 \times 10^3} = 1312.9 \text{ A}$$

$$V_{\text{phase}} = \frac{11 \times 10^3}{\sqrt{3}} = 6351.039 \text{ V}$$

$$\text{Total } \%X = \frac{\text{Full load current}}{\text{Short circuit current}} \times 100 = \frac{1}{6} \times 100$$

$$= 16.67\%$$

External reactance needed $= 16.67 - 6 = 10.67\%$

Let X be the per phase external reactance required in ohms.

$$\%X = \frac{IX}{V} \times 100$$

$$10.67 = \frac{1312.19X \cdot 100}{6351.0393}$$

$$X = \frac{6351.0393 \times 10.67}{1312.19 \times 100} = 0.516428 \text{ ohm}$$

E.11.16. A three-phase line operating at 11 kV and having a resistance of 1.5 ohm and reactance of 6 ohm is connected to a generating station bus bars through a 5-MVA step-up transformer having reactance of 5%. The bus bars are supplied by a 12-MVA generator having 25% reactance. Calculate the short circuit kVA fed into a symmetric fault (1) at the load end of the transformer and (2) at the h.v. terminals of the transformer.

Solution:

Let the base kVA = 12,000 kVA

%X of alternator as base kVA = 25%

%X of transformer as 12,000 kVA base = $\dfrac{12,000}{5000} \times 5 = 12\%$

%X of line = $\dfrac{12,000}{10(11)^2} \times 6 = 59.5\%$

%R of line = $\dfrac{12,000}{10(11)^2} \times 1.5 = 14.876\%$

%X_{Total} = 25 + 12 + 59.5 = 96.5%

%R_{Total} = 14.876%

$$\%Z_{Total} = \sqrt{(96.5)^2 + (14.876)^2} = 97.6398\%$$

Short circuit kVA at the far end or load end $F_2 = \dfrac{12,000 \times 100}{97.6398} = 12290$

If the fault occurs on the h.v. side of the transformer at F_1

%X up to fault $\quad \begin{aligned} F_1 &= \%X_G + \%X_T = 25 + 12 \\ &= 37\% \end{aligned}$

Short circuit kVA fed into the fault

$$= \dfrac{12,000 \times 100}{37} = 32432.43$$

E.11.17. A three-phase generating station has two 15,000 kVA generators connected in parallel each with 15% reactance and a third generator of 10,000 kVA with 20% reactance is also added later in parallel with them. Load is taken as shown from the station bus bars through 6000 kVA, 6% reactance transformers. Determine the maximum fault MVA which the circuit breakers have to interrupt on (1) l.v. side and (2) h.v. side of the system for a symmetrical fault.

Solution:

%X of generator $G_1 = \dfrac{15 \times 15,000}{15,000} = 15\%$

%X of generator $G_2 = 15\%$

%X of generator $G_3 = \dfrac{15 \times 15,000}{15,000} = 30\%$

%X of transformer $T = \dfrac{6 \times 15,000}{6000} = 15\%$

1. If fault occurs at F_1, the reactance is shown in Fig. E.11.17B.

$$\text{The total } \%C \text{ up to fault} \quad = \dfrac{1}{\dfrac{1}{15} + \dfrac{1}{15} + \dfrac{1}{30}}$$

$$= 6\%$$

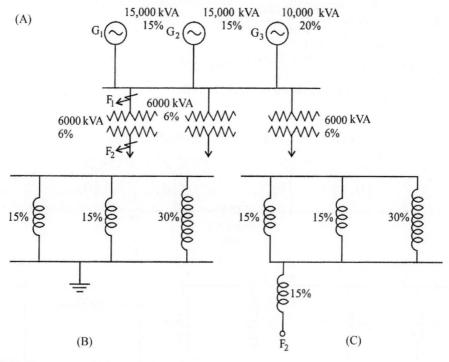

FIGURE E.11.17

System for E.11.17 and its stepwise reduction.

$$\text{Fault MVA} = \frac{15{,}000 \times 100}{6} = 250{,}000 \text{ kVA}$$

$$= 250 \text{ MVA}$$

2. If the fault occurs at F_2, the reactance diagram will be as in Fig. E.11.17C.
The total $\%X$ up to fault $6\% + 15.6\% = 21\%$

$$\text{Fault MVA} = \frac{15{,}000 \times 100}{21 \times 100} = 71.43$$

E.11.18. There are two generators at bus bar A each rated at 12,000 kVA, 12% reactance or
another bus B, two more generators rated at 10,000 kVA with 10% reactance are
connected. The two bus bars are connected through a reactor rated at 5000 kVA with
10% reactance. If a dead short circuit occurs between all the phases on bus bar B, what
is the short circuit MVA fed into the fault?
   Solution:
Let 12,000 kVA be the base kVA
$\%X$ of generator $G_1 = 12\%$
$\%X$ of generator $G_2 = 12\%$

$$\%X \text{ of generator } G_3 = \frac{10 \times 12,000}{10,000} = 12\%$$

$\%X$ of generator $G_4 = 12\%$

$$\%X \text{ of bus bar reactor} = \frac{10 \times 12,000}{5000} = 24\%$$

The reactance diagram is shown in Fig. E.11.18B.

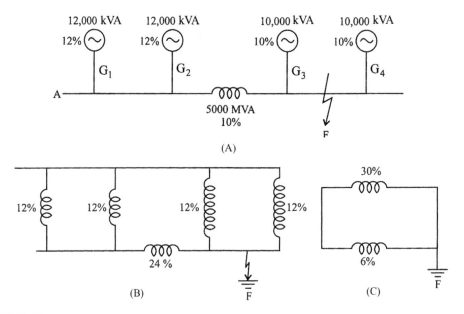

FIGURE E.11.18

System for E.11.18 and its stepwise reduction.

$$\%X \text{ up to fault} = \frac{30 \times 6}{30 + 6} = 50\%$$

$$\text{Fault kVA} = \frac{12,000 \times 100}{6} = 600,000 \text{ kVA}$$

$$= 600 \text{ MVA}$$

E.11.19. A power plant has two generating units rated 3500 kVA and 5000 kVA with percentage reactances 8% and 9%, respectively. The circuit breakers have breaking capacity of 175 MVA. It is planned to extend the system by connecting it to the grid through a transformer rated at 7500 kVA and 7% reactance. Calculate the reactance needed for a reactor to be connected in the bus bar section to prevent the circuit breaker from being overloaded if a short circuit occurs on any outgoing feeder connected to it. The bus bar voltage is 3.3 kV.

Solution:

Let 7500 kVA be the base kVA

%X of generator A $= \dfrac{8 \times 7500}{3500} = 17.1428\%$

%X of generator B $= \dfrac{9 \times 7500}{5000} = 13.5\%$

%X of transformer = 7% (as its own base)

The reactance diagram is shown in Fig. E.11.19B.

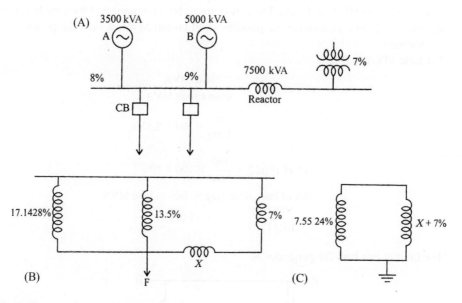

FIGURE E.11.19

System for E.11.19 and its stepwise reduction. *CB*, circuit breaker.

The short circuit kVA should not exceed 175 MVA

Total reactance to fault $= 1 \left/ \left[\dfrac{1}{7.5524} + \dfrac{1}{X+7} \right] \right.$

$\qquad\qquad\qquad = \dfrac{(X+7)(7.5524)}{X+7+7.5524}\% = \dfrac{(X+7)(7.5524)}{X+14.5524}\%$

Short circuit kVA $= 7500 \times 100 \dfrac{X(X+14.5524)}{(X+7)(7.5524)}$

This should not exceed 175 MVA

$$175 \times 10^3 = \dfrac{7500 \times 100(X+14.5524)}{(X+7)(7.5524)}$$

Solving $\quad X = 7.02\%$

Again $\quad \%X = \dfrac{kVA \cdot (X)}{10(kV)^2} = \dfrac{7500 \times (X)}{10 \times (3.3)^2}$

$\therefore \qquad X = \dfrac{7.02 \times 10 \times 3.3^2}{7500} = \mathbf{0.102 \ ohm}$

In each share of the bus bar, a reactance of 0.102 ohm is required to be inserted.

E.11.20. The short circuit MVA at the bus bars for a power plant A is 1200 MVA and for another plant B is 1000 MVA at 33 kV. These two are to be interconnected by a tie line with reactance 1.2 ohm. Determine the possible short circuit MVA at both the plants.

   Solution:

Let base MVA = 100

$$\%X \text{ of plant } 1 = \frac{\text{base MVA}}{\text{short circuit MVA}} \times 100$$

$$= \frac{100}{1200} \times 100 = 8.33\%$$

$$\%X \text{ of plant } 2 = \frac{100}{1000} \times 100 = 10\%$$

$\%X$ of interconnecting tie line on base MVA

$$= \frac{100 \times 10^3}{10 \times (3.3)^2} \times 1.2 = 11.019\%$$

For fault at bus bars for generator A

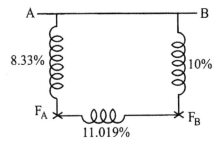

FIGURE E.11.20

Reduced equivalent system for E.11.20.

$$\%X = 1 / \left[\frac{1}{8.33} + \frac{1}{21.019} \right]$$

$$= 5.9657\%$$

$$\text{Short circuit MVA} = \frac{\text{base MVA} \times 100}{\%X}$$

$$= \frac{100 \times 100}{5.96576} = \mathbf{1676.23}$$

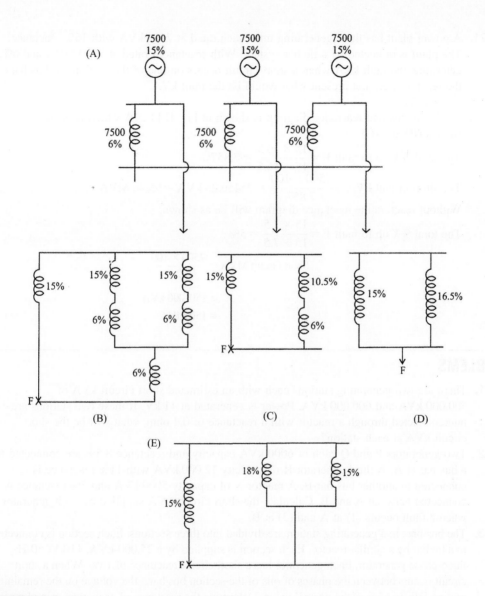

FIGURE E.11.21

System for E.11.21 and its stepwise reduction.

For a fault at the bus bars for plant B

$$\%X = 1 \left/ \left[\frac{1}{19.349} + \frac{1}{10} \right] \right. = 6.59\%$$

$$\text{Short circuit MVA} = \frac{100 \times 100}{6.59} = \mathbf{1517.45}$$

E.11.21. A power plant has three generating units each rated at 7500 kVA with 15% reactance. The plant is protected by a tie bar system. With reactances rated at 7500 MVA and 6%, determine the fault kVA when a short circuit occurs on one of the sections of bus bars. If the reactors were not present what would be the fault kVA.

    Solution:

    The equivalent reactance diagram is shown in Fig. E.11.21A which reduces to figures (B) and (C).

The total %X up to fault $F = \dfrac{15 \times 16.5}{15 + 16.5} = 7.857\%$

The short circuit kVA $= \dfrac{7500 \times 100}{7.857} = 95456.28$ kVA $= 95.46$ MVA

Without reactors the reactance diagram will be as shown.

The total %X up to fault $F = \dfrac{15 \times 7.5}{15 + 7.5} = 5\%$

$$\text{Short circuit MVA} = \frac{7500 \times 100}{5}$$

$$= 150,000 \text{ kVA}$$

$$= 150 \text{ MVA}$$

PROBLEMS

P.11.1. There are two generating stations each with an estimated short circuit kVA of 500,000 kVA and 600,000 kVA. Power is generated at 11 kV. If these two stations are interconnected through a reactor with a reactance of 0.4 ohm, what will be the short circuit kVA at each station?

P.11.2. Two generators P and Q each of 6000 kVA capacity and reactance 8.5% are connected to a bus bar at A. A third generator R of capacity 12,000 kVA with 11% reactance is connected to another bus bar B. A reactor X of capacity 5000 kVA and 5% reactance is connected between A and B. Calculate the short circuit kVA supplied by each generator when a fault occurs (1) at A and (2) at B.

P.11.3. The bus bars in a generating station are divided into three sections. Each section is connected to a tie bar by a similar reactor. Each section is supplied by a 25,000-kVA, 11-kV, 50-Hz, three-phase generator. Each generator has a short circuit reactance of 18%. When a short circuit occurs between the phases of one of the section bus bars, the voltage on the remaining section falls to 65% of the normal value. Determine the reactance of each reactor in ohms.

QUESTIONS

11.1. Explain the importance of per unit system.

11.2. What do you understand by short circuit kVA? Explain.

11.3. Explain the construction and operation of protective reactors.

11.4. How are reactors classified? Explain the merits and demerits of different types of system protection using reactors.

UNBALANCED FAULT ANALYSIS

12

Analysis of power system under unbalanced faults, viz., line-to-ground, line-to-line, and double line-to-ground faults, requires the use of symmetrical components or sequence components of both currents and voltages. It is also necessary to define sequence impedances through which the sequence components of currents flow. The methodology for the analysis of power networks under unbalanced fault conditions is presented in the following.

12.1 SEQUENCE IMPEDANCES

Electrical equipment or components offer impedance to flow of current when potential is applied. The impedance offered to the flow of positive sequence currents is called the positive sequence impedance Z_1. The impedance offered to the flow of negative sequence currents is called the negative sequence impedance Z_2. When zero sequence currents flow through components of power system the impedance offered is called the zero sequence impedance Z_0.

12.2 BALANCED STAR CONNECTED LOAD

Consider the circuit in Fig. 12.1.

A three-phase balanced load with self and mutual impedances Z_s and Z_m drawn currents I_a, I_b, and I_c as shown. Z_n is the impedance in the neutral circuit which is grounded draws and the current in the circuit is I_n.

The line-to-ground voltages are given by

$$\left.\begin{array}{l} V_a = Z_s\,I_a + Z_m\,I_b + Z_m\,I_c + Z_n\,I_n \\ V_b = Z_m\,I_a + Z_s\,I_b + Z_m\,I_c + Z_n\,I_n \\ V_c = Z_m\,I_a + Z_m\,I_b + Z_s\,I_c + Z_n\,I_n \end{array}\right\} \tag{12.1}$$

Since

$$I_a + I_b + I_c = I_n$$

Eliminating I_n from Eq. (12.1)

$$\begin{bmatrix} V_a \\ V_b \\ V_c \end{bmatrix} = \begin{bmatrix} Z_s + Z_n & Z_m + Z_n & Z_m + Z_n \\ Z_m + Z_n & Z_s + Z_n & Z_m + Z_n \\ Z_m + Z_n & Z_m + Z_n & Z_s + Z_n \end{bmatrix} \begin{bmatrix} I_a \\ I_b \\ I_c \end{bmatrix} \tag{12.2}$$

Power Systems Analysis. DOI: http://dx.doi.org/10.1016/B978-0-08-101111-9.00012-4

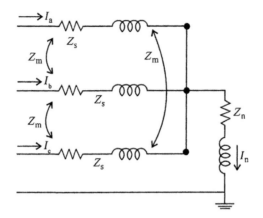

FIGURE 12.1

Balanced star connected load.

Put in compact matrix notation

$$[V_{abc}] = [Z_{abc}][I_{abc}] \tag{12.3}$$

$$V_{abc} = [A] \, V_a^{0,1,2} \tag{12.4}$$

and

$$I_{abc} = [A] \, I_a^{0,1,2} \tag{12.5}$$

Premultiplying Eq. (12.3) by $[A]^{-1}$ and using Eqs. (12.4) and (12.5), we obtain

$$V_a^{0,1,2} = [A]^{-1}[Z_{abc}][A] \, I_a^{0,1,2} \tag{12.6}$$

Defining

$$[Z]^{0,1,2} = [A^{-1}][Z_{abc}][A] \tag{12.7}$$

$$= \frac{1}{3} \begin{bmatrix} 1 & 1 & 1 \\ 1 & a & a^2 \\ 1 & a^2 & 1 \end{bmatrix} \begin{bmatrix} Z_s + Z_n & Z_m + Z_n & Z_m + Z_n \\ Z_m + Z_n & Z_s + Z_n & Z_m + Z_n \\ Z_m + Z_n & Z_m + Z_n & Z_s + Z_n \end{bmatrix} \begin{bmatrix} 1 & 1 & 1 \\ 1 & a^2 & a \\ 1 & a & a^2 \end{bmatrix}$$

$$= \begin{bmatrix} (Z_s + 3Z_n + 2Z_m) & 0 & 0 \\ 0 & Z_s - Z_m & 0 \\ 0 & 0 & Z_s - Z_m \end{bmatrix} \tag{12.8}$$

If there is no mutual coupling

$$[Z^{0,1,2}] = \begin{bmatrix} Z_s + 3Z_n & 0 & 0 \\ 0 & Z_s & 0 \\ 0 & 0 & Z_s \end{bmatrix} \tag{12.9}$$

From the above, it can be concluded that for a balanced load the three sequences are independent, which means that currents flowing in one sequence will produce voltage drops of the same phase sequence only.

12.3 TRANSMISSION LINES

Transmission lines are static components in a power system. Phase sequence has thus no effect on the impedance. The geometry of the lines is fixed whatever may be the phase sequence. Hence, for transmission lines

$$Z_1 = Z_2$$

we can proceed in the same way as for the balanced three-phase load for three-phase transmission lines also (Fig. 12.2).

$$
\left.
\begin{aligned}
V_a - V'_a &= Z_s\, I_a + Z_m\, I_b + Z_m\, I_c \\
V_b - V'_b &= Z_m\, I_a + Z_s\, I_b + Z_m\, I_c \\
V_c - V'_c &= Z_m\, I_a + Z_m\, I_b + Z_s\, I_c
\end{aligned}
\right\}
$$

(12.10) (12.11) and (12.12)

$$
\begin{aligned}
\nu_{abc} &= [V_{abc}][V'_{abc}] \\
&= [Z_{abc}][I_{abc}]
\end{aligned}
$$

(12.13)

$$
[Z]^{0,1,2} = [A^{-1}][Z_{abc}][A]
$$

$$
=
\begin{bmatrix}
Z_s + 2Z_m & 0 & 0 \\
0 & Z_s - Z_m & 0 \\
0 & 0 & Z_s - Z_m
\end{bmatrix}
$$

(12.14)

The zero sequence currents are in phase and flow through the line conductors only if a return conductor is provided. The zero sequence impedance is different from the positive and negative sequence impedances.

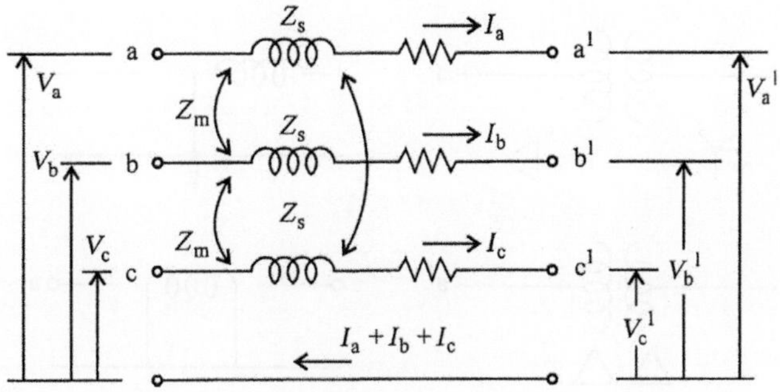

FIGURE 12.2

Three-phase transmission line with mutual impedance.

12.4 SEQUENCE IMPEDANCES OF TRANSFORMER

For analysis, the magnetizing branch is neglected and the transformer is represented by an equivalent series leakage impedance.

Since the transformer is a static device, phase sequence has no effect on the winding reactances. Hence

$$Z_1 = Z_2 = Z_1$$

where Z_1 is the leakage impedance

If zero sequence currents flow then

$$Z_0 = Z_1 = Z_2 = Z_1$$

In star–delta or delta–star transformers the positive sequence line voltage on one side leads the corresponding line voltage on the other side by 30°. It can be proved that the phase shift for the line voltages to be $-30°$ for negative sequence voltages.

The zero sequence impedance and the equivalent circuit for zero sequence currents depend upon the neutral point and its ground connection. The circuit connection for some of the common transformer connection for zero sequence currents is indicated in Fig. 12.3.

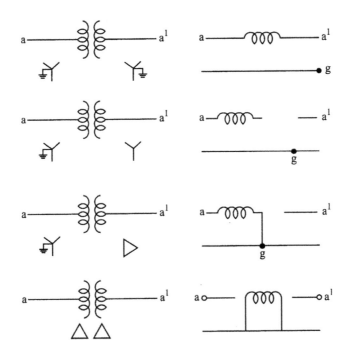

FIGURE 12.3

Zero sequence equivalent circuits.

12.5 SEQUENCE REACTANCES OF SYNCHRONOUS MACHINE

The positive sequence reactance of a synchronous machine may be X_d or X_d' or X_d'' depending upon the condition at which the reactance is calculated with positive sequence voltages applied.

When negative sequence currents are impressed on the stator winding, the net flux rotates at twice the synchronous speed relative to the rotor. The negative sequence reactance is approximately given by

$$X_2 = X_d'' \tag{12.15}$$

The zero sequence currents, when they flow, are identical and the spatial distribution of the mmfs is sinusoidal. The resultant air gap flux due to zero sequence currents is zero. Thus, the zero sequence reactance is approximately the same as the leakage flux

$$X_0 = X_l \tag{12.16}$$

12.6 SEQUENCE NETWORKS OF SYNCHRONOUS MACHINES

Consider an unloaded synchronous generator shown in Fig. 12.4 with a neutral to ground connection through an impedance Z_n. Let a fault occur at its terminals which causes currents I_a, I_b, and I_c to flow through its phases a, b, and c respectively. The generated phase voltages are E_a, E_b, and E_c. Current I_n flows through the neutral impedance Z_n.

12.6.1 POSITIVE SEQUENCE NETWORK

Since the generator phase windings are identical by design and construction, the generated voltages are perfectly balanced. They are equal in magnitude with a mutual phase shift of 120°. Hence, the generated voltages are of positive sequence. Under these conditions a positive sequence current flows in the generator that can be represented as in Fig. 12.5.

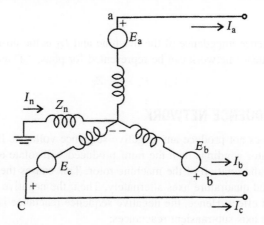

FIGURE 12.4

Three-phase synchronous generator.

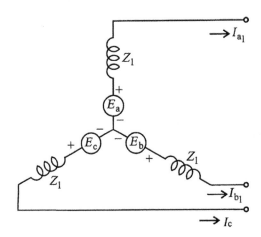

FIGURE 12.5

Positive sequence currents.

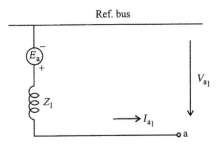

FIGURE 12.6

Positive sequence network.

Z_1 is the positive sequence impedance of the machine and I_{a1} is the positive sequence current in phase a. The positive sequence network can be represented for phase "a" as shown in Fig. 12.6.

$$V_{a1} = E_a - I_{a1} \cdot Z_1 \qquad (12.17)$$

12.6.2 NEGATIVE SEQUENCE NETWORK

Synchronous generator does not produce any negative sequence voltages. If negative sequence currents flow through the stator windings then the mmf produced will rotate at synchronous speed but in a direction opposite to the rotation of the machine rotor. This causes the negative sequence mmf to move past the direct and quadrature axes alternately. Then, the negative sequence mmf sets up a varying armature reaction effect. Hence, the negative sequence reactance is taken as the average of direct axis and quadrature axis subtransient reactances:

$$X_2 = (X_d'' + X_q'')/2 \qquad (12.18)$$

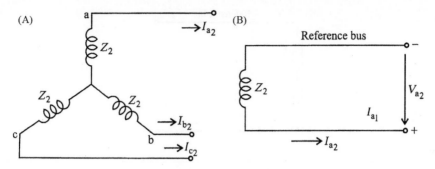

FIGURE 12.7

Negative sequence currents (A) and negative sequence network (B)

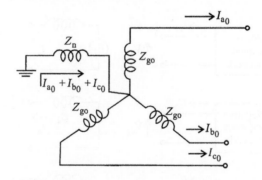

FIGURE 12.8

Zero sequence currents in synchronous machine.

The negative sequence current paths and the negative sequence network are shown in Fig. 12.7.

$$V_{a2} = -Z_2 \, I_{a2} \qquad (12.19)$$

12.6.3 ZERO SEQUENCE NETWORK

Zero sequence currents flowing in the stator windings produce mmfs which are in time phase. Sinusoidal space mmf produced by each of the three stator windings at any instant at a point on the axis of the stator would be zero, when the rotor is not present. However, in the actual machine leakage flux will contribute to zero sequence impedance. Consider the circuit in Fig. 12.8.

Since $I_{a0} = I_{b0} = I_{c0}$

The current flowing through Z_n is $3 \, I_{a0}$.

The zero sequence voltage drop

$$V_{a0} = -3 \, I_{a0} \, Z_n - I_{a0} \, Z_{g0} \qquad (12.20)$$

where Z_{g0} is the zero sequence impedance per phase of the generator

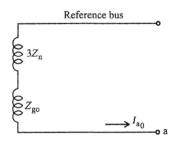

FIGURE 12.9

Zero sequence network.

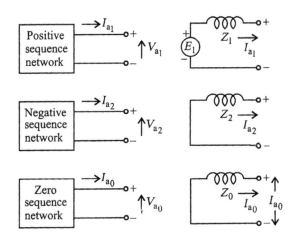

FIGURE 12.10

The three sequence networks.

Hence,

$$Z_0^1 = 3Z_n + Z_{g0} \tag{12.21}$$

so that

$$V_{a0} = -I_{a0}\, Z_0^1 \tag{12.22}$$

The zero sequence network is shown in Fig. 12.9.

Thus, it is possible to represent the sequence networks for a power system differently as different sequence currents flow as summarized in Fig. 12.10.

12.7 UNSYMMETRICAL FAULTS

The unsymmetrical faults generally considered are

- Line-to-ground fault
- Line-to-line fault
- Line-to-line to ground fault

Single line-to-ground fault is the most common type of fault that occurs in practice. Analysis for system voltages and calculation of fault current under the above conditions of operation is discussed now.

12.8 ASSUMPTIONS FOR SYSTEM REPRESENTATION

1. Power system operates under balanced steady-state conditions before the fault occurs. Therefore, the positive, negative, and zero sequence networks are uncoupled before the occurrence of the fault. When an unsymmetrical fault occurs they get interconnected at the point of fault.
2. Prefault load current at the point of fault is generally neglected. Positive sequence voltages of all the three phases are equal to the prefault voltage V_F. Prefault bus voltage in the positive sequence network is V_F.
3. Transformer winding resistances and shunt admittances are neglected.
4. Transmission line series resistances and shunt admittances are neglected.
5. Synchronous machine armature resistance, saliency, and saturation are neglected.
6. All nonrotating impedance loads are neglected.
7. Induction motors are either neglected or represented as synchronous machines.

It is conceptually easier to understand faults at the terminals of an unloaded synchronous generator and obtain results. The same can be extended to a power system and results obtained for faults occurring at any point within the system.

12.9 UNSYMMETRICAL FAULTS ON AN UNLOADED GENERATOR

Single Line-to-Ground Fault:

Consider Fig. 12.11. Let a line-to-ground fault occur on phase a.

We can write under the fault condition the following relations:

$$V_a = 0$$

$$I_b = 0$$

and

$$I_c = 0$$

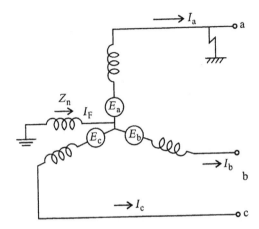

FIGURE 12.11

Line-to-ground fault on unloaded generator.

It is assumed that there is no fault impedance:

$$I_F = I_a + I_b + I_c = I_a = 3I_{a1}$$

Now

$$
\left.
\begin{aligned}
I_{a1} &= \frac{1}{3}(I_a + a\, I_b + a^n\, I_c) \\[4pt]
I_{a2} &= (I_a + a^n I_b + a\, I_c) \\[4pt]
I_{a0} &= \frac{1}{3}(I_a + I_b + I_c)
\end{aligned}
\right\}
\tag{12.23}
$$

Substituting Eq. (12.23) into Eq. (12.22)

$$I_b = I_c = 0 \tag{12.24}$$

$$I_{a1} = I_{a2} = I_{a0} = \frac{1}{3}I_a \tag{12.25}$$

Hence the three sequence networks carry the same current and hence all can be connected in series as shown in Fig. 12.12 satisfying the relation

$$V_a = E_a - I_{a1}\, Z_1 - I_{a2}\, Z_2 - I_{a0}\, Z_0 - I_F\, Z_n \tag{12.26}$$

Since $V_a = 0$

$$E_a = I_{a1}\, Z_1 + I_{a2}\, Z_2 + I_{a0}\, Z_0 + I_F\, Z_n$$

$$E_a = I_{a1}\, Z_1 + I_{a2}\, Z_2 + I_{a0}\, Z_0 + 3I_{a1}\, Z_n$$

$$E_a = I_{a1}[Z_1 + Z_2 + Z_0 + 3Z_n]$$

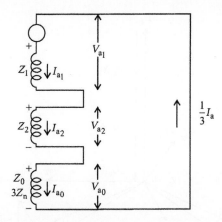

FIGURE 12.12

Sequence network connection.

Hence,

$$I_a = \frac{3 \cdot E_a}{(Z_1 + Z_2 + Z_0 + 3Z_n)} \tag{12.27}$$

The line voltages are now calculated:

$$V_a = 0$$

$$V_b = V_{a0} + a^2 V_{a1} + a V_{a2}$$

$$= (-I_{a0} Z_0) + a^2 (E_a - I_{a1} Z_1) + a(-I_{a2} Z_2)$$

$$= a^2 E_a - I_{a1}(Z_0 + a^2 Z_1 + a Z_2)$$

Substituting the value of I_{a1}

$$V_b = a^2 E_a - \frac{E_a}{(Z_1 + Z_2 + Z_0)} \cdot (Z_0 + a^r Z_1 + a Z_2)$$

$$= E_a \left[a^2 - \frac{Z_0 + a^2 Z_1 + a Z_2}{Z_0 + Z_1 + Z_2} \right] = E_a \left[\frac{a^2 Z_0 + a^2 Z_1 + a^2 Z_2 - Z_0 - a^2 Z_1 - a Z_2}{Z_0 + Z_1 + Z_2} \right]$$

$$\therefore \quad V_b = E_a \left[\frac{(a^2 - a)Z_2 + (a^2 - 1)Z_0}{Z_0 + Z_1 + Z_2} \right] \tag{12.28}$$

$$V_c = V_{a0} + a V_{a1} + a^r V_{a2}$$

$$= (-I_{a0} Z_0) + a(E_a - I_{a1} Z_1) + a^r(-I_{a2} Z_2)$$

Since

$$I_{a1} = I_{a2} = I_{a0}$$

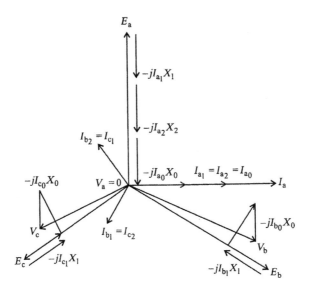

FIGURE 12.13

Phasor diagram for line-to-ground fault.

$$V_c = aE_a - \frac{E_a}{Z_1 + Z_2 + Z_0} \cdot (Z_0 + aZ_1 + a^2 Z_2)$$

$$= E_a \left[a - \frac{(Z_0 + aZ_1 + a^2 Z_2)}{Z_1 + Z_2 + Z_0} \right]$$

$$V_c = E_a \frac{[(a-1)Z_0 + (a-a^2)Z_2]}{Z_0 + Z_1 + Z_2} \qquad (12.29)$$

The phasor diagram for single line-to-ground fault is shown in Fig. 12.13.

12.10 LINE-TO-LINE FAULT

Consider a line-to-line fault across phases b and c as shown in Fig. 12.14.

From Fig. 12.14 it is clear that

$$I_a = 0$$
$$I_b = I_c \qquad (12.30)$$
$$\text{and } V_b = V_c$$

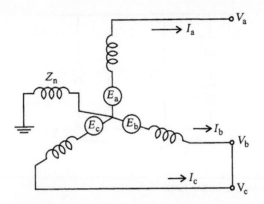

FIGURE 12.14

Double line fault.

Utilizing these relations

$$I_{a1} = \frac{1}{3}(I_a + aI_b + a^2 I_c) = \frac{1}{3}(a^2 - a)I_b$$

$$= j\frac{I_b}{\sqrt{3}} \tag{12.31}$$

$$I_{a2} = \frac{1}{3}(I_a + a^2 I_b + aI_c) = \frac{1}{3}(a^2 - a)I_b$$

$$= -j\frac{I_b}{\sqrt{3}} \tag{12.32}$$

$$I_{a0} = (I_a + I_b + I_c) = \frac{1}{3}(0 + I_b - I_b) = 0 \tag{12.33}$$

Since

$$V_b = V_c$$

$$a^2\, V_{a1} + a\, V_{a2} + V_{a0} = a\, V_{a1} + a^2\, V_{a2} + V_{a0}$$

$$(a^2 - a)V_{a1} = (a^2 - a)V_{a2}$$

$$\therefore \quad V_{a1} = V_{a2} \tag{12.34}$$

The sequence network connection is shown in Fig. 12.15.
From the diagram we obtain

$$E_a - I_{a1} Z_1 = -I_{a2} Z_2$$

$$= I_{a1} Z_2$$

$$E_a = I_{a1}(Z_1 + Z_2)$$

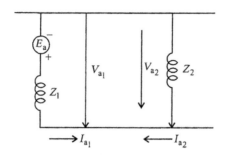

FIGURE 12.15

Sequence network connection for line-to-line fault.

$$I_{a1} = \frac{E_a}{Z_1 + Z_2} \tag{12.35}$$

$$I_b = (a^2 - a)\, I_{a1} = -j\sqrt{3}\, I_{a1} \tag{12.36}$$

$$I_c = (a - a^2)\, I_{a1} = j\sqrt{3}\, I_{a1} \tag{12.37}$$

Also,

$$V_{a1} = \frac{1}{3}(V_a + aV_b + a^2\, V_c) = \frac{1}{3}\left[V_a + (a + a^2)V_b\right]$$

$$V_{a2} = \frac{1}{3}(V_a + a^2V_b + a\, V_c) = \frac{1}{3}\left[V_a + (a^2 + a)V_b\right]$$

Since

$$1 + a + a^2 = 0; \quad a + a^2 = -1$$

Hence

$$V_{a1} = V_{a2} = \frac{1}{3}(V_a + V_b) \tag{12.38}$$

Again

$$I_a = I_{a1} + I_{a2} = I_{a1} - I_{a1} = 0$$

$$I_b = a^2\, I_{a1} + aI_{a2} = (a^2 - a)\, I_{a1}$$

$$= \frac{(a^2 - a)E_a}{Z_1 + Z_2}$$

$$I_c = -I_b = \frac{(a^2 - a)E_a}{Z_1 + Z_2} \tag{12.39}$$

$$V_a = V_{a1} + V_{a2} = E_a - I_{a1} Z_1 + (-I_{a2}Z_2)$$

$$= E_a - \frac{E_a}{Z_1 + Z_2}(Z_1 - Z_2) = E_a\left[1 - \frac{Z_1 - Z_2}{Z_1 + Z_2}\right] \tag{12.40}$$

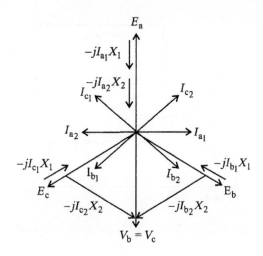

FIGURE 12.16

Phasor diagram for line-to-line fault.

$$V_b = a^2 \ V_{a1} + a \ V_{a2}$$

$$= a^2[E_a - I_{a1} \ Z_1] + a(-I_{a2} \ Z_2)$$

$$= a^2 \ E_a - I_{a1}[a^2 \ Z_1 - a \ Z_2]$$

$$= E_a\left[a^2 - \frac{(a^2 Z_1 - aZ_2)}{Z_1 + Z_2}\right] = E_a\left[\frac{a^n Z_1 + a^2 Z_2 - a^2 Z_1 + aZ_2}{Z_1 + Z_2}\right] \tag{12.41}$$

$$= \frac{E_a Z_2(a + a^2)}{Z_1 + Z_2} = \frac{E_a(-Z_2)}{Z_1 + Z_2}$$

$$V_c = V_b = \frac{E_a(-Z_2)}{(Z_1 + Z_2)} \tag{12.42}$$

The phasor diagram for a double line fault is shown in Fig. 12.16.

12.11 DOUBLE LINE-TO-GROUND FAULT

Consider line-to-line fault on phases b and c also grounded as shown in Fig. 12.17.

From Fig. 12.17.

$$\left.\begin{array}{c} I_a = 0 \\ V_b = V_c = 0 \\ I_b + I_c = I_F \end{array}\right\} \tag{12.43}$$

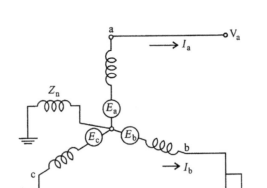

FIGURE 12.17

Double line-to-ground fault.

$$V_{a1} = \frac{1}{3}(V_a + aV_b + a^2 V_c)$$

$$= \frac{1}{3}V_a$$

(12.44)

$$V_{a2} = \frac{1}{3}(V_a + a^2V_b + a V_c)$$

$$= \frac{1}{3}V_a$$

(12.45)

Further

$$V_{a0} = \frac{1}{3}(V_a + V_b + V_c) = \frac{1}{3}V_a$$

Hence

$$V_{a1} = V_{a2} = V_{a0} = \frac{1}{3}V_a$$

(12.46)

But

$$V_{a1} = E_a - I_{a1}Z_1$$
$$V_{a2} = -I_{a2} Z_2$$

and

$$V_{a0} = -I_{a0} Z_0 - I_F Z_n$$
$$= -I_{a0}(Z_0 + 3Z_n) = -I_{a0}(Z_0^1)$$

(12.47)

It may be noted that

$$I_F = I_b + I_c = a^2 I_{a1} + a I_{a2} + I_{a0} + a I_{a1} + a^2 I_{a2} + a I_{a0}$$

$$= (a + a^2)I_{a1}(a + a^2)I_{a2} + 2 I_{a0}$$

$$= -I_{a1} - I_{a2} + 2I_{a0} = -I_{a1} - I_{a2} - I_{a0} + 3I_{a0}$$

$$= -(I_{a1} + I_{a2} + I_{a0}) + 3 I_{a0} = 0 + 3 I_{a0} = 3I_{a0}$$

The sequence network connections are shown in Fig. 12.18:

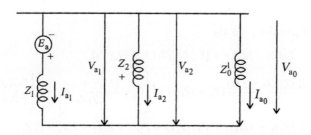

FIGURE 12.18

Sequence network connection for double line-to-ground fault.

$$I_{a1} = \frac{E_a}{Z_1 + \dfrac{Z_2 Z_0^1}{Z_2 + Z_0^1}} \qquad (12.48)$$

$$= \frac{E_a(Z_2 + Z_0^1)}{Z_1 Z_2 + Z_2 Z_0^1 + Z_0^1 Z_1} \qquad (12.49)$$

$$V_{a2} = V_{a1}$$

$$-I_{a1} Z_2 = E_a - I_{a1} Z_1$$

$$I_{a2} = -\left(\frac{E_a - I_{a_1} Z_1}{Z_2}\right)$$

$$= -\left[E_a - \frac{E_a(Z_2 + Z_0^1) \cdot Z_1}{Z_1 Z_2 + Z_2 Z_0^1 + Z_0 Z_1}\right] \cdot \frac{1}{Z_2} \qquad (12.50)$$

$$= \frac{-E_a \cdot Z_0^1}{Z_1 Z_2 + Z_2 Z_0^1 + Z_0^1 Z_1}$$

Similarly

$$-I_{a0} Z_0^1 = -I_{a2} Z_2 \qquad (12.51)$$

$$I_{a0} = -I_{a2} \frac{Z_2}{Z_0^1} = \frac{-E_a Z_2}{Z_1 Z_2 + Z_2 Z_0 + Z_0 Z_1} \qquad (12.52)$$

$$V_a = V_{a1} + V_{a2} + V_{a0}$$

$$= E_a - I_{a1} Z_1 - I_{a2} Z_2 - I_{a0}(Z_0 + 3Z_n)$$

$$= E_a - \frac{E_a(Z_2 + Z_0)}{\Sigma Z_1 Z_2} Z_1 + \frac{E_a Z_0 Z_2}{\Sigma Z_1 Z_2} + \frac{E_a \cdot Z_2(Z_0 + 3Z_n)}{\Sigma Z_1 Z_2} \tag{12.53}$$

$$= E_a \frac{3Z_2 Z_0 + 3Z_2 Z_n}{\Sigma Z_1 Z_2} = 3E_a \left(\frac{Z_2(Z_0 + Z_n)}{Z_1 Z_2 + Z_2 Z_0 + Z_0 Z_1} \right)$$

$$V_b = V_{a0} + a^2 V_{a1} + a V_{a2}$$

$$= -I_{a0}(Z_0 + 3Z_n) + a^2[E_a - I_{a1}Z_1] + a[-I_{a2} Z_2]$$

$$= \frac{E_a(Z_2)(Z_0 + 3Z_n)}{\Sigma Z_1 Z_2}(Z_0 + 3Z_n) + a^2 \left[E_a - \frac{(E_a Z_2 + Z_0)}{\Sigma Z_1 Z_2} \right] + a \left[\frac{E_0 Z_0 Z_2}{\Sigma Z_1 Z_2} \right]$$

$$= \frac{E_a[Z_2 Z_0 + 3Z_2 Z_n] + a^2 E_a[Z_1 Z_2 + Z_2 Z_0 + Z_0 Z_1 - Z_2 Z_1 - Z_0 Z_1]}{\Sigma Z_1 Z_2 + a E_a Z_0 Z_2}$$

$$V_b = \frac{E_a[Z_0 Z_2 + a^2 Z_0 Z_2 + a Z_0 Z_2 + 3Z_2 Z_n]}{\Sigma Z_1 Z_2} \tag{12.54}$$

$$= \frac{E_a[Z_0 Z_2(1 + a + a^2) + 3Z_2 Z_n]}{Z_1 Z_2 + Z_2 Z_0 + Z_0 Z_4} = \frac{3 \cdot E_a \cdot Z_2 Z_n}{Z_1 Z_2 + Z_2 Z_0 + Z_0 Z_1}$$

If

$$Z_n = 0; \quad V_b = 0$$

The phasor diagram for this fault is shown in Fig. 12.19.

12.12 SINGLE LINE-TO-GROUND FAULT WITH FAULT IMPEDANCE

If in Eq. (12.19) the fault is not a dead short circuit but has an impedance Z_F then the fault is represented in Fig. 12.20. Eq. (12.53) will be modified into

$$V_a = E_a - J_{a1} Z_1 - J_{a2} Z_2 - I_{a0} Z_0 - I_F Z_n - I_F Z_F \tag{12.55}$$

Substituting $V_a = 0$ and solving for I_a

$$I_a = \frac{3E_a}{Z_1 + Z_2 + Z_0 + 3(Z_n + Z_F)} \tag{12.56}$$

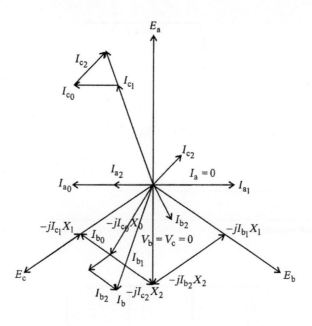

FIGURE 12.19

Phasor diagram for double line ground fault.

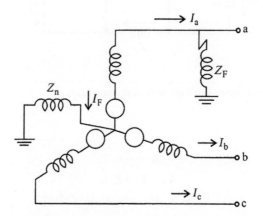

FIGURE 12.20

Line-to-ground fault with fault impedance.

12.13 LINE-TO-LINE FAULT WITH FAULT IMPEDANCE

Consider the circuit in Fig. 12.21 when the fault across the phases b and c has an impedance Z_F.

$$I_a = 0 \qquad (12.57)$$

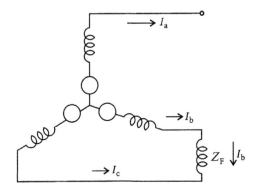

FIGURE 12.21

Line-to-line fault with fault impedance.

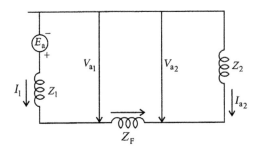

FIGURE 12.22

Sequence network connection for line-to-line fault with fault impedance.

and

$$I_b = -I_c$$

$$V_b - V_c = Z_F I_b \tag{12.58}$$

$$(V_0 + a^2V_1 + aV_2) - (V_0 + aV_1 + a^2V_2)$$

$$= Z_F(I_0 + a^2I_1 + aI_2) \tag{12.59}$$

Substituting Eqs. (12.58) and (12.59) in equation

$$(a^2 - a)\,V_1 - (a^2 - a)\,V_2 = Z_F(a^2 - a)\,I_1$$

i.e.,

$$V_1 - V_2 = Z_F\,I_1 \tag{12.60}$$

The sequence network connection in this case will be as shown in Fig. 12.22.

12.14 DOUBLE LINE-TO-GROUND FAULT WITH FAULT IMPEDANCE

This can be illustrated in Fig. 12.23.

The representative equations are

$$I_a = 0$$

$$V_b = V_c$$

$$V_b = (I_b + I_c)(Z_F + Z_n) \tag{12.61}$$

But

$$I_0 + I_1 + I_2 = 0$$

and also

$$V_0 + aV_1 + a^2 V_2 = V_a + a^2 V_1 + aV_2$$

So that

$$(a^2 - a)V_1 = (a^2 - a) V_2$$

or

$$V_1 = V_2 \tag{12.62}$$

Further,

$$(V_0 + aV_2) = (I_0 + a^2 I_1 + aI^2 + I_0 + aJ_1 + a^2 I_2)Z_F + Z_n$$

Since

$$a^2 + a = -1$$

$$(V_0 - V_1) = (Z_n + Z_F)[2I_0 - I_1 - I_2]\backslash$$

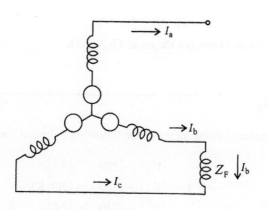

FIGURE 12.23

Double line-to-ground fault with fault impedance.

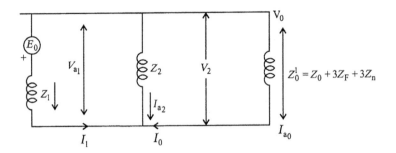

FIGURE 12.24

Sequence network connection for double line-to-ground fault with fault impedance.

But since

$$I_0 = -I_1 - I_2$$

$$V_0 - V_1 = (Z_n + Z_F)(2\,I_0 + I_0) = 3(Z_F + Z_n) \cdot I_0 \tag{12.63}$$

Hence, the fault conditions are given by

$$I_0 + I_1 + I_2 = 0$$

$$V_1 = V_2$$

and

$$V_0 - V_1 = 3(Z_F + Z_n) \cdot I_0$$

$$I_{a1} = \dfrac{E_a}{Z_1 + \dfrac{Z_0^1 Z_2}{Z_0^1 + Z_2}} \tag{12.64}$$

and so on as in case (12.11)

where

$$Z_0^1 = Z_0 + 3Z_F + 3Z_n$$

The sequence network connections are shown in Fig. 12.24.

WORKED EXAMPLES

E.12.1. Calculate the sequence components of the following balanced line-to-network voltages.

$$\bar{V} = \begin{bmatrix} V_{an} \\ V_{bn} \\ V_{cn} \end{bmatrix} = \begin{bmatrix} 220 & \underline{|0°} \\ 220 & \underline{|-120°} \\ 220 & \underline{|+120°} \end{bmatrix} KV$$

Solution:

$$V_0 = \frac{1}{3}(V_{an} + V_{bn} + V_{cn})$$
$$= \frac{1}{3}\left[200\underline{|0°} + 200\underline{|-120°} + 220\underline{|+120°}\right]$$
$$= 0$$

$$V_1 = \frac{1}{3}[V_{an} + aV_{bn} + a^2V_{cn}]$$
$$= \frac{1}{3}\left[220\underline{|0°} + 220\underline{|(-120° + 120°)} + 220\underline{|(120° + 240°)}\right]$$
$$= 220\underline{|0°}\,\text{KV}$$

$$V_2 = \frac{1}{3}[V_{an} + a^2V_{bn} + a^2V_{cn}]$$
$$= \frac{1}{3}\left[220\underline{|0°} + 220\underline{|-120° + 240°} + 220\underline{|120° + 120°)}\right]$$
$$= \frac{1}{3}[220 + 220\,120° + 220\,240°]$$
$$= 0$$

Note: Balanced three-phase voltages do not contain negative sequence components.

E.12.2. Prove that neutral current can flow only if zero sequence currents are present.

Solution:

$$I_a = I_{a1} + I_{a2} + I_{a0}$$
$$I_b = a^2I_{a1} + aI_{a2} + I_{a0}$$
$$I_c = aI_{a1} + a^2I_{a2} + I_{a0}$$

If zero sequence currents are not present
then

$$I_{a0} = 0$$

In that case

$$I_a + I_b + I_c = I_{a1} + I_{a2} + a^2I_{a1} + aI_{a2} + aI_{a1} + a^2I_{a2}$$
$$= (I_{a1} + aI_{a1} + a^2I_{a1}) + (I_{a2} + a^2I_{a2} + aI_{a2})$$
$$= 0 + 0 = 0$$

The neutral cement $I_n = I_R = I_Y + I_B = 0$. Hence, neutral cements will flow only in case of zero sequence components of currents that exist in the network.

E.12.3. Given the negative sequence cements

$$\bar{I} = \begin{bmatrix} I_a \\ I_b \\ I_c \end{bmatrix} = \begin{bmatrix} 100 & 0° \\ 100 & 120° \\ 100 & -120° \end{bmatrix}$$

Obtain their sequence components

Solution:

$$I_0 = \frac{1}{3}[I_a + I_b + I_c]$$
$$= \frac{1}{3}[100|\underline{0°} + 100|\underline{120°} + 100|\underline{-120°}] = 0\,\text{A}$$

$$I_1 = \frac{1}{3}[I_a + aI_b + a^r I_c]$$
$$= \frac{1}{3}[100|\underline{0°} + 100|\underline{120° + 120°} + 100|\underline{-120° + 240°}]$$
$$= \frac{1}{3}[100|\underline{0°} + 100|\underline{240°} + 100|\underline{120°}]$$
$$= 0\,\text{A}$$

$$I_2 = \frac{1}{3}[I_a + a^2 I_b + aI_c]$$
$$= \frac{1}{3}[100|\underline{0°} + 100|\underline{120° + 240°} + 100|\underline{-120° + 240°}]$$
$$= \frac{1}{3}[100|\underline{0°} + 100|\underline{0°} + 100|\underline{-0°}]$$
$$= 100\,\text{A}$$

Note: Balanced currents of any sequence, positive or negative, do not contain currents of the other sequences.

E.12.4. Find the symmetrical components for the given three-phase currents.

$$I_a = 10|\underline{0°}$$
$$I_b = 10|\underline{-90°}$$
$$I_c = 15|\underline{135°}$$

Solution:

$$
\begin{bmatrix} I_0 \\ I_1 \\ I_2 \end{bmatrix} = \frac{1}{3} \begin{bmatrix} 1 & 1 & 1 \\ 1 & a & a^2 \\ 1 & a^r & a \end{bmatrix} \begin{bmatrix} 10 & \underline{|0°} \\ 10 & \underline{|-90°} \\ 10 & \underline{|135°} \end{bmatrix}
$$

$$I_0 = \frac{1}{3}[100° + 10 - 90° + 15,135°]$$

$$= \frac{1}{3}\left[10(1 + j0.0) + 10(0 - j1.0) + 15(-0.707 + j0.707)\right]$$

$$= \frac{1}{3}\left[10 - j10 - 10.605 + j10.605\right]$$

$$= \frac{1}{3}\left[0.605 + j0.605\right] = \left[0.8555 \underline{|135°}\right]$$

$$= 0.285 \underline{|135°} \text{ A}$$

$$I_1 = \frac{1}{3}\left[10\underline{|0°} + 10 - \underline{|90° + 120°} + 15\underline{|-135° + 240°}\right]$$

$$= \frac{1}{3}\left[10(1 + j0.0 + 10\underline{|30°} + 15\underline{|15°}\right]$$

$$= \frac{1}{3}\left[10 + 10(0.866 + j0.5) + 15(0.9639 + j0.2588\right]$$

$$= \frac{1}{3}\left[33.1485 + j8.849\right] = \frac{1}{3}[34.309298]\underline{|15°}$$

$$= 11.436\underline{|15°} \text{ A}$$

$$I_2 = \frac{1}{3}\left[10\underline{|0°} + 10\underline{|240° - 90°} + 15\underline{|135° + 120°}\right]$$

$$= \frac{1}{3}\left[10 + (1 + j0) + 10(-0.866 + j0.5) + 15(-0.2588 - j0.9659)\right]$$

$$= \frac{1}{3}\left[-2.542 - j9.4808\right]$$

$$= 3.2744\underline{|105°} \text{ A}$$

E.12.5. In a fault study problem the following currents are measured:

$$I_R = 0$$

$$I_Y = 10 \text{ A}$$

$$I_B = -10 \text{ A}$$

Find the symmetrical components
Solution:

$$IR_1 = \frac{1}{3}\left[I_R + a\,I_Y + a^2\,I_B\right]$$

$$= \frac{1}{3}\left[0 - a(10) + a^2(-10)\right] = \frac{10}{\sqrt{3}} \text{ A}$$

$$IR_2 = [I_R + a^2\,Z_Y + a\,J_B]$$

$$= \frac{1}{3}(a^2 \cdot 10 + a(-10)] = -\frac{10}{\sqrt{3}}$$

$$IR_0 = \frac{1}{3}(I_R + I_Y + I_B)$$

$$= \frac{1}{3}(10 - 10) = 0$$

E.12.6. Draw the zero sequence network for the system shown in Fig. E.12.6.

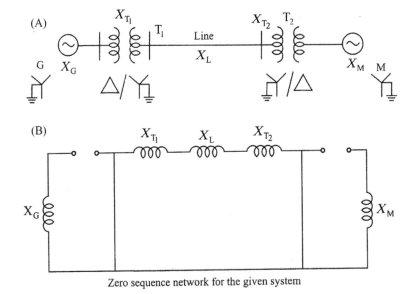

FIGURE E.12.6

System for E12.6 (A) and its zero sequence network connection (B).

Solution:

The zero sequence network is shown in Fig. E.12.6.

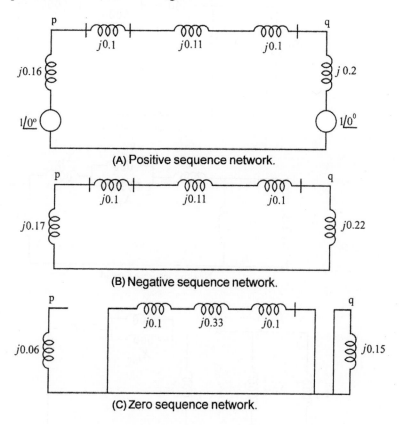

(A) Positive sequence network.

(B) Negative sequence network.

(C) Zero sequence network.

E.12.7. Draw the sequence networks for the system shown in Fig. E.12.7.

FIGURE E.12.7

Network for E12.7.

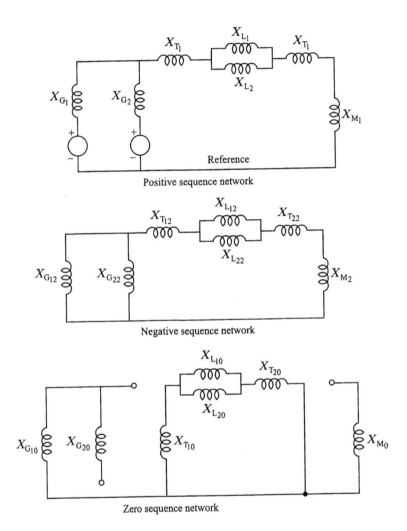

Positive sequence network

Negative sequence network

Zero sequence network

E.12.8. Consider the system shown in Fig. E.12.8. Phase b is open due to conductor break.
Calculate the sequence currents and the neutral current.

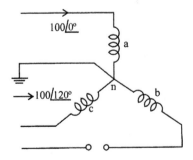

FIGURE E.12.8

System for E 12.8.

$$I_a = 100\underline{|0°}\,\text{A}$$
$$I_b = 100\underline{|120°}\,\text{A}$$

Solution:

$$\overline{I} = \begin{bmatrix} I_a \\ I_b \\ I_c \end{bmatrix} = \begin{bmatrix} 100 & \underline{|0°} \\ 0 & \\ 100 & \underline{|120°} \end{bmatrix} \text{A}$$

$$I_0 = \frac{1}{3}\left[100\underline{|0°} + 0 + 100\underline{|120°}\right]$$
$$= \frac{1}{3}\left[100(1 + j0) + 0 + 100(-0.5 - j0.866)\right]$$
$$= \frac{100}{3}\left[0.5 + j0.866) - 33.3\underline{|60°}\,\text{A}\right]$$

$$I_1 = \frac{1}{3}\left[100\underline{|0°} + 0 + 100\underline{|120° + 240°}\right]$$
$$= \frac{1}{3}\left[100\underline{|0°} + 100\underline{|0°}] = \frac{200}{3} = 66.66\,\text{A}\right]$$

$$I_2 = \frac{1}{3}\left[100\underline{|0°} + 0 + 100\underline{|120° + 120°}\right]$$
$$= \frac{1}{3}\left[100[1 + j0 - 0.5 - j0.866]\right]$$
$$= \frac{100}{3} = [-0.5 - j0.866] = 33.33\underline{|-60°}\,\text{A}\right]$$

Neutral current:

$$I_n = I_0 + I_1 + I_2$$
$$= 100\underline{|0°} + 0 + 100120°$$
$$= 100\left[1 + j0 - 0.5 + j0.866\right]$$
$$= 100\underline{|60°}\,\text{A}$$

Also

$$I_n = 3I_0 = 3(33.33(60°)) = 100\underline{|60°}\,\text{A}$$

E.12.9. Calculate the subtransient fault current in each phase for a dead short circuit on one phase to ground at bus "q" for the system shown in Fig. E.12.9.

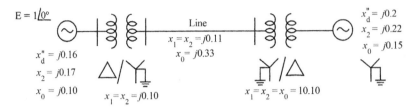

FIGURE E.12.9

Network with data for E12.9 and the sequence networks and their interconnection.

All the reactances are given in p.u. on the generator base.

Solution:

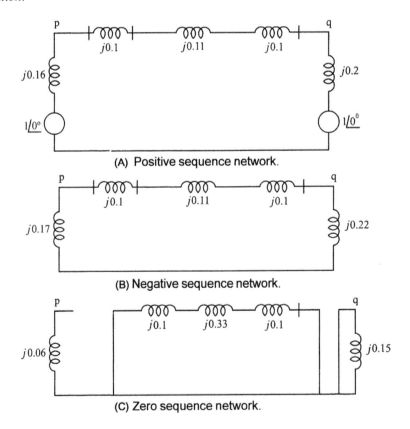

The three sequence networks are shown in Fig. E.12.9(A−C). For a line-to-ground fault on phase a, the sequence networks are connected as in Fig. E.12.9(D) at bus "q".

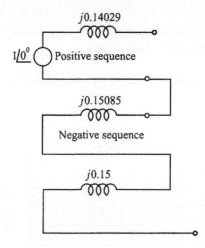

The equivalent positive sequence network reactance X_p is given from Fig. (A)

$$\frac{1}{X_p} = \frac{1}{0.47} + \frac{1}{0.2}$$

$$X_p = 0.14029$$

The equivalent negative sequence reactance X_n is given from Fig. E.12.9(B)

$$\frac{1}{X_n} = \frac{1}{0.48} + \frac{1}{0.22} \quad \text{or} \quad X_n = 0.01508$$

The zero sequence network impedance is $j0.15$. The connection of the three sequence networks is shown in Fig. E.12.9(D).

$$I_0 = I_1 = I_2 = \frac{1\underline{/0^\circ}}{j0.14029 + j0.150857 + j0.15}$$

$$= \frac{1\,0^\circ}{j0.44147} = -j2.2668 \,\text{p.u.}$$

E.12.10. In the system given in example (E.12.9) if a line-to-line fault occurs calculate the sequence components of the fault current.

Solution:

The sequence network connection for a line-to-line fault is shown in Fig. E.12.10.

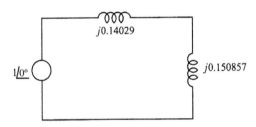

FIGURE E.12.10

Reduced reactance diagram.

From the figure

$$I_1 = I_2 = \frac{1\underline{|0°}}{j0.1409 + j0.150857} + \frac{1\underline{|0°}}{j0.291147}$$

$$= -j3.43469 \, \text{p.u.}$$

E.12.11. If the line-to-line fault in example E.12.9 takes place involving ground with no fault impedance determine the sequence components of the fault current and the neutral fault current.

Solution:

The sequence network connection is shown in Fig. E.12.11.

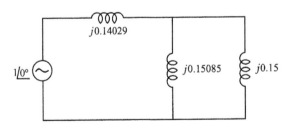

FIGURE E.12.11

Sequence network for E12.11.

$$I_1 = \frac{1\underline{|0°}}{j0.14029 + \dfrac{j(0.150857)(j0.15)}{j0.150857 + j0.15}}$$

$$= \frac{1\underline{|0°}}{j0.14029 + j0.0752134} = \frac{1\underline{|0°}}{j0.2155034}$$

$$= -j4.64 \,\text{p.u.}$$

$$I_2 = -j(4.64)\left(\frac{j0.15}{j0.300857}\right) = -j2.31339 \,\text{p.u.}$$

$$I_0 = -j(4.64)\left(\frac{j0.150857}{j0.300857}\right) = -j2.326608 \,\text{p.u.}$$

The neutral fault current $= 3\,j_0 = 3(-j2.326608) = -j6.9798 \,\text{p.u.}$

E.12.12. A dead earth fault occurs on one conductor of a three-phase cable supplied by a 5000 kVA, three-phase generator with earthed neutral. The sequence impedances of the altemator are given by

$$Z_1 = (0.4 + j4)\,\Omega; \quad Z_2 = (0.3 + j0.6)\,\Omega$$

and

$$Z_0 = (0 + j0.45)\,\Omega \text{ per phase}$$

The sequence impedance of the line up to the point of fault are $(0.2 + j0.3)\,\Omega$, $(0.2 + j0.3)\,\Omega$, $(0.2 + j0.3)\,\Omega$, and $(3 + j1)\,\Omega$. Find the fault current and the sequence components of the fault current. Also find the line-to-earth voltages on the infaulted lines. The generator line voltage is 6.6 kV.

Solution:
Total positive sequence impedance is $Z_1 = (0.4 + j4) + (0.2 + j0.3) = (0.6 + j4.3)\,\Omega$.
Total negative sequence impedance to fault is $Z_0 = (0.3 + j0.6) + (0.2 + j0.3) = (0.5 + j0.9)\,\Omega$
Total zero sequence impedance to fault is

$$Z_0 = (0 + j0.45) + (3 + j1.0) = (3 + j1.45)\,\Omega$$

$$Z_1 + Z_2 + Z_3 = (0.6 + j4.3) + (0.5 + j0.9) + (3.0 + j1.45)$$
$$= (4.1 + j6.65)\,\Omega$$

$$I_{a1} = I_{a0} = I_{a2} = \frac{6.6 \times 1000}{\sqrt{3}} - \frac{1}{(4.1 + j6.65)} = \frac{3810.62}{7.81233} \,\text{A}$$

$$= 487.77 - 58°.344 \,\text{A}$$

$$= (255.98 - j415.198)\,\text{A}$$

$$I_a = 3 \times 487.77\underline{|-58°.344}$$

$$= 1463.31\,\text{A}\underline{|-58°.344}$$

E.12.13. A 20 MVA, 6.6 kV star connected generator has positive, negative, and zero sequence reactances of 30%, 25%, and 7%, respectively. A reactor with 5% reactance based on the rating of the generator is placed in the neutral to ground connection. A line-to-line fault occurs at the terminals of the generator when it is operating at rated voltage. Find the initial symmetrical line-to-ground r.m.s. fault current. Find also the line-to-line voltage.

Solution:

$$Z_1 = j0.3; \quad Z_2 = j0.25$$

$$Z_0 = j0.07 + 3 \times j0.05 = j0.22$$

$$I_{a1} = \frac{1\lfloor 0° }{j(0.3) + j(0.25)} = \frac{1}{j0.55} = -j1.818 \, \text{p.u.}$$

$$= -j1.818 \times \frac{20 \times 1000}{\sqrt{3} \times 6.6} = -j3180 \, \text{A}$$

$$= -I_{a1}$$

$I_{a0} = 0$ as there is no ground path

$$V_a = E_a - I_{a_1} Z_1 - I_{a_2} Z_2$$
$$V_a = 1 - (-j1.818)(j0.3) + (j1.818)(i0.25)$$
$$= 1 - (1.818 \times 0.3 - 1.818 \times 0.25)$$
$$= 1 - 0.5454 + 0.4545 = 0.9091 \, \text{p.u}$$
$$= 0.9091 \times 3180 = 2890.9 \, \text{V}$$

$$V_b = a^2 E - (a^2 I_{a1} Z_1 + a I_{a2} Z_2)$$
$$= (-0.5 - j0.866) \cdot 1 + j(-j1.818)(j0.3)$$
$$= (-j0.866 - 0.5 + j0.94463)$$
$$= (-0.5 + j0.078 \, 6328) \times 3180$$
$$= (-1590 - j250) = 1921.63$$

$$V_c = V_b = 1921.63 \, \text{V}$$

E.12.14. A balanced three-phase load with an impedance of $(6 + j8)$ Ω per phase, connected in star is having in parallel a delta connected capacitor bank with each phase reactance of 27 Ω. The star point is connected to ground through an impedance of $0 + j5$ Ω. Calculate the sequence impedance of the load.

Solution:

The load is shown in Fig. E.12.14.

Converting the delta connected capacitor tank into star

$$C_A/\text{phase} = 27 \, \Omega$$

$$C_Y/\text{phase} = 27 = a \, \Omega$$

The positive sequence network is shown in Fig. E.12.14(A)

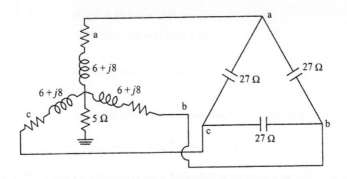

FIGURE E.12.14

System for E12.14.

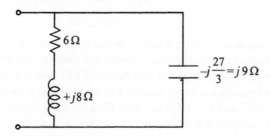

The negative sequence network is also the same as the positive sequence network

$$Z_1 = Z_2 = Z_{star} \left\| \frac{Z}{3} delta \right.$$

$$= \frac{(6 + j8)(-j9)}{6 + j8 - j9} = \frac{72 - j54}{6 - j1} = \frac{90\lfloor 36°.87}{6.082\lfloor 9°.46}$$

$$= 14.7977 \lfloor 27°.41 \, \Omega$$

The zero sequence network is shown in figure

$$Z_0 = Z_{star} + 3 Z_n = 6 + j8 + 3(j5)$$
$$= (6 + j23)\ \Omega = 23.77\ 80°.53$$

PROBLEMS

P.12.1. Determine the symmetrical components for the three-phase currents

$$I_R = 15\angle 0°, \quad I_Y = 15\underline{|230°} \quad \text{and} \quad I_B = 15\underline{|130°}\ A$$

P.12.2. The voltages at the terminals of a balanced load consisting of three 12 Ω resistors connected in star are

$$V_{RY} = 120\angle 0°\ V$$
$$V_{YB} = 96.96\angle - 121.44°\ V$$
$$V_{BR} = 108\angle 130°\ V$$

Assuming that there is no connection to the neutral of the load determine the symmetrical components of the line currents and also the line currents from them.

P.12.3. A 50 Hz turbo generator is rated at 500 MVA 25 kV. It is star connected and solidly grounded. It is operating at rated voltage and is on no-load. Its reactances are $x_d'' = x_1 = x_2 = 0.17$ and $x_0 = 0.06$ per unit. Find the subtransient line current for a single line-to-ground fault when it is disconnected from the system.

P.12.4. Find the subtransient line current for a line-to-line fault on two phases for the generator in problem (P12.3)

P.12.5. A 125 MVA, 22 kV turbo generator having $x_d'' = x_1 = x_2 = 22\%$ and $x_0 = 6\%$ has a current limiting reactor of 0.16 Ω in the neutral, while it is operating on no-load at rated voltage. A double line-to-ground fault occurs on two phases. Find the initial symmetrical r.m.s. fault current to the ground.

QUESTIONS

12.1. What are symmetrical components? Explain.

12.2. What is the utility of symmetrical components?

12.3. Derive an expression for power in a three-phase circuit in terms of symmetrical components.

12.4. What are sequence impedances? Obtain expression for sequence impedances in a balanced static three-phase circuit.

12.5. What is the influence of transformer connections in single-phase transformers connected for three-phase operation?

12.6. Explain the sequence networks for synchronous generator.

12.7. Derive an expression for the fault current for a single line-to-ground fault on an unloaded generator.

12.8. Derive an expression for the fault current for a double line fault on an unloaded generator.

12.9. Derive an expression for the fault current for a double line-to-ground fault on an unloaded generator.

12.10. Draw the sequence network connections for single line-to-ground fault, double line fault and double line-to-ground fault conditions.

12.11. Draw the phasor diagrams for
   i. Single line-to-ground fault,
   ii. Double line fault, and
   iii. Double line-to-ground fault
        Conditions on unloaded generator.

12.12. Explain the effect of prefault currents.

12.13. What is the effect of fault impedance? Explain.

POWER SYSTEM STABILITY

13

13.1 ELEMENTARY CONCEPTS

Maintaining synchronism between the various elements of a power system has become an important task in power system operation as systems expanded with increasing interconnection of generating stations and load centers. The electromechanical dynamic behavior of the prime mover-generator-excitation systems, various types of motors, and other types of loads with widely varying dynamic characteristics can be analyzed through somewhat oversimplified methods for understanding the processes involved. There are three modes of behavior generally identified for the power system under dynamic condition. They are

1. Steady state stability
2. Transient stability
3. Dynamic stability.

Stability is the ability of a dynamic system to remain in the same operating state even after a disturbance that occurs in the system.

Stability when used with reference to a power system is that attribute of the system or part of the system, which enables it to develop restoring forces between the elements thereof, equal to or greater than the disturbing force so as to restore a state of equilibrium between the elements.

A power system is said to be steady state stable for a specific steady state operating condition, if it returns to the same steady state operating condition following a disturbance. Such disturbances are generally small in nature.

A stability limit is the maximum power flow possible through some particular point in the system when the entire system or part of the system to which the stability limit refers is operating with stability.

Larger disturbances may change the operating state significantly, but still into an acceptable steady state. Such a state is called a transient state.

The third aspect of stability viz. Dynamic stability is generally associated with excitation system response and supplementary control signals involving excitation system. It is also concerned with small disturbances lasting for a long time. This will be dealt later.

Instability refers to a condition involving loss of "synchronism" which is also the same as "falling out of the step" with respect to the rest of the system.

Power Systems Analysis. DOI: http://dx.doi.org/10.1016/B978-0-08-101111-9.00013-6
Copyright © 2017 BSP Books Pvt. Ltd. Published by Elsevier Ltd. All rights reserved.

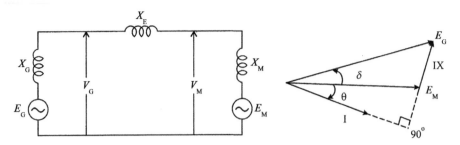

FIGURE 13.1

Generator load system through external load reactance and its phasor diagram.

13.2 ILLUSTRATION OF STEADY STATE STABILITY CONCEPT

Consider the synchronous generator–motor system shown in Fig. 13.1. The generator and motor have reactances X_G and X_M, respectively. They are connected through a line of reactance X_e. The various voltages are indicated.

From the Fig. 13.1

$$E_G = E_M + j \times I$$

$$I = \frac{E_G - E_M}{jX_{12}}$$

where

$$X_{12} = X_G + X_e + X_M$$

Power delivered to motor by the generator is

$$P = \text{Re}[E\ I^*]$$

$$= \text{Re}[E_G \angle \delta] \frac{[E_G - \delta - E_M\ 0°]}{X_{12} - 90°}$$

$$= \frac{E_G^2}{X_{12}} \cos 90° - \frac{E_G\ E_M}{X_{12}} \cos(90 + \delta)$$

$$P = \frac{E_G\ E_M}{X_{12}} \sin \delta \qquad (13.1)$$

P is a maximum when $\delta = 90°$

$$P_{\max} = \frac{E_G\ E_M}{X_{12}} \qquad (13.2)$$

The graph of P versus δ is called power angle curve and is shown in Fig. 13.2. The system will be stable so long $dP/d\delta$ is positive. Theoretically, if the load power is increased in very small

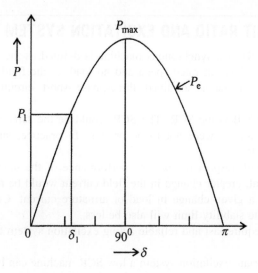

FIGURE 13.2

Power angle characteristics.

increments from $\delta = 0$ to $\pi/2$, the system will be stable. At $\delta = \pi/2$, the steady state stability limit will be reached. $P_{\max}$ is dependent on E_G, E_M, and X. Thus we obtain the following possibilities for increasing the value of $P_{\max}$ indicated in the next section.

13.3 METHODS FOR IMPROCESSING STEADY STATE STABILITY LIMIT

1. Use of higher excitation voltages, thereby increasing the value of E_G.
2. Reducing the reactance between the generator and the motor. The reactance $X = X_G + X_M + X_e$ is called the transfer reactance between the two machines and this has to be brought down to the possible extent.

13.4 SYNCHRONIZING POWER COEFFICIENT

We have

$$P = \frac{E_G E_M}{X_{12}} \sin \delta$$

The quantity

$$\frac{dP}{d\delta} = \frac{E_G E_M}{X_{12}} \cos \delta \qquad (13.3)$$

is called synchronizing power coefficient or stiffness.

For stable operation $dP/d\delta$, the synchronizing coefficient must be positive.

13.5 SHORT CIRCUIT RATIO AND EXCITATION SYSTEM

The short circuit ratio (SCR) of a synchronous machine is defined as the ratio of the field current required to produce rated voltage at rated speed and no load, to the field current required to produce rated armature current under sustained three-phase short circuit. This is illustrated in Fig. 13.3.

From the Fig. 13.3 OA/OB is the SCR. The SCR would be the same as the reciprocal of synchronous reactance, if saturation were not to be present. In practice, since saturation exists, its effect is to increase the SCR.

SCR has importance with respect to both the performance of the machine and its cost. If the SCR is less than the normal, greater change in the field current would be required to maintain constant terminal voltage for a given change in load or armature current. Correspondingly it can be inferred that the steady state stability limit will also be less.

Low SCR generators require fast and reliable acting excitation system than a machine of higher SCR.

By choosing an appropriate excitation system a low SCR machine can be satisfactorily operated. Low SCR machine means smaller size, less weight, and less cost for the machine.

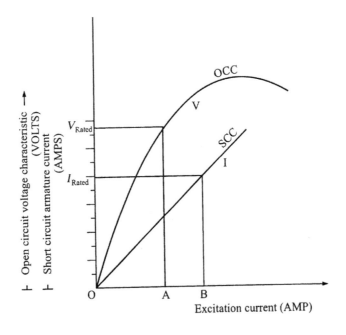

FIGURE 13.3

Short circuit ratio (SCR).

A modern fast-acting excitation system can assist in the utilization of relatively smaller and cheaper generating set with the same steady state stability limit as a bigger and expensive machine.

13.6 TRANSIENT STABILITY

Steady state stability studies often involve a single machine or the equivalent to a few machines connected to an infinite bus undergoing small disturbances. The study includes the behavior of the machine under small incremental changes in operating conditions about an operating point on small variation in parameters.

When the disturbances are relatively larger or faults occur on the system, the system enters transient state. Transient stability of the system involves nonlinear models. Transient internal voltage E_i' and transient reactances X_d' are used in calculations.

The first swing of the machine (or machines) that occur in a shorter time generally does not include the effect of excitation system and load−frequency control system. The first swing transient stability is a simple study involving a time space not exceeding 1 second. If the machine remains stable in the first second, it is presumed that it is transient stable for that disturbances. However, where disturbances are larger and require study over a longer period beyond 1 second, multiswing studies are performed taking into effect the excitation and turbine−generator controls. The inclusion of any control system or supplementary control depends upon the nature of the disturbances and the objective of the study.

13.7 STABILITY OF A SINGLE MACHINE CONNECTED TO INFINITE BUS

Consider a synchronous motor connected to an infinite bus. Initially the motor is supplying a mechanical load P_{m0} while operating at a power angle δ_0. The speed is the synchronous speed ω_s. Neglecting losses power input is equal to the mechanical load supplied. If the load on the motor is suddenly increased to P_{m1}, this sudden load demand will be met by the motor by giving up its stored kinetic energy and the motor, therefore, slows down. The torque angle δ increases from δ_0 to δ_1 when the electrical power supplied equals the mechanical power demand at b as shown in Fig. 13.4. Since, the motor is decelerating, the speed, however, is less than N_s at b. Hence, the torque angle δ_1 increases further to δ_2 where the electrical power P_e is greater than P_{m1}, but $N = N_s$ at point c. At this point c further increase of δ is arrested as $P_e > P_{m1}$ and $N = N_s$. The torque angle starts decreasing till δ_1 is reached at b but due to the fact that till point b is reached P_e is still greater than P_{m1}, speed is more than N_s. Hence, δ decreases further till point a is reacted where $N = N_s$ but $P_{m1} > P_e$. The cycle of oscillation continues. But, due to the damping in the system that includes friction and losses, the rotor is brought to the new operating point b with speed $N = N_s$.

In Fig. 13.4 area "abd" represents deceleration and area bce acceleration. The motor will reach the stable operating point b only if the accelerating energy A_1 represented by bce equals the decelerating energy A_2 represented by area abd.

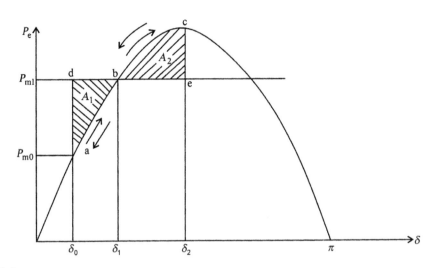

FIGURE 13.4

Stability of synchronous motor connected to infinite bus.

13.8 THE SWING EQUATION

The interconnection between electrical and mechanical side of the synchronous machine is provided by the dynamic equation for the acceleration or deceleration of the combined-prime mover (turbine)—synchronous machine rotor. This is usually called swing equation.

The net torque acting on the rotor of a synchronous machine is

$$T = \frac{WR^2}{g}\alpha \qquad (13.4)$$

where

T = algebraic sum of all torques in Kg-m.
a = Mechanical angular acceleration
WR^2 = Moment of inertia in kg-m^2

Electrical angle

$$\vartheta_e = \vartheta_m \cdot \frac{P}{Z} \qquad (13.5)$$

where ϑ_m is mechanical angle and P is the number of poles.

The frequency

$$f = \frac{PN}{120} \qquad (13.6)$$

where N is the rpm.

$$f = \frac{P}{2}\left(\frac{rpm}{60}\right)$$

$$\frac{60f}{rpm} = \frac{P}{2}$$

$$\vartheta_e = \left(\frac{60f}{rpm}\right)\vartheta_m \tag{13.7}$$

The electrical angular position d in radians of the rotor with respect to a synchronously rotating reference axis is

$$\delta = \vartheta_e - \omega_0 t \tag{13.8}$$

where

w_0 = rated synchronous speed in rad/s
   and
t = time in seconds (Note: $\delta + \omega_0 t = \vartheta_e$)

The angular acceleration taking the second derivative of Eq. (13.8) is given by

$$\frac{d^2\delta}{dt^2} = \frac{d^2\vartheta_e}{dt^2}$$

From Eq. (13.7) differentiating twice

$$\frac{d^2\vartheta_e}{dt^2} = \left(\frac{60f}{rpm}\right)\frac{d^2\vartheta_m}{dt^2}$$

$$\therefore$$

$$\frac{d^2\vartheta_m}{dt^2} = \alpha = \left(\frac{rpm}{60f}\right)\frac{d^2\vartheta_e}{dt^2}$$

From Eq. (13.4)

$$T = \frac{WR^2}{g}\left(\frac{rpm}{60f}\right)\frac{d^2\vartheta_e}{dt^2} = \frac{WR^2}{g}\left(\frac{rpm}{60f}\right)\frac{d^2\delta}{dt^2} \tag{13.9}$$

Let the base torque be defined as

$$T_{Base} = \frac{Base\ kVA}{2\pi\left(\frac{rpm}{60}\right)} \tag{13.10}$$

Torque in per unit, $\quad T_{p.u.} = \dfrac{T}{T_{Base}} = \dfrac{WR^2}{g}\left(\dfrac{rpm}{60f}\right)\dfrac{d^2\delta}{dt^2}\cdot\dfrac{2\pi\left(\dfrac{rpm}{60}\right)}{base\ kVA} \tag{13.11}$

Kinetic energy, $\quad$ K.E. $= \dfrac{1}{2}\dfrac{WR^2}{g}\,\omega_0^2 \tag{13.12}$

where

$$\omega_0 = 2\pi\frac{rpm}{60}$$

Defining

$$H = \frac{\text{kinetic energy at rated speed}}{\text{base kVA}}$$

$$= \underbrace{\frac{1}{2}\frac{WR^2}{g}\left(2\pi\frac{rpm}{60}\right)^2}_{\text{KE at rated speed}}\frac{1}{\text{base kVA}}$$

$$T = \frac{H}{\pi f}\cdot\frac{d^2\delta}{dt^2} \tag{13.13}$$

The torque acting on the rotor of a generator includes the mechanical input torque from the prime mover, torque due to rotational losses (i.e., friction, wind age, and core loss), electrical output torque, and damping torques due to prime mover, generator and power system.

The electrical and mechanical torques acting on the rotor of a motor are of opposite sign as are the result of the electrical input and mechanical load output. We may neglect the damping and rotational losses, so that the accelerating torque.

$$T_a = T_m - T_e$$

where T_e is the air-gap electrical torque and T_m the mechanical shaft torque.

$$\frac{H}{\pi f}\frac{d^2\delta}{dt^2} = T_m - T_e \tag{13.14}$$

(i.e.,)

$$\frac{d^2\delta}{dt^2} = \frac{\pi f}{H}(T_m - T_e) \tag{13.15}$$

Torque in per unit is equal to power in per unit if speed deviations are neglected. Then

$$\frac{d^2\delta}{dt^2} = \frac{\pi f}{H}(P_m - P_e) \tag{13.16}$$

The Eqs. (13.15) and (13.16) are called swing equations.

It may be noted, that, since $\delta = \vartheta - \omega_0 t$

$$\frac{d\delta}{dt} = \frac{d\vartheta}{dt} - \omega_0$$

Since the rated synchronous speed in rad/s is $2\pi f$

$$\frac{d\vartheta}{dt} = \frac{d\delta}{dt} + \omega_0$$

we may put the equation in another way.

$$\text{K. E.} = \frac{1}{2}I\omega^2 \text{ J}$$

The moment of inertia I may be expressed in Joule$-$(second)^2 per (radian)^2 since ω is in radian per second. The stored energy of an electrical machine is more usually expressed in mega joules and angles in degrees. Angular momentum M is thus described by mega joule$-$second per electrical degree.

$$M = I \cdot \omega$$

where ω is the synchronous speed of the machine and M is called inertia constant. In practice ω is not synchronous speed while the machine swings and hence M is not strictly a constant.

The quantity H defined earlier as inertia constant has the units in mega joules.

$$H = \frac{\text{Stored energy in mega joules}}{\text{Machine rating in mega volt ampers(G)}} \qquad (13.17)$$

But Stored energy $= \frac{1}{2} I \omega^2 = \frac{1}{2} M \omega$
In electrical degrees

$$\omega = 360 f (= 2\pi f) \qquad (13.18)$$

$$\text{GH} = \frac{1}{2} M (360 f) = \frac{1}{2} M 2\pi f = M \pi f$$

$$M = \frac{\text{GH}}{\pi f} \text{MJ} - \text{s/elec degree} \qquad (13.19)$$

In the per unit systems

$$M = \frac{H}{\pi f} \qquad (13.20)$$

So that

$$\frac{\mathrm{d}^2 \delta}{\mathrm{d} t^2} = \frac{\pi f}{H} (P_{\mathrm{m}} - P_{\mathrm{e}}) \qquad (13.21)$$

which may be written also as

$$M \frac{\mathrm{d}^2 \delta}{\mathrm{d} t^2} = P_{\mathrm{m}} - P_{\mathrm{e}} \qquad (13.22)$$

where

$$M = \frac{H}{\pi f}$$

This is another form of swing equation.
Further

$$P_{\mathrm{e}} = \frac{EV}{X} \sin \delta$$

So that

$$M \frac{\mathrm{d}^2 \delta}{\mathrm{d} t^2} = P_{\mathrm{m}} - \frac{EV}{X} \sin \delta \qquad (13.23)$$

with usual notation.

13.9 EQUAL AREA CRITERION AND SWING EQUATION

Equal area criterion is applicable to single machine connected to infinite bus. It is not directly applicable to multimachine system. However, the criterion helps in understanding the factors that influence transient stability.

The swing equation connected to infinite bus is given by

$$\frac{H}{\pi f}\frac{d^2\delta}{dt^2} = P_m - P_e = P_a \tag{13.24}$$

or

$$\frac{2H}{w_s}\frac{d^2\delta}{dt^2} = P_m - P_e = P_a \tag{13.25}$$

Also

$$M\frac{d^2\delta}{dt^2} = P_a \tag{13.26}$$

As t increases δ increases to a maximum value δ_{max} where $\frac{d\delta}{dt} = 0$,

Multiplying Eq. (13.26) on both sides by $2\frac{d\delta}{dt}$ we obtain

$$2\frac{d^2\delta}{dt^2}\frac{d\delta}{dt} = \frac{P_a}{M}\cdot 2\frac{d\delta}{dt}$$

Integrating both sides

$$\left(\frac{d\delta}{dt}\right)^2 = \frac{2}{M}\int P_a\, d\delta$$

$$\frac{d\delta}{dt} = \sqrt{\frac{2}{M}\int_{\delta_0}^{\delta} P_a\, d\delta}$$

δ_0 is the initial rotor angle from where the rotor starts swinging due to the disturbance.

For stability

$$\text{For stability,} \quad \frac{d\delta}{dt} = 0$$

Hence,

$$\sqrt{\frac{2}{M}\int_{\delta_0}^{\delta} P_a\, d\delta} = 0$$

i.e.,

$$\int_{\delta_0}^{\delta} P_a\, d\delta - \int_{\delta_0}^{\delta} (P_m - P_e)\, d\delta = 0$$

The system is stable, if we could locate a point c on the power angle curve such that areas A_1 and A_2 are equal. Equal area criterion states that whenever, a disturbance occurs, the accelerating and decelerating energies involved in swinging of the rotor of the synchronous machine must be equal so that a stable operating point (such as b) could be located.

$$A_1 - A_2 = 0 \text{ means that,}$$

$$\int_{\delta_0}^{\delta_1} (P_{m1} - P_e)d\delta - \int_{\delta_1}^{\delta_2} (P_e - P_m)\, d\delta = 0$$

But

$$P_e = P_{max} \sin \delta$$

$$\int_{\delta_0}^{\delta_1} (P_{m_1} - P_{max} \sin \delta)d\delta - \int_{\delta_1}^{\delta_2} (P_{max} \sin \delta - P_{m_1})\, d\delta = 0$$

$$P_{m1}(\delta_1 - \delta_0) + P_{max}(\cos \delta_1 - \cos \delta_0)$$

$$P_{max}(\cos \delta_1 - \cos \delta_2) + P_{m1}(\delta_2 - \delta_1) = 0$$

i.e.,

$$P_{m1}[\delta_2 - \delta_0] = P_{max}[\cos \delta_0 - \cos \delta_2]$$

$$\cos \delta_0 - \cos \delta_2 = [\delta_2 - \delta_0]$$

But

$$\frac{P_{m_1}}{P_{max}} = \frac{P_{max} \sin \delta_1}{P_{max}} = \sin \delta_1$$

Hence

$$(\cos \delta_0 - \cos \delta_2) = \sin \delta_1[\delta_2 - \delta_0] \tag{13.27}$$

The above is a transcendental equation and hence cannot be solved using normal algebraic methods.

13.10 **TRANSIENT STABILITY LIMIT**

Now consider that the change in P_m is larger than the change shown in Fig. 13.5. This is illustrated in Fig. 13.6.

In the case $A_1 > A_2$, i.e., we fail to locate an area A_2 that is equal to area A_1. Then, as stated the machine will lose its stability since the speed cannot be restored to N_s.

Between these two cases of stable and unstable operating cases, there must be a limiting case where A_2 is just equal to A_1 as shown in Fig. 13.7. Any further increase in P_{m1} will cause A_2 to be less than A_1. $P_{m1} - P_{m0}$ in Fig. 13.7 is the maximum load change that the machine can sustain synchronism and is thus the transient stability limit.

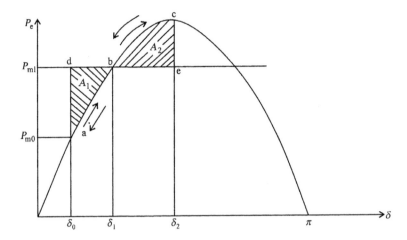

FIGURE 13.5

Equal area criterion.

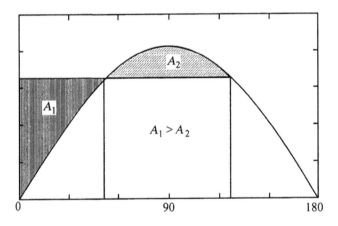

FIGURE 13.6

Unstable system ($A_1 > A_2$).

13.11 FREQUENCY OF OSCILLATIONS

Consider a small change in the operating angle δ_0 due to a transient disturbance by $\Delta\delta$. Corresponding to this we can write

$$\delta = \delta_0 + \Delta\delta$$

and

$$P_e = P_e^o + \Delta P_e$$

where ΔP_e is the change in power and P_e^o, the initial power at δ_0

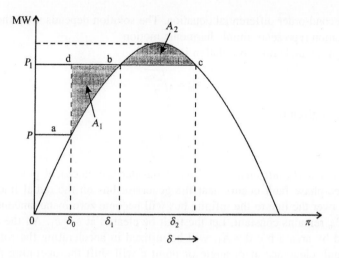

FIGURE 13.7

Transient stability limit.

$$(P_e + \Delta P_e) = P_{max} \sin \delta_0 + (P_{max} \cos \delta_0)\Delta\delta$$

Also,

$$P_m = P_e^o = P_{max} \sin \delta_0$$

Hence,

$$(P_m - P_e^o + \Delta P_e)$$
$$= P_{max} \sin \delta_0 - [P_{max} \sin \delta_0 - (P_{max} \cos \delta_0)\Delta\delta]$$
$$= (P_{max} \cos \delta_0)\Delta\delta$$

$d\frac{P_e}{d\delta}$ is the synchronizing coefficient S.
The swing equation is

$$\frac{2H}{\omega}\frac{d^2\delta_0}{dt^2} = P_a = P_m - P_e^o$$

Again,

$$\frac{2H}{\omega}\frac{d^2(\delta_0 + \Delta\delta)}{dt^2} = P_m - (P_e^o + \Delta P_e)$$

Hence,

$$\frac{2H}{\omega}\frac{d^2(\Delta\delta)}{dt^2} = -P_{max}(\cos \delta_0) \cdot \Delta\delta = -S^o \cdot \Delta\delta$$

where S^o is the synchronizing coefficient at P_e^o.
Therefore,

$$\frac{d^2(\Delta\delta)}{dt^2} + \left(\frac{\omega S^o}{2H}\right) \quad \Delta\delta = 0$$

which is a linear second-order differential equation. The solution depends upon the sign of δ_0. If δ_0 is positive, the equation represents simple harmonic motion.

The frequency of the undamped oscillation is

$$\omega_m = \sqrt{\frac{\omega\delta_0}{2\pi}} \qquad (13.28)$$

The frequency f is given by

$$f = \frac{1}{2\pi}\sqrt{\frac{\omega\delta_0}{2\pi}} \qquad (13.29)$$

Transient stability and fault clearance time: Consider the electrical power system shown in Fig. 13.8. If a three-phase fault occurs near the generator bus on the radial line connected to it, power transmitted over the line to the infinite bus will become zero instantaneously. The mechanical input power P_m remains constant. Let the fault be cleared at $\delta = \delta_1$. All the mechanical input energy represented by area a b c d $= A_1$, will be utilized in accelerating the rotor from δ_0 to δ_1. (see Fig. 13.9). Fault clearance at δ_1 angle or point c will shift the operating point from c to e instantaneously on the $P-\delta$ curve. At point f, an area $A_2 = $ d e f g is obtained which is equal to

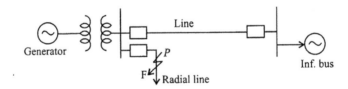

FIGURE 13.8

Power system with fault.

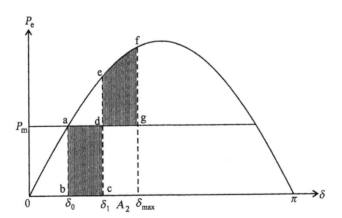

FIGURE 13.9

Equal area criterion applied when fault is cleared at δ_1.

A_1. The rotor comes back from f and finally settles down at "a" where $P_m = P_e$. δ_1 is called the clearing angle and the corresponding time t_1 is called the clearing time for the fault from the inception of it at δ_0.

13.12 CRITICAL CLEARING TIME AND CRITICAL CLEARING ANGLE

If, in the previous case, the clearing time is increased from t_1 to t_c such that δ_1 is δ_c as shown in Fig. 13.10, where A_1 is just equal to A_2. Then, any further increase in the fault clearing time t_1 beyond t_c, would not be able to enclose an area A_2 equal to A_1. This is shown in Fig. 13.11.

beyond δ_c, A_2 starts decreasing. A fault clearance cannot be delayed Beyond t_c. This limiting Fault clearance angle δ_c is called critical clearing angle and the corresponding time to clear the fault is called critical clearing time t_c (Fig. 13.11).

From Fig. 13.11

$$\delta_{max} = \pi - \delta_0$$

$$P_m = P_{max} \sin \delta_0$$

$$A_1 = \int_{\delta_0}^{\delta_c} (P_m - 0)d\delta = P_m[\delta_c - \delta_0]$$

$$A_2 = \int_{\delta_c}^{\delta_{max}} (P_{max} \sin \delta - P_m)d\delta$$

$$= P_{max}(\cos \delta_c - \cos \delta_{max}) - P_m(\delta_{max} - \delta_c)$$

$A_1 = A_2$ gives

$$\cos \delta_c - \cos \delta_m = \frac{P_m}{P_{max}}[\delta_{max} - \delta_0]$$

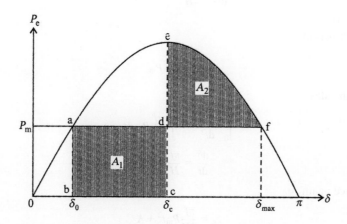

FIGURE 13.10

Equal area criterion when $A_1 = A_2$ at $\delta = \delta_c$.

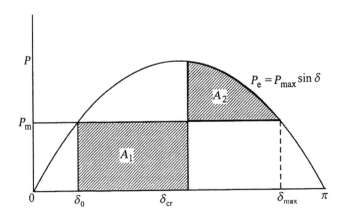

FIGURE 13.11

Equal area criterion when fault is cleared beyond δ_c.

$$\cos \delta_c = \frac{P_m}{P_{max}}[(\pi - \delta_0) - \delta_0] + \cos(\pi - \delta_0)$$

$$= \frac{P_m}{P_{max}}[(\pi - 2\delta_0)] - [\cos \delta_0]$$

$$\delta_c = \cos^{-1}\left[\frac{P_m}{P_{max}}(\pi - 2\delta_0) - (\cos \delta_0)\right] \tag{13.30}$$

During the period of fault the swing equation is given by

$$\frac{d^2\delta}{dt^2} = \frac{\pi f}{H}(P_m - P_e).$$

But, since $P_e = 0$
During the fault period

$$\frac{d^2\delta}{dt^2} = \frac{\pi f}{H}P_m$$

Integrating both sides

$$\int_0^t \frac{d^2\delta \, dt}{dt^2} = \int_0^t \frac{\pi f}{H}P_m \, dt$$

$$\frac{d\delta}{dt} = \frac{\pi f}{H}P_m t$$

and integrating once again

$$\delta_c = \frac{\pi f}{2H}P_m t^2 + K$$

At $t = 0$; $\delta = \delta_0$, Hence $K = \delta_0$

Hence

$$\delta_c = \frac{\pi f}{2H} P_m t^2 + \delta_0 \tag{13.31}$$

Hence the critical cleaning time

$$t_c = \sqrt{\frac{2H(\delta_0 - \delta_c)}{P_m\,\pi\cdot f}}\ \text{second} \tag{13.32}$$

For a given clearing time, the clearing angle is decreased by increasing the inertia constant. The inertia constant depends upon the design and the designs are standardized and hence cannot be manipulated for the sake of stability. However, it will be interesting to compare the effect on hydro and steam generators. Taking the average value of H for hydro machines as 3 and for steam turbines as 6 and since the value of t_c varies as the square root of H

$$\frac{t_c\ \text{hydro}}{t_c\ \text{steam}} = \sqrt{\frac{H_a}{H_s}} = \sqrt{\frac{3}{6}} = 0.707$$

The critical clearing time for hydro turbine is about 70% of the time for a steam turbine generator.

13.13 FAULT ON A DOUBLE-CIRCUIT LINE

Consider a single generator or generating station supplying power to a load or an infinite bus through a double-circuit line as shown in Fig. 13.12.

The electrical power transmitted is given by $P_{e12} = \frac{EV}{x'_d + x_{12}} \sin\delta$ where $\frac{1}{x_{12}} = \frac{1}{x_1} + \frac{1}{x_2}$ and x'_d is the transient reactance of the generator. Now, if a fault occurs on line 2, e.g., then the two circuit breakers on either side will open and disconnect the line 2. Since, $x_1 > x_{12}$ (two lines in parallel), the $P-\delta$ curve for one line in operation is given by

$$P_{e1} = \frac{EV}{x'_d + x_1} \sin\delta$$

will be below the $P-\delta$ curve P_{e12} as shown in Fig. 13.13. The operating point shifts from a to b on $P-\delta$ curve P_{e1} and the rotor accelerates to point c where $\delta = \delta_1$. Since the rotor speed is not

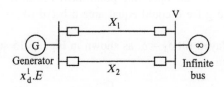

FIGURE 13.12

Double-circuit line and fault.

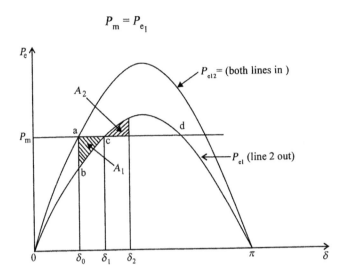

FIGURE 13.13

Fault on double-circuit line, faulted line removed.

synchronous, the rotor decelerates till the point d is reached at $\delta = \delta_2$ so that area A_1 (=area a b c) is equal to area A_2 (=area c d e). The rotor will finally settle down at point c due to damping. At point c

$$P_m = P_{e1}$$

13.14 TRANSIENT STABILITY WHEN POWER IS TRANSMITTED DURING THE FAULT

Consider the case during the fault period some load power is supplied to the load or to the infinite bus. If the $P-\delta$ curve during the fault is represented by curve 3 in Fig. 13.14.

Upon the occurrence of fault, the operating point moves from a to b on to the fault curve 3. When the fault is cleared at $\delta = \delta_1$, the operating point moves from b to c along the curve P_{e3} and then shifts to point e. If area d g f e d could equal area a b c d ($A_2 = A_1$) then the system will be stable.

If the fault clearance is delayed till $\delta_1 = \delta_c$ as shown in Fig. 13.15 such that area a b c d (A_1) is just equal to and e d f (A_2) then

$$\int_{\delta_0}^{\delta_c} (P_{max} \sin \delta - P_m)d\delta = \int_{\delta_c}^{\delta_{max}} (P_{max} \sin \delta - P_m)\, d\delta$$

It is clear from the Fig. 13.14 that $\delta_{max} = \pi - \delta_0 = \pi - \sin^{-1} \dfrac{P_m}{P_{max\,2}}$

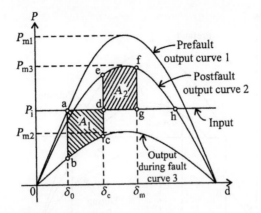

FIGURE 13.14

Fault on a double-circuit line, fault line removed with output during fault.

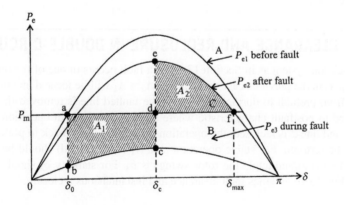

FIGURE 13.15

Critical clearing angle-power transmitted during fault.

Integrating

$$(P_m \cdot \delta + P_{max} \cos \delta) \Big|_{\delta_0}^{\delta_c} + (P_{max\,2} \cos \delta - P_m \cdot \delta) \Big|_{\delta_c}^{\delta_{max}} = 0$$

$$P_m(\delta_c - \delta_0) + P_{max}\,3(\cos \delta_c - \cos \delta_0) + P_m(\delta_{max} - \delta_c)$$

$$+ P_{max\,2}(\cos \delta_{max} - \cos \delta_c) = 0$$

$$\cos \delta_c \frac{P_m(\delta_{max} - \delta_0) - P_{max\,3} \cos \delta_0 + P_{max\,2} \cos \delta_{max}}{P_{max\,2} - P_{max\,3}} \tag{13.33}$$

The angles are all in radians.

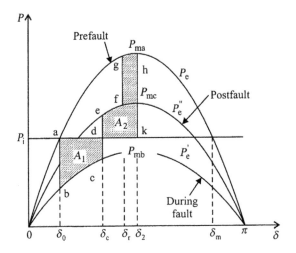

FIGURE 13.16

Fault clearance and reclosing.

13.15 FAULT CLEARANCE AND RECLOSURE IN DOUBLE-CIRCUIT SYSTEM

Consider a double-circuit system as in Section 13.12. If a fault occurs in one of the lines while supplying a power of P_{m0}; as in the previous case then an area $A_2 = A_1$ will be located and the operating characteristic changes from prefault to during the fault. If the faulted line is removed then power transfer will be again shifted to postfault characteristic where the line 1 only is in operation. Subsequently if the fault is cleared and line 2 is reclosed, the operation once again shifts back to prefault characteristic and normalcy will be restored. For stable operation area A_1 (=area abcd) should be equal to area A_2 (=area defghk). The maximum angle the rotor swings is δ_3. For stability δ_2 should be less than δ_m. The illustration in Fig. 13.16 assumes fault clearance and instantaneous reclosure.

13.16 FIRST SWING STABILITY

Under transient conditions the macine reactances X_g and X_m of X_{12} in Eq. 13.2 will be the transient reactances of the synchronous machines involved and the voltages are values behind transient reactances. One should remember that these reactances are proportional to the field flux linkages. When a fault occurs, the flux linkages do not change initially (theorem of constant flux linkages). Subsequently they decay at a rate determined by the short circuit time constant T_d^1. This time constant is minimum for a three-phase short circuit at the terminals of the macine and is slightly more for other type of faults depending upon the severity and location of the fault.

   In case of a fault cleared within a short time, machine may remain stable surviving the first swing of the rotor. But, in case the fault remains uncleared for longer time because of the continued decrease of the field flux linkages the macine or machines may pull out of step after the second or subsequent swings.

13.17 SOLUTION TO SWING EQUATION STEP-BY-STEP METHOD

Solution to swing equation gives the change in δ with time. Uninhibited increase in the value of δ will cause instability. Hence, it is desired to solve the swing equation to see that the value of δ starts decreasing after an initial period of increase, so that at some later point in time, the machine reaches the stable state. Generally 8, 5, 3, or 2 cycles are the times suggested for circuit breaker interruption after the fault occurs. A variety of numerical step-by-step methods are available for solution to swing equation. The plot of δ versus t in seconds is called the swing curve.

There are several methods available for the numerical evaluation of second-order differential equations. We evaluate step-by-step the dependent variable for small increments of independent variables, i.e., time. But elaborate computations are possible only on a computer. For hand computation, a step-by-step procedure is presented at first. To calculate the rotor angular δ, the following assumptions are made:

1. The accelerating power P_a computed at the beginning of an interval is assumed constant from the middle of the preceding interval to the middle of the interval under consideration.
2. During any interval, the angular velocity is assumed constant at the value computed for the middle of the interval.

Since δ is changing continuously both the assumptions are strictly speaking not correct. P_a and ω are both functions of δ.

When Δt is made very small, the calculated values become more accurate. Let the time intervals be Δt.

Consider $(n-2)$, $(n-1)$, and nth intervals. The accelerating power P_a is computed at the end of these intervals and plotted at circles in Fig. 13.17A.

Note that these are the beginnings for the next intervals viz., $(n-1)$, n, and $(n+1)$. P_a is kept constant between the mid-points of the intervals.

Likewise, w_r, the difference between w and w_s is kept constant throughout the interval at the value calculated at the mid-point. The angular speed therefore is assumed to change between $(n-3/2)$ and $(n-1/2)$ ordinates.

we know that $\Delta\omega = \frac{d\omega}{dt} \cdot \Delta t$

Hence

$$\omega_{r(n-1)} - \omega_{r(n-3/2)} = \frac{d^2\delta}{dt^2} \cdot \Delta t = \frac{180f}{H} P_{a(n-1)} \cdot \Delta t \tag{13.34}$$

Again change in δ

$$\Delta\delta = \frac{d\delta}{dt} \cdot \Delta t$$

i.e.,

$$\Delta\delta_{n-1} = \delta_{n-1} - \delta_{n-2} = \omega_{r(n-3/2)} \cdot \Delta t \tag{13.35}$$

for $(n-1)$th interval
and

$$\Delta\delta = \delta_n - \delta_{n-1} = \omega_{r(n-1/2)} \cdot \Delta t \tag{13.36}$$

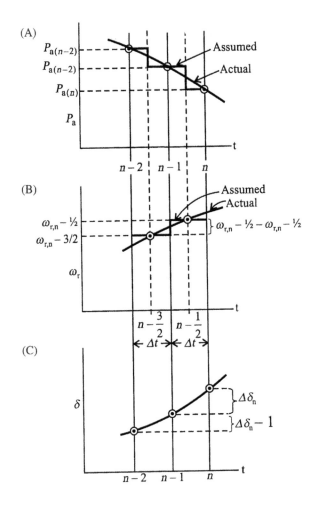

FIGURE 13.17

Plotting swing curve.

From the two Eqs. (13.16) and (13.15) we obtain

$$\Delta\delta_n = \delta_{n-1} + = \frac{180f}{H}\Delta t^2 \cdot P_{a(n-1)} \tag{13.37}$$

Thus the plot of δ with time increasing after a transient disturbance has occurred or a fault takes place can be plotted as shown in Fig. 13.17C.

Eq. 13.37 forms the basis for calculating the swing curve. The accelerating power P_a has to be computed at the beginning of each new interval and the process is continued till enough number of points are obtained to plot the swing curve or till the desired time instant is reached. Smaller the value of Δt, more accurate are the results obtained. In general $\Delta t = 0.05$ second is a good choice for time increment.

When a fault occurs or load changes, there would be a discontinuity at that instant in the accelerating power P_a. For example, before a fault, $P_a = 0$ and after the fault it has a certain value governed by $P_a = P_m - P_e$. Since the accelerating power has two values, at the beginning of the interval, the average of the two values has to be taken (see Example 13.5).

13.18 FACTORS AFFECTING TRANSIENT STABILITY

The reactive stability of a generating unit is determined by (1) the angular swing of the machine relative to the system (infinite bus) during and after the occurrence of a fault and (2) the critical clearing time. These factors are dependent on the inertia constant (H) of the machine and also the direct axis transient reactance X^1_d. From Eq. 13.37 it can be noted that smaller the value of H constant larger the angular swing during any time interval.

If P_{max} is relatively less than for a given shaft power P_m, δ_0 gets increased, δ_{max} gets decreased, and the difference between δ_0 and δ_{cr} also gets reduced. Hence, with reduction in P_{max} the swing of the machine rotor from its initial position gets restricted to a lower value before it reaches its critical clearing angle. Any consideration of decrease of H and increase of X^1_d of the machine is detrimental to maintain stability.

Transient stability is very much affected by the type of the fault. A three-phase dead short circuit is the most severe fault; the fault severity decreasing with two-phase fault and single line-to-ground fault in that order.

If the fault is farther from the generator the severity will be less in the case of a fault occurring at the terminals of the generator.

Power transferred during fault also plays a major role. When, part of the power generated is transferred to the load, the accelerating power is reduced to that extent. This can easily be understood from the curves of Figs. 13.13, 13.14, and 13.15.

Theoretically an increase in the value of inertia constant M reduces the angle through which the rotor swings farther during a fault. However, this is not a practical proposition since, increasing M means, increasing the dimensions of the machine, which is uneconomical. The dimensions of the machine are determined by the output desired from the machine and stability cannot be the criterion. Also, increasing M may interfere with speed governing system. Thus looking at the swing equations

$$M\frac{d^2\delta}{dt^2} = P_a = P_m - P_e = P_m - \frac{EV}{X_{12}}\sin\delta \tag{13.38}$$

The possible methods that may improve the transient stability are:

1. Increase of system voltages, and use of automatic voltage regulators.
2. Use of quick response excitation systems.
3. Compensation for transfer reactance X_{12} so that P_e increases and $P_m - P_e = P_a$ reduces.
4. Use of high-speed circuit breakers reduces the fault duration time and hence the accelerating power.

When faults occur the system voltage drops. Support to the system voltages by automatic voltage controllers and fast-acting excitation systems will improve the power transfer during the fault and reduce the rotor swing.

Reduction in transfer reactance is possible only when parallel lines are used in place of single line or by use of bundle conductors. Other theoretical methods such as reducing the space between the conductors and increasing the size of the conductors are not practicable and are uneconomical.

Quick opening of circuit breakers and single pole reclosing is helpful. Since majority of the faults are line-to-ground faults, selective single pole opening and reclosing will ensure transfer of power during the fault and improve stability. These methods are briefly mentioned.

13.18.1 EFFECT OF VOLTAGE REGULATOR

If the excitation system is controlled by an automatic voltage regulator, then the regulator controls the field flux linkages in such a manner that the field flux linkages decrease more slowly initially and in fact increase later. Hence, if the machine survives the first swing of the rotor, then with the voltage regulator take over control of excitation, the machine will remain stable during subsequent swings. However, if the disturbance is more severe, the machine may lose synchronism on one the first few swings.

Even after the fault is cleared, when the machines are swinging, the armature current will still have a demagnetizing effect. Of course, this effect is much less than that during the fault period. The excitation system, thus can reduce the decrease of flux linkages or increase even when high-speed relaying and fault clearance are used. In fact the assistance of the excitation system is more after the fault is cleared in sustaining the internal voltage build up. It can be said that the faster the excitation system responds to control reduction in voltage, the more effective it will be in improving stability. It has been proved by Concordia that with increase of switching time, the exciter time constant must decrease for maintaining constant flux linkages so that the same stability limit is obtained.

Fast switching: When a fault occurs, the voltages fall considerably. This in turn causes less power transfer. The accelerating power thereby increases. The time required for the removal of a fault is the total time required for the relay to respond and the breaker to operate. It is possible now to achieve this in less than two cycles. From equal area criterion it can be observed that decreasing the accelerating area and increasing the deceleration area will improve transient stability. High-speed circuit breakers are capable of achieving this.

Auto reclosing: Majority of faults are transient in nature and are self-clearing. Most of the modern circuit breakers are auto reclosing. Opening of the line through relay actuated circuit breaker will remove the fault in most of the cases. The line cannot be instantaneously reclosed as the line insulation has to fully regain its dielectric strength. In some cases, the reclosure may have to be repeated for a second time and if required even for a third time.

Single switching: Majority of the power system faults are single line-to-ground faults. It is possible to design and arrange the protective relaying scheme with circuit breakers so that only the faulted phase is isolated in the event. Deenergizing the faulted line and reclosing must be accomplished in the shortest possible time as the three-phase system cannot be sustained along with one-phase open. If it takes longer time, then all the three phases must be tripped.

Breaking resistors: Consider the system shown in Fig. 13.18 with breaking resistors located at the generator bus. For a fault on a line connected to the generator bus, tripping and reclosure of the line is a routine matter as indicated earlier. This process must clear most of the faults. The interruption to power flow is only for a very short duration. If resistors are connected to the generator bus

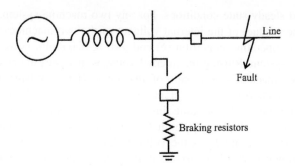

FIGURE 13.18

Breaking resistors.

as shown in Fig. 13.18, acceleration of the generator rotor can further be reduced, as power transferred during the fault duration is increased by this way. The resistors may be switched into the circuit immediately after the breaker opens and again switched out after reclosing.

Fast valving: It has been stated that whenever a fault occurs, the affected circuit is switched out through appropriate relaying. During the fault period the generator rotor accelerates. Reducing the power from the turbine P_T instantaneously, the accelerating power P_a and the subsequent rotor angle advancement could be minimized. The turbine power P_T is the net power obtained from all the stages of the turbine. Generally, high pressure, intermediate and low-pressure stages or sections exists. Also, between the high-pressure and intermediate stages steam is passed through a reheat boiler. Along with it a bypass path is also provided. Sudden closure followed by slow opening of such valves. Turbine power can be considerably reduced for a short time period. The duration of power curtailment may be of the order of a second. This is called fast valving.

Switching of series capacitors: A capacitor inserted in a line and normally short circuited by a breaker and opened during fault periods can improve transient stability substantially. Control schemes for switching of capacitors are nevertheless more complicated and involve higher mathematical logic.

Load shedding: It is already mentioned that breaking resistors reduce angular acceleration of the generator rotor. Load shedding or removal of load or loads at appropriate locations may improve transient stability. This is more applicable if the occurrence of a fault and subsequent isolation of lines causes power deficiency, in an area.

13.19 EXCITATION SYSTEM AND THE STABILITY PROBLEM

Consider Eq. (13.38)

$$M \frac{\mathrm{d}^2 \delta}{\mathrm{d}t^2} = P_\mathrm{m} \frac{Ev}{x_{12}} \sin \delta$$

In order to reduce the accelerating power, increased speed of exciter response will be one of the choices before us. This, in fact, was the first means suggested for improving power system stability under both transient and steady state conditions. For any two machine system, the power transmitted is proportional to the product of the internal voltages of the two machines, divided by the reactance. This is true at any operating condition (δ) and also at maximum power condition. It is true even for a multimachine system. Raising the internal voltages increases the power that can be transmitted between any two machines or groups of machines. Hence, raising the internal voltages increases the stability limits. An increase in internal voltages generally is accompanied by an increase of load on the machines. At the same time the input P does not increase in general.

Consider a synchronous machine with terminal voltage V_t. The voltage due to excitation acting along the quadrature axis is E_q and E_q^1 is the voltage along this axis. The direct axis rotor angle with respect to a synchronously revolving axis is δ. If a load change occurs and the field current, I_f, is not changed then the various quantities mentioned change with the real power delivered P as shown in Fig. 13.19A.

In case the field current I_f is changed such that the transient flux linkages along the q-axis E_q^1 proportional to the field flux linkages is maintained constant the power transfer could be increased by 30%−60% greater than case (a) and the quantities for this case are plotted in Fig. 13.19B.

If the field current I_f is changed along with P simultaneously so that V_t is maintained constant, then it is possible to increase power delivery by 50%−80% more than case (a). This is shown in Fig. 13.19C.

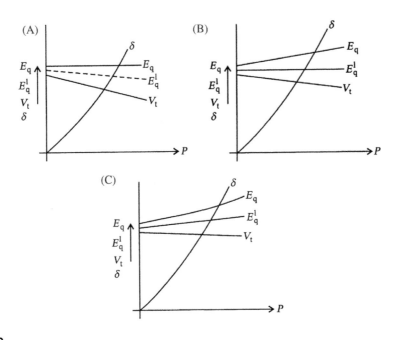

FIGURE 13.19

Effect of excitation on power transfer capability in a synchronous machine. (A) Synchronous machine—load changes but field excitation constant. (B) Synchronous machine—Field current varied to maintain field flux linkages constant. (C) Synchronous machine—Excitation changed to maintain terminal voltage constant.

It can be concluded from the above, that excitation control has a great role to play in power system stability and the speed with which this control is achieved is very important in this context.

Note that

$$P_{max} = \frac{E \cdot V}{X}$$

and increase of E matters in increasing P_{max}.

In Russia and other countries, control signals utilizing the derivatives of output current and terminal voltage deviation have been used for controlling the voltage in addition to propositional control signals. Such a situation is termed 'forced excitation' or "forced field control." Not only the first derivatives of ΔI and ΔV are used, but also higher derivatives have been used for voltage control on load changes.

These controllers have not much control on the first swing stability, but have effect on the operation subsequent swings.

This way of system control for satisfactory operation under changing load conditions using excitation control comes under the purview of dynamic stability.

13.20 DYNAMIC STABILITY

"Stable operation with voltage regulators at values of power beyond the stability limit obtainable with hand control of the excitation is called dynamic stability" (Kimbark).

The excitation system has to increase internal voltages such that the synchronizing power increases more rapidly than the increase of angle δ decreases the synchronizing power. $\left[\frac{dp}{d\delta} = \frac{Ev}{x_{12}} \cos \delta \text{ must be positive for stable operation} \right]$. The benefit in the case of machines connected with large external reactances, however, is relatively less.

The reactance x_{12} between two machines consists mainly of the reactances of the two machines, transformer reactances and line reactance. The transient reactances of large machines generally lie within a narrow range and are characteristic values that do change much in normal designs. A low value of reactance generally needs a larger size for the machine and under rating is required. This is quite uneconomical.

In the steady state, power is calculated with saturated synchronous reactances and the voltages behind these reactances. Power is a maximum when δ, the phase angle between the voltages reaches 90°. This power limit is then limited by the transfer reactance x_{12}. The voltage regulators cannot raise the power limit nearer to the theoretical limit. This is because, while the regulator acts faster, the field circuits of the exciter and the main generator arc are sluggish. Hence, the voltages cannot be built up faster, countering the demagnetizing flux linkages so as to raise the electrical power.

13.20.1 POWER SYSTEM STABILIZER

An voltage regulator in the forward path of the exciter–generator system will introduce a damping torque and under heavy load conditions this damping torque may become negative. This is a situation where dynamic instability may occur and a cause for concern. It is also observed that the several time constants in the forward path of excitation control loop introduce large phase lag at low frequencies just above the natural frequency of the excitation system.

To overcome these effects and to improve the damping, compensating networks are introduced to produce torque in phase with the speed.

Such a network is called "Power System Stabilizer" (PSS).

13.21 SMALL DISTURBANCE ANALYSIS

It is already explained that steady state analysis refers to the capability of the power system to remain in synchronism when subjected to small disturbances. To examine steady state stability of a power system, it is possible to obtain a linear model and apply classical stability analysis. For the analysis voltage regulator, speed governor, and other control devices will not be included. The motion of the system is assumed free so that its characteristic equation can be examined. From Eq. (13.21)

$$M\frac{d^2\delta}{dt^2} = P_m - P_e \tag{13.39}$$

where

$$M = \frac{H}{\pi f_0} \text{ and } P_e = P_{max} \sin \delta$$

Let the system operate initially at an angle δ_0. Consider a load increment that causes an angular deviation of $\Delta\delta$, from the initial operating angle δ_0 then,

$$\delta = \delta_0 + \Delta\delta$$

Eq. (13.39) becomes

$$M\frac{d^2(\delta_0 + \Delta\delta)}{dt^2} = P_m - P_{max} \sin[\delta_0 + \Delta\delta]$$

Expanding on both sides and remembering that $\Delta\delta$ is small so that $\cos \Delta\delta \approx 1$ and $\sin \Delta\delta \approx \Delta\delta$.

$$M\frac{d^2\delta_0}{dt^2} + M\frac{d^2\Delta\delta}{dt^2} = P_m - P_{max} \sin \delta_0 - P_{max} \cos \delta_0 \Delta\delta \tag{13.40}$$

The change in accelerating energy is given by

$$M\frac{d^2\delta_0}{dt^2} + M\frac{d^2\Delta\delta}{dt^2} - \left(M\frac{d^2\delta_0}{dt^2}\right) = P_m - P_{max} \sin \delta_0$$

$$- P_{max} \cos \delta_0\Delta\delta - (P_m - P_{max} \sin \delta_0) \tag{13.41}$$

i.e.,

$$M\frac{d^2\Delta\delta}{dt^r} + P_{max} \cos \delta_0 \Delta\delta = 0 \tag{13.42}$$

Remembering that $P_{max} \cos \delta_0$ is the synchronizing coefficient P_s, the equation of motion becomes

$$M \frac{d^2 \Delta \delta}{dt^r} + P_s \Delta \delta = 0 \qquad (13.43)$$

During a disturbance, the rotor speed differs from the speed of the rotating magnetic field and this produces induction motor action and the rotor experiences a damping torque. The damping power is proportional to the difference in angular velocities.

The damping power is expressed by

$$P_d = D \frac{d\delta}{dt} \qquad (13.44)$$

The coefficient D can be determined easily by performing a test or from design data. The equation of motion then becomes

$$M \frac{d^2 \Delta \delta}{dt^2} + D \frac{d\Delta \delta}{dt} + P_s \Delta \delta = 0 \qquad (13.45)$$

or,

$$\frac{d^2 \Delta \delta}{dt^2} + \left(\frac{D}{M}\right) \frac{d\Delta \delta}{dt} + \left(\frac{P_s}{M}\right) \Delta \delta = 0 \qquad (13.46)$$

Damping ratio ξ, natural frequency of oscillation ω_n and damped frequency of oscillation ω_d can be obtained by comparing Eq. 13.46 with the standard second-order equation.

$$\frac{d^2 \Delta \delta}{dt^2} + 2\xi \omega_n \frac{d\Delta \delta}{dt} + \omega_n^2 \Delta \delta = 0 \qquad (13.47)$$

$\xi = \frac{D}{2} \sqrt{\frac{1}{MP_s}}$ which under normal operating condition is less than one. The roots of the characteristic equation are given with usual notation by

$$S_1, S_2 = -\xi \omega_n \pm j \omega_n \sqrt{1 - \xi^2} \qquad (13.48)$$

$$= -\xi \omega_n \pm j \omega_d \qquad (13.49)$$

The roots are complex conjugate.

If P_s is positive, the real parts are negative for positive damping. In this case, the response is bounded and the system is stable.

State Space Solution:

Let

$$x_1 = \Delta \delta$$

$$\dot{x}_1 = x_2 = \frac{d}{dt} \Delta \delta = \Delta \omega$$

$$\dot{x}_2 = \frac{d^2 \Delta \delta}{dt^2} = -\omega_n^2 \Delta \delta - 2\xi \omega_m \frac{d\Delta \delta}{dt}$$

$$\dot{x}_2 = -\omega_n^2 x_1 - 2 \omega_m \xi x_2$$

Thus, the state space modal becomes

$$\begin{bmatrix} x_1^{\cdot} \\ x_2^{\cdot} \end{bmatrix} = \begin{bmatrix} 0 & 1 \\ -\omega_n^2 & -2\xi\omega_n \end{bmatrix} \begin{bmatrix} x_1 \\ x_2 \end{bmatrix} \tag{13.50}$$

$$[SI - A] = \begin{bmatrix} S & 0 \\ 0 & S \end{bmatrix} - \begin{bmatrix} 0 & 1 \\ -\omega_n^2 & -2\xi\omega_n \end{bmatrix}$$
$$= \begin{bmatrix} S & 0 \\ \omega_n^2 & 2\xi\omega_n \end{bmatrix} \tag{13.51}$$

$$[SI - A]^{-1} = \frac{\begin{bmatrix} S + 2\xi\omega_n & 1 \\ -\omega_n^2 & S \end{bmatrix}}{(S^2 + 2\xi\omega_n S + \omega_n^2)} \tag{13.52}$$

$$X(s) = [SI - A]^{-1} X(0) \tag{13.53}$$

For a small perturbation of the rotor by $\Delta\delta_0$

$$x_1(0) = \Delta\delta_0 \text{ and } x_2(0)\frac{dx_1}{dt} = \Delta\omega_o = 0$$

Hence,

$$\Delta\delta(S) = \frac{(S + 2\xi\omega_n)\Delta\delta_0}{S^2 + 2\xi\omega_n S + \omega_n^2} \tag{13.54}$$

Obtaining the inverse Laplace transform

$$\Delta\delta = \frac{\Delta\delta_0}{\sqrt{1 - \xi^2}} e^{-\xi\omega_n t} \sin(\omega_d t + \theta) \tag{13.55}$$

where

$$\theta = \cos^{-1}\xi \tag{13.56}$$

The rotor angle after the disturbance at any time t is given by

$$\delta = \delta_0 + \frac{\Delta\delta_0}{\sqrt{1 - \xi^2}} e^{-\xi\omega_n t} \sin(\omega_d t + \theta) \tag{13.57}$$

The response time constant

$$\tau = \frac{1}{\xi\omega_n} = \frac{2H}{\pi f_o D} \tag{13.58}$$

If H is increased, ω_n and ξ decrease thereby the time required to settle to new load angle increases.

If P_s, the synchronizing coefficient is increased, ω_n increases and ξ decreases.

13.22 NODE ELIMINATION METHODS

In all stability studies, buses that are excited by internal voltages of the machines only are considered. Hence, load buses are eliminated. As an example consider the system shown in Fig. 13.20.

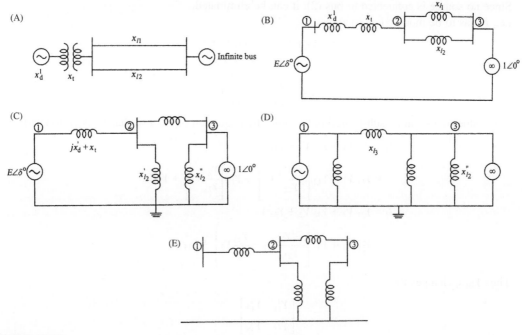

FIGURE 13.20

Double line power system. (A) Reactance diagram for the system and its stepwise reduction.

The transfer reactance between the two buses (1) and (3) is given by

$$X_{13} = j + x_t + x_{l1l2}$$

where

$$\frac{1}{x_{l_1 l_2}} = \frac{1}{x_{l_1}} + \frac{1}{x_{l_2}}$$

If a fault occurs on all the phases on one of the two parallel lines, say, line 2, then the reactance diagram will become as shown in Fig. 13.20C.

Since no source is connected to bus (2), it can be eliminated. The three reactances between buses (1), (2), (3), and (g) become a star network, which can be converted into a delta network using the standard formulas. The network will be modified into Fig. 13.20(D).

X_{13}^1 is the transfer reactance between buses (1) and (3).

Consider the same example with delta network reproduced as in Fig. 13.20E.
For a three bus system, the nodal equations are

$$\begin{bmatrix} I_1 \\ I_2 \\ I_3 \end{bmatrix} = \begin{bmatrix} Y_{11} & Y_{12} & Y_{13} \\ Y_{21} & Y_{22} & Y_{23} \\ Y_{31} & Y_{32} & Y_{33} \end{bmatrix} \begin{bmatrix} V_1 \\ V_2 \\ V_3 \end{bmatrix}$$

Since no source is connected to bus (2), it can be eliminated.
i.e., I_2 has to be mode equal to zero

$$Y_{21} V_1 + Y_{22} V_2 + Y_{23} V_3 = 0$$

Hence

$$V_2 = -\frac{Y_{21}}{Y_{22}} V_1 - \frac{Y_{23}}{Y_{22}} V_3$$

This value of V_2 can be substituted in the other two equation of (. . .) so that V_2 is eliminated

$$I_1 = Y_{11} V_1 + Y_{12} V_2 + Y_{13} V_3$$

$$= Y_{11} V_1 + Y_{12}\left[\frac{-Y_{21}}{Y_{22}} V_1\right] + Y_{13}\left[\frac{Y_{23}}{Y_{22}} V_3\right] + Y_{13} V_3$$

$$I_3 = Y_{31} V_1 + Y_{32} V_2 + Y_{33} V_3$$

$$= Y_{31} V_1 + Y_{32}\left[\frac{-Y_{21}}{Y_{22}} V_1 - \frac{Y_{23}}{Y_{22}} V_3\right] + Y_{33} V_3$$

Thus Y_{BUS} changes to

$$\begin{bmatrix} Y_{11}^1 & Y_{12}^1 \\ Y_{31}^1 & Y_{33}^1 \end{bmatrix}$$

where

$$Y_{11}^1 = Y_{11} - Y_{12}\frac{Y_{21}}{Y_{22}} \text{ and } Y_{13}^1 = Y_{31}^1 = Y_{13} - \frac{Y_{23} Y_{12}}{Y_{22}}$$

$$Y_{33}^1 = Y_{33} - \frac{Y_{32} \cdot Y_{23}}{Y_{22}}$$

13.23 OTHER METHODS FOR SOLUTION OF SWING EQUATION

In addition to step-by-step method described in Section 13.17, there are various other efficient methods to find the solution to swing equation. The simplest method is Euler's method and the most widely used method for greater accuracy is the Fourth-Order-Ringe-Kutta method.

In the following Euler's method modified so that the method is more suitable for solution of swing equation is briefly described.

13.23.1 MODIFIED EULER'S METHOD

Consider the first order differential equation

$$\frac{dx}{dt} = f(x)$$

Consider an instant t and its neighborhood point $t + \Delta t$. At the beginning of the interval Δt, $\frac{dx_t}{dt} = f(x_t)$ and at the end of the interval Δt

$$x_{t+\Delta t} = x_t + \Delta x = x_t + \left(\frac{dx_t}{dt}\right) \cdot \Delta t$$

This is Euler's method to estimate a new point $x_{t+\Delta t}$ from x_t and from its slope $\frac{dx_t}{dt}$. The method assumes constant slope over the interval Δt. Δw improvement can be obtained by calculating the slope at the beginning and also at the end of the interval and using the average of these two slopes.

Let

$$\bar{x} = x_t + \left(\frac{dx_t}{dt}\right)^{\Delta t}$$

be calculated as a first estimate.

The slope at $\bar{x}$ is calculated from

$$\frac{d\bar{x}}{dt} = f(\bar{x})$$

New value of the variable is computed as

$$x_{t+\Delta t} = x_t + \frac{\left(\dfrac{dx_t}{dt} + \dfrac{d\bar{x}}{dt}\right)}{2} \cdot \Delta t$$

The above procedure can be applied to compute δ and ω, the rotor angle, and angular velocity.

$$\frac{d\delta t}{dt} = \omega_t - \omega_s \quad \text{and} \quad \frac{d\omega t}{dt} = \frac{P_a}{M}$$

$$\bar{\delta} = \delta t + \left(\frac{d\delta t}{dt}\right) \cdot \Delta t$$

$$\bar{\omega} = \omega t + \left(\frac{d\omega t}{dt}\right) \cdot \Delta t$$

The slopes are computed from

$$\frac{d\bar{\delta}}{dt} = \bar{\omega} - \omega_s$$

and

$$\frac{d\bar{\omega}}{dt} = \frac{\bar{P}_a}{M}$$

Then,

$$\delta_{t+\Delta t} = \delta t + \frac{\left(\dfrac{d\delta t}{dt} + \dfrac{d\bar{\delta}}{dt}\right)}{2} \cdot \Delta t$$

and

$$\omega_{t+\Delta t} = \omega t + \frac{\left(\dfrac{d\omega t}{dt} + \dfrac{d\bar{\omega}}{dt}\right)}{2}\Delta t$$

The calculations are to be continued till the desired time period is covered.

WORKED EXAMPLES

E.13.1. A 4-pole, 50 Hz, 11 kV turbo generator is rated 75 MW and 0.86 power factor lagging. The machine rotor has a moment of inertia of 9000 Kg-m². Find the inertia constant in MJ/MVA and M constant or momentum in MJ/elec degree?

Solution:

$$\omega = 2\pi f = 100\,\pi \text{ rad/s}$$

$$\text{Kinetic energy} = \frac{1}{2}I\omega^2 = \frac{1}{2} \times 9000 + (100\pi)^2$$

$$= 443.682 \times 106 \text{ J}$$
$$= 443.682 \text{ MJ}$$

$$\text{MVA rating of the machine} = \frac{75}{0.86} = 87.2093$$

$$H = \frac{MJ}{MVA} = \frac{443.682}{87.2093} = 8.08755$$

$$M = \frac{GH}{180f} = \frac{87.2093 \times 5.08755}{180 \times 50}$$

$$= 0.0492979 \text{ MJS}/0 \text{ dc}$$

E.13.2. Two generators rated at 4-pole, 50 Hz, 50 MW 0.85 p.f. (lag) with moment of inertia 28,000 kg-m² and 2-pole, 50 Hz, 75 MW, 0.82 p.f. (lag) with moment of inertia 15,000 kg-m² are connected by a transmission line. Find the inertia constant of each machine and the inertia constant of single equivalent machine connected to infinite bus. Take 100 MVA base.

Solution:

For machine *I*

$$\text{K.E.} = \frac{1}{2} \times 28,000 \times (100\,\pi)^2 = 1380.344 \times 10^6 \text{ J}$$

$$\text{MVA} = \frac{50}{0.85} = 58.8235$$

$$H_1 = \frac{1380.344}{58.8235} = 23.46586 \text{ MJ/MVA}$$

$$M_1 = \frac{58.8235 \times 23.46586}{180 \times 50} = \frac{1380.344}{180 \times 50}$$

$$= 0.15337 \text{ MJS/degree elect.}$$

For the second machine

$$\text{K.E.} = \frac{1}{2} \times 15,000 \frac{1}{2} \times (100\,\pi)^2 = 739,470,000 \text{ J}$$

$$= 739.470 \text{ MJ}$$

$$\text{MVA} = \frac{75}{0.82} = 91.4634$$

$$H_2 = \frac{739.470}{91.4634} = 8.0848$$

$$M_2 = \frac{91.4634 \times 8.0848}{180 \times 50} = 0.082163 \text{ MJS/degree elec}$$

$$\frac{1}{M} = \frac{1}{M_1} = \frac{1}{M_2}$$

$$\therefore \quad M = \frac{M_1 M_2}{M_1 + M_2} = \frac{0.082163 \times 0.15337}{0.082163 + 0.15337}$$

$$= \frac{0.0126}{0.235533} = 0.0535 \text{ MJS/elec.degree}$$

$$GH = 180 \times 50 \times M = 180 \times 50 \times 0.0535$$
$$= 481.5 \text{ MJ}$$

on 100 MVA base, inertia constant.

$$H = \frac{481.5}{100} = 4.815 \text{ MJ/MVA}$$

E.13.3. A 4-pole synchronous generator rated 110 MVA, 12.5 KV, 50 HZ has an inertia constant of 5.5 MJ/MVA

1. Determine the stored energy in the rotor at synchronous speed.
2. When the generator is supplying a load of 75 MW, the input is increased by 10 MW. Determine the rotor acceleration, neglecting losses.
3. If the rotor acceleration in (2) is maintained for 8 cycles, find the change in the torque angle and the rotor speed in rpm at the end of 8 cycles?

Solution:

1. Stored energy = $GH = 110 \times 5.5 = 605$ MJ where G = Machine rating
2. P_a = The accelerating power = 10 MW

$$10 \text{ MW} = M \frac{d^2\delta}{dt^2} = \frac{GH}{180f} \frac{d^2\delta}{dt^2}$$

$$\frac{605}{180 \times 50} \frac{d^2\delta}{dt^2} = 10$$

$$0.0672 \frac{d^2\delta}{dt^2} = 10 \quad \text{or} \quad \frac{d^2\delta}{dt^2} = \frac{10}{0.0672} = 148.81$$

$\alpha = 148.81$ elec degrees/s^2

3. 8 cycles = 0.16 s

Change in $\delta = \frac{1}{2} \times 148.81 \times (0.16)^2$

Rotor speed at the end of 8 cycles

$$= \frac{120f}{P} \cdot (\delta) \times t = \frac{120 \times 50}{4} \times 1.90476 \, \delta \times 0.16$$

$$= 457.144 \text{ rpm}$$

E.13.4. Power is supplied by a generator to a motor over a transmission line as shown in Fig. E.13.4(A). To the motor bus a capacitor of 0.8 p.u. reactance per phase is connected through a switch. Determine the steady state power limit with and without the capacitor in the circuit.

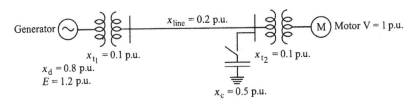

FIGURE E.13.4A

System for E13.4 (A) and its stepwise reduction (B,C,D).

Steady state power limit without the capacitor

$$P_{\text{max 1}} = \frac{1.2 \times 1}{0.8 + 0.1 + 0.2 + 0.8 + 0.1} = \frac{1.2}{2.0} = 0.6 \text{ p.u.}$$

With the capacitor in the circuit, the following circuit is obtained:

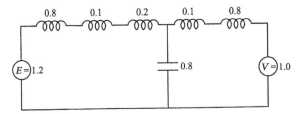

FIGURE E.13.4B

Simplifying

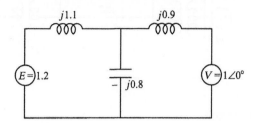

FIGURE E.13.4C

Converting the star to delta network, the transfer reactance between the two nodes X_{12}.

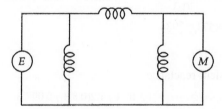

FIGURE E.13.4D

$$X_{12} = \frac{(j1.1)(j0.9) + (j0.9)(-j0.8) + (-j0.8 \times j1.1)}{-j0.8}$$

$$= \frac{-0.99 + 0.72 + 0.88}{-j0.8} = \frac{-0.99 + 1.6}{-j0.8} = \frac{j0.61}{0.8}$$

$$= j0.7625 \text{ p.u.}$$

$$\text{Steady state power limit} = \frac{1.2 \times 1}{0.7625} = 1.5738 \text{ p.u.}$$

E.13.5. A generator rated 75 MVA is delivering 0.8 p.u. power to a motor through a transmission line of reactance j 0.2 p.u. The terminal voltage of the generator is 1.0 p.u. and that of the motor is also 1.0 p.u. Determine the generator e.m.f behind transient reactance. Find also the maximum power that can be transferred?
Solution:
When the power transferred is 0.8 p.u.

$$0.8 = \frac{1.0 \times 1.0 \sin \theta}{(0.1 + 0.2)} = \frac{1}{0.3} \sin \theta$$

$$\text{Sin } \theta = 0.8 \times 0.3 = 0.24$$

$$\theta = 13.^\circ 8865$$

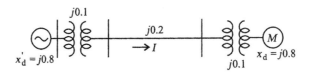

FIGURE E.13.5

System for E13.5.

Current supplied to motor

$$I = \frac{1\angle 13.°8865 - 1\angle 0°}{j0.3} = \frac{(0.9708 + j0.24) - 1}{j0.3}$$

$$= \frac{-0.0292 + j0.24}{j0.3} = j0.0973 + 0.8 = 0.8571\lfloor\tan^{-1}0.1216$$

$$I = 0.8571\lfloor 6.°934$$

Voltage behind transient reactance

$$= 1\angle 0° + j\,1.2(0.8 + j\,0.0973)$$

$$= 1 + j\,0.96 - 0.11676$$

$$= 0.88324 + j\,0.96$$

$$= 1.0496\ 47°.8$$

$$P_{max} = \frac{EV}{X} = \frac{1.0496 \times 1}{1.2} = 0.8747\ \text{p.u.}$$

E.13.6. Determine the power angle characteristic for the system shown in Fig. E.13.6(A). The generator is operating at a terminal voltage of 1.05 p.u. and the infinite bus is at 1.0 p.u. voltage. The generator is supplying 0.8 p.u. power to the infinite bus.

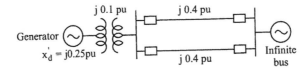

FIGURE E.13.6A

System for E13.6 (A) and its reactance diagram (B).

Solution:
The reactance diagram is drawn as in Fig. E.13.6 (B).

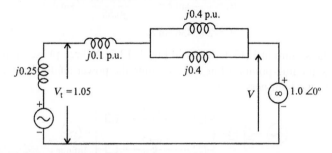

FIGURE E.13.6B

The transfer reactance between V_t and V is $= j0.1 + \dfrac{j0.4}{2} = j0.3$ p.u.

we have $\dfrac{V_t V}{X} \sin \delta = \dfrac{(1.05)(1.0)}{0.3} \sin \delta = 0.8$

Solving for δ, $\sin \delta = 0.22857$ and $\delta = 13°.21$

The terminal voltage is $1.05 \underline{|13°.21}$

$$1.022216 + j0.24$$

The current supplied by the generator to the infinite bus

$$
\begin{aligned}
I &= \frac{1.022216 + j0.24 - (1 + j0)}{j0.3} \\
&= \frac{(0.022216 + j0.24)}{j0.3} = 0.8 - j0.074 \\
&= 1.08977 \underline{|5.°28482}\, \text{p.u.}
\end{aligned}
$$

The transient internal voltage in the generator

$$
\begin{aligned}
E^1 &= (0.8 - j0.074)\, j0.25 + 1.22216 + j0.24 \\
&= j0.2 + 0.0185 + 1.02216 + j0.24 \\
&= 1.040 + j0.44 \\
&= 1.1299 \underline{|22°.932}
\end{aligned}
$$

The total transfer reactance between E^1 and V

$$= j0.25 + j0.1 + \frac{j0.4}{2} = j0.55 \text{ p.u.}$$

The power angle characteristic is given by

$$P_e = \frac{E^1 V}{X} \sin \delta = \frac{(1.1299) \times (1.0)}{j0.55} \sin \delta$$

$$P_e = 2.05436 \sin \delta$$

E.13.7. Consider the system in E.13.1 shown in Fig. E.13.7. A three-phase fault occurs at point P as shown at the mid-point on line 2. Determine the power angle characteristic for the system with the fault persisting.

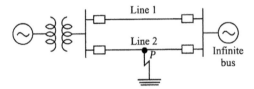

FIGURE E.13.7A

System for E13.7 (A), its reactance diagram (B), and admittance diagram (C).

Solution:
The reactance diagram is shown in Fig. E.13.7B.

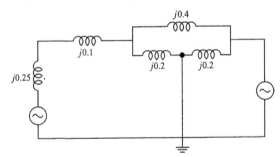

FIGURE E.13.7B

The admittance diagram is shown in Fig. E.13.7C.

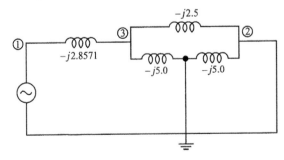

FIGURE E.13.7C

The buses are numbered and the bus admittance matrix is obtained.
Node 3 or bus 3 has no connection to any source directly, it can be eliminated.

	1	2	3
1	$-j1.85271$	0.0	$j2.8555271$
2	0.0	$-j7.5$	$j2.5$
3	$j2.8271$	$j2.5$	$-j10.3571$

$$Y_{11(\text{modified})} = Y_{11(\text{old})} - \frac{Y_{13}Y_{31}}{Y_{33}}$$

$$= -j\,2.8571 - \frac{(2.8527)(2.85271)}{(-10.3571)} = -2.07137$$

$$Y_{12(\text{modified})} = 0 - \frac{(2.85271)(2.5)}{(-10.3571)} = 0.6896$$

$$Y_{22(\text{modified})} = Y_{22(\text{old})} - \frac{Y_{32}Y_{23}}{Y_{33}}$$

$$= -7.5 - \frac{(2.5)(2.5)}{(-10.3571)} = -6.896549$$

The modified bus admittance matrix between the two sources is
The transfer admittance between the two sources is 0.6896 and the transfer
reactance = 1.45

	1	2
1	-2.07137	0.06896
2	0.6896	-6.59655

$$P_2 = \frac{1.05 \times 1}{1.45} \sin \delta \text{ p.u.}$$

or

$$P_e = 0.7241 \sin \delta \text{ p.u.}$$

E.13.8. For the system considered in E.13.6 if the *H* constant is given by 6 MJ/MVA obtain the
swing equation.
Solution:

The swing equation is $\dfrac{H}{\pi f} \dfrac{d^2 \delta}{dt^2} = P_m - P_e = P_a$, the accelerating power

If δ is in electrical radians

$$\frac{d^2 \delta}{dt^2} = \frac{180 \times f}{H} P_a = \frac{180 \times 50}{6} P_a = 1500\, P_a$$

E.13.9. In E13.7 if the three-phase fault is cleared on line 2 by operating the circuit breakers on both sides of the line, determine the postfault power angle characteristic.

Solution:

The net transfer reactance between E' and V_a with only line 1 operating is

$$j0.25 + j0.1 + j0.4 = j0.75 \text{ p.u.}$$

$$P_e = \frac{(1.05)(1.0)}{j0.75} \sin \delta = 1.4 \sin \delta$$

E.13.10. Determine the swing equation for the condition in E.13.9 when 0.8 p.u. power is delivered.

Given

$$M = \frac{1}{1500}$$

Solution:

$$\frac{180f}{H} = \frac{180 \times 50}{6} = 1500$$

$$\frac{1}{1500} \frac{d^2\delta}{dt^2} = 0.8 - 1.4 \sin \delta \text{ is the swing equation}$$

where δ in electrical degrees.

E.13.11. Consider example E.13.6 with the swing equation

$$P_e = 2.05 \sin \delta$$

If the machine is operating at 28° and is subjected to a small transient disturbance, determine the frequency of oscillation and also its period.

Given

$$H = 5.5 \text{ MJ/MVA}$$

$$P_e = 2.05 \sin 28° = 0.9624167$$

Solution:

$$\frac{dP_e}{d\delta} = 2.05 \cos 28° = 1.7659$$

The angular frequency of oscillation $= \omega_n$

$$\omega_n = \sqrt{\frac{\omega S^0}{2H}} = \sqrt{\frac{2\pi \times 50 \times 1.7659}{2 \times 5.5}}$$

$$= 7.099888 = 8 \text{ elec rad/s}$$

$$f_n = \frac{1}{2\pi} \times 8 = \frac{4}{\pi} = 1.2739 \text{ Hz}$$

$$\text{Period of oscillation} = T = \frac{1}{f_n} = \frac{1}{1.2739} = T = 0.785 \text{ s}$$

E.13.12. The power angle characteristic for a synchronous generator supplying infinite bus is given by

$$P_e = 1.25 \sin \delta$$

The H constant is 5 seconds and initially it is delivering a load of 0.5 p.u. Determine the critical angle.

Solution:

$$\cos \delta_c = \frac{P_{mo}}{P_{max}}[(\pi - 2\delta_0)] + \cos(\pi - \delta_0)$$

$$\frac{P_{mo}}{P_{max}} = \frac{0.5}{1.25} = 0.4 = \sin d\delta_0; \quad \delta_0 = 23°.578$$

$$\cos \delta_0 = 0.9165$$

$$\delta_0 \text{ in radians} = 0.4113$$

$$2\delta_0 = 0.8226$$

$$\pi - 2\delta_0 = 2.7287$$

$$\frac{P_{mo}}{P_{max}}(\pi - 2\delta_0) = 1.09148$$

$$\cos \delta_c = 1.09148 - 0.9165 = 0.17498$$

$$\delta_c = 79°.9215$$

E.13.13. Consider the system shown in Fig. E.13.13.

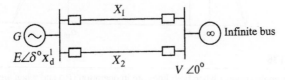

FIGURE E.13.13A

System for E13.13. (A) Power angle diagram.

$$x_d^l = 0.25 \text{ p.u.}$$

$$|E| = 1.25 \text{ p.u. and } |V| = 1.0 \text{ p.u.}; \quad X_1 = X_2 = 0.4 \text{ p.u.}$$

Initially the system is operating stable while delivering a load of 1.25 p.u. Determine the stability of the system when one of the lines is switched off due to a fault.

Solution:

When both the lines are working

$$P_{e\,max} = \frac{1.25 \times 1}{0.25 + 0.2} = \frac{1.25}{0.45} = 2.778 \text{ p.u.}$$

When one line is switched off

$$P_{e\,max}^l = \frac{1.25 \times 1}{0.25 + 0.4} = \frac{1.25}{0.65} = 1.923 \text{ p.u.}$$

$$P_{e0} = 2.778 \sin \delta_0 = 1.25 \text{ p.u.}$$

$$\sin \delta_0 = 0.45$$

$$\delta_0 = 26°.7437 = 0.4665 \text{ rad}$$

At point C

$$P_e^l = 1.923 \quad \sin \delta_1 = 1.25$$
$$\sin \delta_1 = 0.65$$
$$\delta_1 \quad = 40°.5416$$
$$= 0.7072 \text{ rad}$$

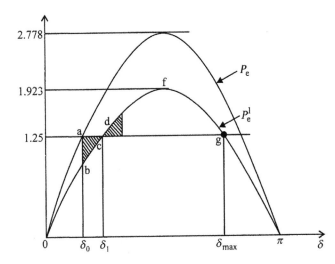

FIGURE E.13.13B

(B) The power angle curves f when both lines are in operation (P_e) and when are line is switched off (P_e^l).

$$A_1 = \text{area abc} = \int_{\delta_0}^{\delta_1} (P_2 - P_e^l) d\delta = \int_{0.4665}^{0.7072} (1.25 - 1.923 \sin \delta) d\delta$$

$$= 1.25 \Big|_{0.4665}^{0.7072} + 1.923 \cos \delta \Big|_{26°.7437}^{40°.5416}$$

$$= 0.300625 + (-0.255759) = 0.0450$$

Maximum area available $= \text{area c d f g c} = A_{2 \text{ max}}$

$$A_{2\text{max}} = \int_{\delta_1}^{\delta_{\text{max}}} (P_e^l - P_i) d\delta = \int_{0.7072}^{\pi - 0.7072} (1.923 \sin \delta - 1.25) d\delta$$

$$= -1.923 \cos \delta \Big|_{40.°5416}^{139°.46} - 1.25(2.4328 - 0.7072)$$

$$= 0.7599 - 1.25 \times 1.7256$$

$$= 0.7599 - 2.157 = -1.3971 \gg A_1$$

The system is stable

[Note: area A_1 is below $P_2 = 1.25$ line and area A_2 is above $P_2 = 1.25$ line; hence the negative sign]

E.13.14. Determine the maximum value of the rotor swing in the example E.13.13.

Solution:

Maximum value of the rotor swing is given by condition

$$A_1 = A_2$$

$$A_1 = 0.044866$$

$$A_2 = \int_{\delta_1}^{\delta_2} (-1.25 + 1.923 \sin \delta)d\delta$$

$$= (-1.25 \, \delta_2 + 1.25 \times 0.7072) - 1.923(\cos \delta_2 - 0.76)$$

i.e.,

$$= +1.923 \cos \delta_2 + 1.25 \, \delta_2 = 2.34548 - 0.0450$$

i.e.,

$$= 1.923 \cos \delta_2 + 1.25 \, \delta_2 = 2.30048$$

By trial and error $\delta_2 = 55°.5$

E.13.15. The M constant for a power system is 3×10^{-4} S^2/elec. degree

The prefault, during the fault and postfault power angle characteristics are given by

$$P_{e1} = 2.45 \sin \delta$$

$$P_{e2} = 0.8 \sin \delta$$

and

$$P_{e3} = 2.00 \sin \delta, \quad \text{respectively}$$

choosing a time interval of 0.05 second to obtain the swing curve for a sustained fault on the system. The prefault power transfer is 0.9 p.u.

Solution:

$$P_{e1} = 0.9 = 2.45 \sin \delta_0$$

The initial power angle
$$\delta_0 = \sin^{-1}\left(\frac{0.9}{2.45}\right)$$

$$= 21.55°$$

At $t = 0_-$ just before the occurrence of fault.

$$P_{max} = 2.45$$

$$\sin \delta_0 = \sin 21°.55 = 0.3673$$

$$P_e = P_{max} \sin \delta_0 = 0.3673 \times 2.45 = 0.9$$

$$P_a = 0$$

At $t = 0_+$, just after the occurrence of fault

$$P_{max} = 0.8; \quad \sin \delta_0 = 0.6373 \text{ and hence}$$

$$P_e = 0.3673 \times 0.8 = 0.2938$$

$$P_a, \text{ the acclerating power} = 0.9 - P_e$$

$$= 0.9 - 0.2938 = 0.606$$

Hence, the average accelerating power at $t = 0_{ave}$

$$= \frac{0 + 0.606}{2} = 0.303$$

$$\frac{(\Delta t)^2}{M} P_a = \frac{(0.05 \times 0.05)}{3 \times 10^{-4}} = 8.33 \ P_a = 8.33 \times 0.303 = 2°.524$$

$$\Delta \delta = 2°.524 \text{ and } \delta^0 = 21°.55.$$

The calculations are tabulated up to $t = 0.4$ second.

Table E.13.15 Table of Results for E13.15

S.No.	t (s)	P_{max} (p.u.)	Sin δ	$P_e = P_{max} \sin \delta$	$P_a = 0.9 - P_e$	$(\Delta t)^2/M \cdot P_a$ $= 8.33 \times P_a$	$\Delta \delta$	δ
1	0 −	2.45	0.3673	0.9	0	—	—	21.55°
	0+	0.8	0.3673	0.2938	0.606	—	—	21.55°
	0ave		0.3673	—	0.303	2.524	2°.524	24°.075
2	0.05	0.8	0.4079	0.3263	0.5737	4.7786	7°.3	24°.075
3	0.10	0.8	0.5207	0.4166	0.4834	4.027	11°.327°	31.3766
4	0.15	0.8	0.6782	0.5426	0.3574	2.977	14°.304	42°.7036
5	0.20	0.8	0.8357	0.6709	0.2290	1.9081	16°.212	57°.00
6	0.25	0.8	0.9574	0.7659	0.1341	1.1170	17°.329	73°.2121
7	0.30	0.8	0.9999	0.7999	0.1000	0.8330	18°.1623	90.5411
8	0.35	0.8	0.9472	0.7578	0.1422	1.1847	19°.347	108.70
9	0.40	0.8	0.7875	0.6300	0.2700	2.2500	21°.596	128.047
								149°.097

From the table it can be seen that the angle δ increases continuously indicating instability.

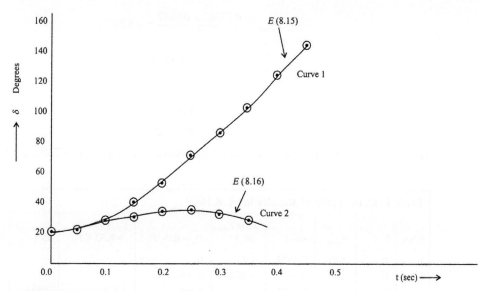

FIGURE E.13.15

Swing curve.

E.13.16. If the fault in the previous example E.13.14 is cleared at the end of 2.5 cycles determine the swing curve and examine the stability of the system.
Solution:
As before

$$\frac{(\Delta t^2)}{M} P_a = 8.33 \, P_a$$

$$\text{Time to clear the fault} = \frac{2.5 \text{ cycles second}}{50 \text{ cycles}}$$

$$= 0.05 \text{ second}$$

In this the calculations performed in the previous example E8.14 hold good for O_{ave}. However, since the fault is cleared at 0.05 second, there will be two values for P_{a1} one for $P_{e2} = 0.8 \sin \delta$ and another for $P_{e3} = 2.00 \sin \delta$.
At $t = 0.5-$ (just before the fault is cleared)

$$P_{max} = 0.5; \; \sin \delta = 0.4079, \text{ and}$$

$$P_e = P_{max} \sin d\delta = 0.3263, \text{ so that } P_a = 0.9 - P_e = 0.57367$$

giving as before

$$\delta = 24°.075$$

But, at $t = 0.5 +$ (just after the fault is cleared) P_{max} becomes 2.0 p.u. at the same δ and $P_e = P_{max} \sin \delta = 0.8158$. This gives a value for $P_a = 0.9 - 0.815 \, \delta = 0.0842$. Then for $t = 0.05$ are the average accelerating power at the instant of fault clearance becomes

$$P_{\text{a ave}} = \frac{0.57367 + 0.0842}{2} = 0.8289$$

$$\frac{(\Delta t)^2}{M} \cdot P_{\text{a}} = 8.33 \times 0.3289 = 2°.74$$

and

$$\Delta\delta = 5.264$$

$$\delta = 5.264 + 24.075 = 29°.339$$

These calculated results and further calculated results are tabulated in Table E.13.16.

Table E.13.16 Table of Results for E.8.16

S.No.	t	$P_{\max}$	$\sin\delta$	$P_e = P_{\max}$ $\sin\delta$	$P_a = 0.9 - P_e$	$(\Delta t)^2/M \cdot P_a$ $= 8.33 \times P_a$	$\Delta\delta$	δ
1	$0-$	2.45	0.3673	0.9	0	—	—	21.55°
	0_+	0.8	0.3673	0.2938	0.606	—	—	21.55°
	0_{ave}		0.3673	—	0.303	2.524	2.524	24.075
2	0.05_-	0.8	0.4079	0.3263	0.5737	—	—	—
	0.05_+	2.0	0.4079	0.858	0.0842	—	—	—
	0.05_{ave}		0.4079	—	0.3289	2.740	5.264	29.339
3	0.10	2.0	0.49	0.98	−0.08	−0.6664	4.5976	33.9367
4	0.15	2.0	0.558	1.1165	−0.2165	−1.8038	2.7937	36.730
5	0.20	2.0	0.598	1.1196	−0.296	−2.4664	0.3273	37.05
6	0.25	2.0	0.6028	1.2056	−0.3056	−2.545	−2.2182	34°.83
7	0.30	2.0	0.5711	1.1423	−0.2423	−2.018	−4.2366	30°.5933

The fact that the increase of angle δ, started decreasing indicates stability of the system.

E.13.17. A synchronous generator represented by a voltage source of 1.1 p.u. in series with a transient reactance of $j0.15$ p.u. and an inertia constant $H = 4$ seconds is connected to an infinite bus through a transmission line. The line has a series reactance of $j0.40$ p.u. while the infinite bus is represented by a voltage source of 1.0 p.u.

FIGURE E.13.17A

System for E13.17 and swing curves at different fault clearing times (B, C).

The generator is transmitting an active power of 1.0 p.u. when a three-phase fault occurs at its terminals. Determine the critical clearing time and critical clearing angle. Plot the swing curve for a sustained fault.

Solution:

$$P = \frac{EV}{X} \sin \delta_0; \quad 1.0 \frac{1.1 \times 1.0}{(0.45 + 0.15)} \sin \delta_0; \quad \delta_0 = 30°$$

$$\delta_c = \cos^{-1}[(\pi - 2\delta_0)\sin \delta_0 - \cos \delta_0]$$
$$= \cos^{-1}[(180° - 2 \times 30°)\sin 30° - \cos 30°]$$
$$= \cos^{-1}\left[\frac{\pi}{3} - 0.866\right] \quad = \cos^{-1}[1.807]$$
$$= 79°.59$$

Critical clearing angle $= 79°.59$

Critical clearing time $= \sqrt{\dfrac{2H}{P_m}\dfrac{(\delta_c - \delta_0)}{\pi f}}$

$$\delta_c - \delta_0 = 79°.59 - 30° = 49.59° = \frac{49.59 \times 3.14}{180}\text{rad}$$

$$= 0.86507 \text{ rad}$$

$$t_c = \sqrt{\frac{2 \times 4 \times 0.86507}{1 \times 3.14 \times 50}} = 0.2099 \text{ second}$$

Calculation for the swing curve

$$\Delta\delta_n = \delta_{n-1} + \left(\frac{180f}{H}\right)\Delta t^2 \, P_{a(n-1)}$$

Let

$$\Delta t = 0.05 \text{ second}$$

$$\delta_{n-1} = 30°$$

$$\frac{180f}{H} = \frac{180 \times 50}{4} = 2250$$

$$M = \frac{H}{180f} = \frac{1}{2250} = 4.44 \times 10^{-4}$$

$$\frac{(\Delta t)^2}{M} P_a = \frac{(0.05 \times 0.05)}{(4.44 \times 10^{-4})} P_a = 5.63 \, P_a$$

Accelerating power before the occurrence of the fault $= P_{a-} = 2 \sin \delta_0 - 1.0 = 0$
Accelerating power immediately after the occurrence of the fault.

$$P_{a+} = 2 \sin \delta_0 - 0 = 1 \text{ p.u.}$$

Average accelerating power $= \frac{0+1}{2} = 0.5$ p.u. Change in the angle during 0.05 second after fault occurrence.

$$\Delta\delta_1 = 5.63 \times 0.5 = 2°.81$$

$$\delta_1 = 30° + 2°.81 = 32°.81$$

The results are plotted in Fig. E.13.17.

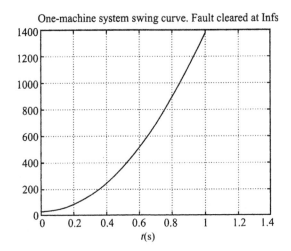

One-machine system swing curve. Fault cleared at Infs

FIGURE E.13.17B

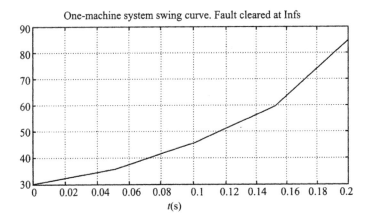

One-machine system swing curve. Fault cleared at Infs

FIGURE E.13.17C

The system is unstable.

E.13.18. In example E13.17, if the fault is cleared in 100 ms, obtain the swing curve.

Solution:

The swing curve is obtained using MATLAB and plotted in Fig. E.13.18.

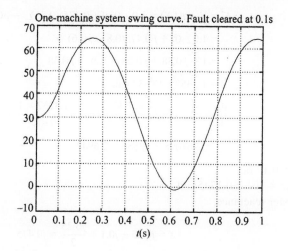

One-machine system swing curve. Fault cleared at 0.1s

FIGURE E.13.18

Swing curve for E13.18.

The system is stable.

E.13.19. A 50 Hz synchronous generator delivering a real power of 0.60 p.u. at 0.8 p.f. lagging to the infinite bus operating at 1.0 p.u. voltage has an inertia constant $H = 6.5$ MJ/MVA and a transient reactance of 0.25 p.u. as shown in figure. All the reactances are in p.u. on a common base. If a small disturbance in load results in change of load angle d by $10°$, if the breakers open and then quickly reclose, determine the motion of rotor up to 2 seconds.

Solution:

The transfer reactance between generator and load bus

$$X = i\,0.25 + j\,0.1 + \frac{i\,0.25}{2} = j\,0.475 \text{ p.u.}$$

Load current supplied to infinite bus.

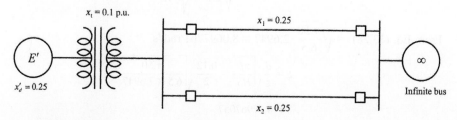

FIGURE E.13.19

System for E13.19.

The voltage of the generator behind excitation.

$$E' = V + j(I\ X) = 1.0\angle 0° + (j0.475)(0.75\angle -36° -87)$$

$$= 1.0\angle 0° + j\,0.475(0.8 - j\,0.6)$$

$$= 1 + j\,0.38 + j\,0.28 = 1.28 + j\,0.38$$

$$= 1.3352\angle 18°.37 \text{ p.u.}$$

$$\frac{dp}{d\delta} = P_{max}\cos\delta_0 = \frac{E'V}{X}\cos\delta_0 = \frac{1.3352 \times 1}{0.475}\cos 18°.37$$

$$= 2.81098 \times 09586 = 2.6947$$

Let $= d = 0.12$ rad
Given $\Delta\delta = 10° = 0.1745$ rad
The transfer reactance

$$X = j0.25 + j0.1 + \frac{j0.25}{2} = j0.475$$

$$\text{Current } I = \frac{S_x}{V_x} = \frac{0.75\angle -36°.87}{1\angle 0°} = 0.75\angle -36°.87$$

The voltage of the generator behind excitation

$$E^1 = V + j(Ix) = 1.0\angle 0° + (j0.475)(0.75\angle -36°.87)$$
$$= 1.0\angle 0° + j0.475(0.8 - j0.6)$$
$$= 1 + j0.38 + j0.28 = 1.28 + j0.38$$
$$= 1.3352\angle 18°.37 \text{ p.u}$$

$$\frac{dP}{d\delta} = P_{max}\cos\delta_0 = \frac{E^1V}{X}\cos\delta_0 = \frac{1.3352 \times 1}{0.475}\cos\angle\delta°.37$$

$$= 2.6947$$

$$\omega_m = \sqrt{\frac{ws°}{2H}} = \sqrt{\frac{\pi f° s°}{H}}$$

From Eq. (13.28) $= \sqrt{\frac{\pi.50}{6.5}\cdot 2.6947} = 8.0697761 \text{ rad/s}$

$$\xi = \frac{d}{z}\sqrt{\frac{\pi f°}{HS°}} = \frac{0.12}{2}\sqrt{\frac{\pi.50}{6.5 \times 7.6947}}$$

$$= 0.17967057$$

$$2\xi\omega_n = 2 \times 8.0697761 \times 0.17967057 \text{ u.2.9}$$

$$\omega_m^2 = 8.069776^2 = 65.1212847$$

$$\omega_d = \omega_m \sqrt{1 - \xi^2} = 8.069776 \sqrt{1 - 0.17968^2}$$
$$= 8.06976 \sqrt{0.967715}$$
$$= 8.06976 \times 0.9837251$$
$$= 7.9384409 \text{ rad/s}$$

$$\theta = \cos^{-1} \xi = \cos^{-1}(0.17967857) = 88.4900467$$

$$\delta = \delta_0 = \frac{\Delta \delta_o}{\sqrt{1 - \xi^2}} e^{-\xi \omega_m t} \sin(\omega_d' + \theta)$$

$$= 18.372 + \frac{10}{0.9891251} e^{-1.45t} \sin(7.93\delta t + \delta\delta.5^\circ)$$

$$= 18.3TZ + 10.1654415 e^{-1.45t} \sin(7.93\delta t + \delta\delta.5^\circ)$$

The solution to the equation is plotted in Fig. E.13.19 with δ versus t in seconds.

PROBLEMS

P.13.1. A 2-pole, 50 Hz, 11 kV synchronous generator with a rating of 120 MW and 0.87 lagging power factor has a moment of inertia of 12,000 kg-m^2. Calculate the constants H and M.

P.13.2. A 4-pole synchronous generator supplies over a short line a load of 60 MW to a load bus. If the maximum steady state capacity of the transmission line is 110 MW, determine the maximum sudden increase in the load that can be tolerated by the system without losing stability.

P.13.3. The prefault power angle characteristic for a generator infinite bus system is given by

$$P_{e1} = 1.62 \sin \delta$$

and the initial load supplied is 1 p.u. During the fault power angle characteristic is given by

$$P_{e2} = 0.9 \sin \delta$$

Determine the critical clearing angle and the clearing time.

P.13.4. Consider the system operating at 50 Hz.
If a three-phase fault occurs across the generator terminals plot the swing curve.
Plot also the swing curve, if the fault is cleared in 0.05 second.

QUESTIONS

13.1. Explain the terms
1. Steady state stability
2. Transient stability
3. Dynamic stability

13.2. Discuss the various methods of improving steady state stability.

13.3. Discuss the various methods of improving transient stability.

13.4. Explain the terms (1) critical clearing angle, and (2) critical clearing time.

13.5. Derive an expression for the critical clearing angle for a power system consisting of a single machine supplying to an infinite bus, for a sudden load increment.

13.6. A double-circuit line feeds an infinite bus from a power station. If a fault occurs in one of the lines and the line is switched off, derive an expression for the critical clearing angle.

13.7. Explain the equal area criterion.

13.8. What are the various applications of equal area criterion? Explain.

13.9. State and derive the swing equations

13.10. Discuss the method of solution for swing equation.

Index

Printed and bound by CPI Group (UK) Ltd, Croydon, CR0 4YY

08/05/2025

01864860-0001